HONG KONG AS A GLOBAL METROPOLIS

Hong Kong has remained the global metropolis for Asia since its founding in the 1840s following the Opium Wars between Britain and China. David Meyer traces its vibrant history from the arrival of the foreign trading firms, when it was established as one of the leading Asian business centers, to its celebrated handover to China in 1997. Throughout this period, Hong Kong has been prominent as a pivotal meeting-place of the Chinese and foreign social networks of capital, and as such has been China's window onto the world economy, dominating other financial centers such as Singapore and Tokyo. Looking into the future, the author presents an optimistic view of Hong Kong in the twenty-first century, challenging those who predict its decline under Chinese rule. This accessible and broad-ranging look at the story of Hong Kong's success will interest anyone concerned with its past, present, and future.

DAVID R. MEYER is Professor of Sociology and Urban Studies at Brown University.

Cambridge Studies in Historical Geography 30

Series editors
ALAN R. H. BAKER, RICHARD DENNIS, DERYCK HOLDSWORTH

Cambridge Studies in Historical Geography encourages exploration of the philosophies, methodologies and techniques of historical geography and publishes the results of new research within all branches of the subject. It endeavours to secure the marriage of traditional scholarship with innovative approaches to problems and to sources, aiming in this way to provide a focus for the discipline and to contribute towards its development. The series is an international forum for publication in historical geography which also promotes contact with workers in cognate disciplines.

For a full list of titles in the series, please see end of book.

HONG KONG AS A GLOBAL METROPOLIS

DAVID R. MEYER

CAMBRIDGE UNIVERSITY PRESS
Cambridge, New York, Melbourne, Madrid, Cape Town, Singapore, São Paulo

Cambridge University Press
The Edinburgh Building, Cambridge CB2 2RU, UK

Published in the United States of America by Cambridge University Press, New York

www.cambridge.org
Information on this title: www.cambridge.org/9780521643443

First published 2000
This digitally printed first paperback version 2006

A catalogue record for this publication is available from the British Library

Library of Congress Cataloguing in Publication data

Meyer, David R.
Hong Kong as a global metropolis / David R. Meyer.
p. cm. – (Cambridge studies in historical geography 30)
Includes bibliographical references and index.
ISBN 0 521 64344 9 (hardback)
1. Hong Kong (China) – History. 2. Hong Kong (China) – Economic
conditions. I. Title. II. Series.
DS796.H757M48 2000
951.25–dc21 99-32285 CIP

ISBN-13 978-0-521-64344-3 hardback
ISBN-10 0-521-64344-9 hardback

ISBN-13 978-0-521-02690-1 paperback
ISBN-10 0-521-02690-3 paperback

Contents

Figures

Maps

Maps prepared by Donna Souza.

Tables

Preface

Hong Kong brims with energy, glitter, and excitement. From the heights of Victoria Peak, the cityscape of skyscrapers, apartment towers, forested hills, and harbor forms a spectacular panorama, and at night the sparkling lightshow of the soaring buildings is second to none in the world. This glamorous city houses an extraordinarily talented people that made this research project a sheer delight. The citizens on the street, the storeowners, clerks, taxi-drivers, and my tailor graciously offered assistance whenever requested, provided directions, and gave me their opinions on the economy and politics. Officials across a wide range of government agencies and business organizations kindly met with me to answer questions, directed me to data sources, and provided materials. University faculty in Hong Kong willingly shared their knowledge, offered suggestions, and provided forums to debate ideas. Numerous business executives generously gave me an hour of their time to interview them about their strategic views of the economy and politics of Hong Kong and Asia, and they explained how they behaved as decision-makers in control of commodity and financial capital. Although I cannot cite them personally because they stated their views in confidence, their insights form critical components of the interpretations and explanations offered in this book; I am indebted to them.

Closer to home, thanks are due to units of Brown University which offered financial and logistical support, including the Population Studies and Training Center, the Watson Institute for International Studies, the Graduate School, the Rockefeller Library, the Department of Sociology, and the Urban Studies Program. Tal Halpern served as my able research assistant for a time, and Donna Souza used her talents with GIS to create the maps.

Finally, thanks go to two individuals who were part of this research throughout. Jim Handrich, principal of the High School of the Hong Kong International School, and a friend for almost forty years, graciously provided lodging for my many trips to Hong Kong, offered warm hospitality, and included me in social occasions that gave me a chance to see another side to Hong Kong. Judy, wife and friend for a lifetime, kindly listened more than she wished to my commentary about Hong Kong and joined me on several trips.

1

Enigma

> Her Majesty's plenipotentiary has now to announce the conclusion of preliminary arrangements between the Imperial commissioner and himself involving the following conditions: The cession of the island and harbour of Hongkong to the British crown.[1]

Hong Kong remains wrapped in an enigma. Its intermediaries of capital, who include traders, financiers, and corporate managers, have made Hong Kong the pivot of decision-making about the exchange of capital within Asia and between that region and the rest of the world. Yet, for 150 years, this tiny island and adjacent peninsula could not even lay claim to status as a city-state. When Britain declared sovereignty over Hong Kong in 1841, after taking it from China under the terms of the Treaty of Nanking that settled the Opium War, the government and merchants had to build a town. The British viewed Hong Kong as their emporium of trade in the Far East, but they did not aspire to transform it into a commercial-military power similar to the earlier aggressive city-states of Genoa and Venice. From the start, Hong Kong and Asia remained peripheral to a British foreign policy focused on Europe, and up to 1860, the meager fleet on the China station seldom numbered more than six ships. Britain devoted greater attention to avoiding being drawn into the interior of China than to expanding trade.[2]

British governors of Hong Kong supported the traders and financiers and worked closely with them. Yet, for all the attention paid to British, and to a lesser extent, other "foreign" traders and financiers, city residents overwhelmingly consisted of Chinese, many of them also traders and financiers. Britain signed a treaty with China in 1898 that leased the New Territories, an area north of the Kowloon Peninsula, for ninety-nine years, and the governments set the return date for 1997, setting off a ticking clock that ended with

[1] Notification of Captain Elliot, January 20, 1841; quoted in Morse, *The international relations of the Chinese empire*, vol. I, p. 271.

[2] Endacott, *A history of Hong Kong*, pp. 25–50; Graham, *The China station*.

the monumental peaceful transfer of a global metropolis between nations. Nevertheless, seen from Britain's perspective, Hong Kong still resided outside the mainstream of foreign policy. Even during World War II, Britain conceded Hong Kong to Japan and concentrated its defensive resources in Singapore. At the end of the war, the foreign and Chinese traders and financiers quickly regrouped in Hong Kong, reconstituting it as the pivot of the Asian networks of capital. When Britain finally realized that it could not retain control of Hong Kong and signed the Joint Declaration with China in 1984 that set the return to China for July 1, 1997, Britain did not seriously consult Hong Kong's residents and refused to commit extensive political and economic resources to contest the transfer.[3] As the transfer date loomed and Hong Kong would gain a territorial hinterland coterminous with its sovereign power, skeptics depicted pro-democracy movements, China's assertions that it would not tolerate challenges to its authority, and emigration of professionals as signs that Hong Kong would decline; and shortly after the transfer proceeded smoothly, economic travails in Asia impacted Hong Kong and again called its viability into question.[4]

An interpretation of Hong Kong as the global metropolis for Asia must explain the enigma of its expansion without a sovereign territorial hinterland even as a British colonial policy kept it at the periphery of concern. It soared after 1945 even as advances in telecommunications raised the specter of seamless capital exchanges without the need for face-to-face communication, and improved air travel seemingly allowed firms to manage trade, finance, and corporate organizations from almost any city. And with increasing uncertainty surrounding its return to China in 1997, competitors such as Tokyo and Singapore failed to dethrone Hong Kong as the dominant Asian venue for decision-making about the exchange of capital. Suggestions that Hong Kong may become one of the greatest global metropolises must reckon both with the capacity of its intermediaries to remain dominant in Asia, even as economic turmoil threatens, and with its status as a "capitalist" bastion ruled by a "socialist" state.

Traders and financiers in Hong Kong always operated in multitiered national, world-regional, and global economies. Recognition of that business scope provides one key for unlocking the enigma of Hong Kong as the global metropolis for Asia. Since the Canton days of the early nineteenth century, foreign merchant traders operated as agents of powerful global firms headquartered in London, New York, and Boston, among other metropolises. Their arrival in Asia represented an extension of expansive colonial states in Europe, and Britain, the leading extractor of concessions from China after the

[3] Endacott, *A history of Hong Kong*, pp. 260–69; Welsh, *A borrowed place*, pp. 374–440, 502–36.

[4] For examples of skepticism, see: Kraar, "The death of Hong Kong"; Theroux, "Letter from Hong Kong." For the economic crisis in Asia and its impact on Hong Kong, see Guyot, "Fears rise in Hong Kong over credit"; Pesek, "Dis-oriented markets."

Opium War of the early 1840s, was the strongest imperial power based on its industrial might. The mad rush of great foreign trading firms to Hong Kong in the 1840s instantaneously established it as an arm of global firms in leading metropolises, and the simultaneous arrival of numerous Chinese trading firms boldly indicated that this metropolis operated more than as an outpost of foreign capital.

The trading firms and financial institutions that followed, nevertheless, always remained embedded in the political economy of Asia. Peasant impoverishment continually thwarted attempts of foreign trade and financial firms to find markets for their home-country products and capital, and this impoverishment impacted the firms' specialization, capitalization, and capacity to compete for control of the exchange of commodity and financial capital. As peasants started to rise out of poverty in the late twentieth century, traders and financiers transformed their businesses, but foreign firms confronted anew the reality of Asian economic exchange. Chinese traders and financiers dominated the exchange of commodity and financial capital at unspecialized, less-capitalized levels, and those intermediaries leveraged that dominance into higher levels of specialization and capitalization as economic growth and development of Asian countries accelerated.

Chinese and foreign traders and financiers always operated as two social networks of capital, a Chinese and a foreign network, and those networks intersected at key metropolises. The term "social networks" emphasizes that intermediary decision-making about the exchange of capital rests on bonds that extend beyond pure market calculations of profit and loss to include deeper, wider social relations. Those relations are essential to build trust and monitor malfeasant behavior, thus reducing the risks of exchange. Social networks provided the means for economic exchange within Asia and between Asia and the developed world of Europe and North America, and Hong Kong operated as the pivotal meeting-place of the Chinese and foreign social networks of capital in Asia. The first step in interpreting Hong Kong as the global metropolis for Asia requires a specification of behavioral principles that intermediaries use to control the exchange of commodity and financial capital. Then, these principles frame the interpretation of the changes in the social networks of capital in Asia that revolve around Hong Kong from antecedents in the Canton days around 1800 to the present. Rather than viewing the Chinese business networks as exceptional, these principles portray both the Chinese and foreign networks as pieces cut from the same cloth.

These behavioral principles also provide a lens with which to evaluate the skeptics' claim that Chinese government control weakens Hong Kong as the global metropolis for Asia. This claim dismisses too readily both Hong Kong's status as the pivot of the Chinese and foreign social networks of capital and China's commitment to preserving the Hong Kong jewel as its window to the world economy. The arrival of "red chips," mainland Chinese firms with government connections, infiltrates *guanxi* (connections) into the business environment and this undermines Hong Kong, the skeptics argue; yet, increasingly, mainland Chinese work in foreign firms and foreigners work in mainland firms. The influx of red chips and private mainland firms strengthens Hong Kong as the meeting-place of the Chinese and foreign social networks of capital. As economic crises swirl in Asia, the critics' predictions that China would undermine Hong Kong are contradicted by the unwavering support that China expresses for its international business center. That affirmation will reinforce Hong Kong's advance towards the level of London and New York during the twenty-first century as it becomes the global metropolis for one of the largest economies in the world.

2

Intermediaries of capital

> As it is the power of exchanging that gives occasion to the division of labour, so the extent of this division must always be limited by the extent of that power, or, in other words, by the extent of the market.[1]

Insights of the social theorists

Social theorists who observed the world economy during the early history of Hong Kong provided clues to understanding the behavior of traders, financiers, corporate managers, and other intermediaries of capital. From the late eighteenth century to the early twentieth century, theorists such as Emile Durkheim, Karl Marx, Adam Smith, Herbert Spencer, Ferdinand Tonnies, and Max Weber witnessed a transformation of the world economy as revolutionary as that of the late twentieth century.[2] Railroads dramatically lowered the cost and raised the volume and speed of commodity and passenger movement over land, and steamboats and steamships did the same for waterborne transport. The telegraph bound cities within nations from the 1840s, and by the 1880s, a global network had emerged. This provided almost instantaneous communication of information and separated information transmission from physical movements of passengers and commodities. Industrial growth, first in Western Europe, then in the United States, and finally in Japan, generated swelling volumes of commodities for shipment, and burgeoning factories drew on widening source areas for raw material inputs and forged increasingly elaborate linkages of intermediate goods. This astounding rise in social complexity fascinated the theorists; their clues to explaining it rested in the causes and consequences of the division of labor.[3]

1 Smith, *An inquiry into the nature and causes of the wealth of nations*, p. 8.
2 Durkheim, *The division of labor in society*; Marx, *Capital*; Smith, *An inquiry into the nature and causes of the wealth of nations*; Spencer, *The principles of sociology*; Tonnies, *Community and society (Gemeinschaft und Gesellschaft)*; Weber, *Economy and society*.
3 Rueschemeyer, *Power and the division of labor*.

They had divergent social, economic, and political views, but the social theorists shared three fundamental points: local economies made a transition from self-sufficiency to integration with other local economies as the division of labor advanced; the increase in exchange among local economies supported this metamorphosis; and differentiation (specialization) and integration were mutually reinforcing processes. To grapple with this complexity, they posed an ideal state, the local self-sufficient economy, that existed only in remote, exotic places. A web of intertwined economic relations bound the residents: isolation from external information kept technological innovation low; production technologies stayed primitive; most goods were exchanged face to face because inadequate transportation media made transactions over greater distances impossible; locally produced goods limited population size; the small population kept demand low and the labor force tiny; and this constricted labor specialization and economies of scale. In sum, the theorists had articulated the state of impoverishment.[4]

To break out of this state, residents of a local economy had to exchange with other local economies. Adam Smith articulated the brilliant insight that the growth of exchange unleashed and molded the possibilities of the division of labor, but citations of his statement trivialized it to an aphorism: "the division of labor is limited by the extent of the market."[5] This aphorism focuses on a body-count of consumers that form the market and shifts attention to the supply side and the production economies of firms. The growth of the market enhances possibilities for the division of labor and economies of scale of firms; these translate into lower production costs, such as in the pin factory that Smith immortalized. The aphorism, however, circumvents attention to broader effects of increased exchange that the social theorists recognized; exchange enhances flows of information and that raises awareness of new sources of demand and supply and promotes technological innovation. Greater exchange also stimulates demand for improvements in transportation and communication, and this permits greater specialization. Residents of local economies do not have the time, information, and capital directly to forge exchange linkages to other local economies. Many theorists took another critical step; they identified agents, or intermediaries, of exchange and termed their headquarters the "metropolis."[6] These actors, who included wholesalers and financiers, destroyed the stagnation of self-sufficient local

[4] Meyer, "The division of labor and the market areas of manufacturing firms," pp. 433–38.

[5] For the full statement of Adam Smith, see the quotation at the head of this chapter; Smith, *An inquiry into the nature and causes of the wealth of nations*, p. 8. The shortened aphorism appears in numerous places; for one of the most famous, see Stigler, "The division of labor is limited by the extent of the market."

[6] Smith, *An inquiry into the nature and causes of the wealth of nations*, book 2; Weber, *Economy and society*, vol. I, pp. 156–59, vol. II, pp. 1216–17; Tonnies, *Community and society (Gemeinschaft und Gesellschaft)*, pp. 82, 227; Marx, *Capital*.

economies through non-local exchange of commodity and financial capital. Intermediaries specialize in controlling and coordinating exchange among local economies; as the division of labor advances, greater complexity of exchange requires more sophisticated intermediaries. The theorists did not elaborate their ideas about intermediaries and metropolises, but subsequent studies provide a basis for an explanation of the growth and change in Hong Kong as the global metropolis for Asia.[7]

Intermediaries confront dilemmas

Control of exchange

To control the exchange of commodity and financial capital, intermediaries must acquire public and specialized business information about their international demand and supply.[8] Because intermediary profitability often depends on being the first to make exchanges, delays in receipt and transmission of information are costly. Printed and electronic mass media, unrestricted spoken information, and official government sources provide public information, but all intermediaries have similar access to this information. Instead, specialized business information communicated face to face, in written forms (mail and journals), and through telecommunications (telegraph, satellite, and fiber optics) constitutes the critical information for intermediary decision-making. This complex information requires synthesis, analysis, and interpretation, and these processes have large fixed-cost components because intermediaries need highly skilled people, trained in quantitative and financial analysis, and information processing capabilities, such as computers, models, and software. Intermediaries often rely on face-to-face contact for close coordination, negotiation, communication of complex information, and transmission of confidential information. Short-distance travel requires little time and money, but these costs escalate rapidly for longer-distance travel; therefore, intermediaries agglomerate at origins and destinations for efficient face-to-face contact. Those who engage in larger-scale, more complex exchange over wider territories typically require greater amounts of capital to fund information acquisition and processing.

The amount of capital required to underwrite exchange must rise along with swelling volumes of exchange. Even intermediaries that exchange capital without acquiring ownership of it need larger capital bases as exchange

[7] The framework draws partly on Meyer, "A dynamic model of the integration of frontier urban places into the United States system of cities"; Meyer, "The world system of cities"; Meyer, "The formation of a global financial center," pp. 98–99; Meyer, "Change in the world system of metropolises," pp. 398–406.

[8] This discussion of information draws on Pred, *City-systems in advanced economies*, pp. 19–22.

expands because they must fund costs associated with transactions before they receive payment for services. For example, a commodity broker of wheat who operates on commission must fund storage, transportation, and insurance costs of moving wheat between seller and buyer. As business with a buyer expands, the broker may buy on the basis of an order or even advance money to the buyer for the purchase. Intermediaries who take ownership of capital as a customary business practice without a binding or firm commitment from a buyer must augment their capital bases to fund larger ownership positions. For example, an investment bank that speculates in currency movements must raise its capital as larger positions are taken even though these positions are often financed with borrowed money, because the capital base of the firm directly impacts the capacity to borrow.

Crossing boundaries

The exchange of commodity and financial capital across international boundaries confronts intermediaries with two distinctive problems: they must physically move themselves and commodities and they must transfer the control of capital.[9] Total cost rises with distance, but typically less than proportionally to the increase in distance, because fixed terminal costs at origin and destination are spread over longer distances; therefore, the fixed portion of the cost per kilometer declines. Transportation media are organized hierarchically; smaller carriers bring passengers and commodities to nodes for aggregation onto the services of larger carriers before long-distance transportation. Because intermediaries use passenger travel for contact with other intermediaries and for gathering information for exchange, centers of intermediary activity require high-quality passenger services. Physical movements of passengers and commodities need not directly trace linkages that intermediaries forge to transfer control of capital. After the introduction of the telegraph (1840s), information could move independently of the transportation of passengers and commodities. Oil traders, for example, buy and sell oil on different international commodity exchanges, whereas physical movements of oil connect producing nations with consuming ones. This focus on transfers of control of capital, rather than simply their physical exchanges, directs attention to the fundamental activities of intermediaries. To implement exchanges of capital, they must either take ownership before transferring capital between buyer and seller or provide services that enable exchanges to occur. An exporter, for example, buys textiles from a factory in one nation and sells them

[9] This discussion of physical movement and the transfer of the control of capital is adapted from Meyer, "A dynamic model of the integration of frontier urban places into the United States system of cities." The discussion focuses on international exchange, but the concepts also fit intranational exchange.

to an importer or to a retail chain in another nation. Similarly, a bank collects money from depositors in one nation, pays them interest, and lends this money to a factory in another nation. In this case, the bank controls the financial exchange without actually taking ownership of capital, but the bank remains liable for the security of the capital.

The control of the process of exchange across boundaries constitutes the core business of intermediaries; they exert that control whether capital exists as physical or symbolic assets. Intermediaries may directly exchange physical assets, such as precious metals (gold, silver) and commodities (grain, manufactures), or create symbolic measures of the assets such as option contracts to buy or sell. Similarly, they can exchange currencies (dollars, yen), stocks, and bonds that represent stores of value, or transform them into even more abstract symbols such as derivatives. Intermediaries can transform assets such as loans and real estate portfolios into marketable securities that trade. The shift of emphasis from physical to symbolic assets forces intermediaries to invest more resources in telecommunications and information processing capabilities. Assets ultimately exist as claims on ownership; exchange reduces to a transfer of information.

Regardless of the form of assets, exchange of capital across international boundaries always enmeshes intermediaries in political negotiation and conflict.[10] Because intermediaries need approval to operate as importers and exporters of capital, government officials have leverage to extract political and financial support from them. Intermediaries may lobby their government to enforce sanctions against foreign competitors, or, when sanctions from other nations impact them, they may request support from their government. Governmental actions pose risks for intermediaries, but these episodes pale compared with the risks they confront on every exchange of capital across political boundaries. Partners to an agreement to buy or sell may fail to complete transactions, payments may not be made, and credits or loans may remain unpaid. All businesses face these risks, but most have only a small share of their capital at risk away from their business at any time. Because intermediaries allocate most of their capital to underwriting exchange, the bulk of it remains outside their immediate control either in transit or at a distant site as a store of value such as in commodities, investments, or loans. They cannot continuously monitor partners to exchange transactions, and the costs to guarantee or enforce contracts are prohibitive; those efforts slow the exchange process, reducing the rate of return on capital. Enforcement of contracts across political boundaries remains difficult because exchange partners reside outside the authority of the home nation. Sanctions against exchange

[10] Block, "The roles of the state in the economy"; Corbridge, Thrift, and Martin, *Money, power and space*.

partners for malfeasance depend on the willingness of the other nation to take enforcement actions; yet, malefactors can lobby national leaders for support, thus deflecting or stopping sanctions.

Intermediaries must reduce this ongoing risk of exchange across boundaries; exchange with friends offers a plausible solution. Social theorists from the eighteenth-century Scottish Enlightenment, including Adam Smith and David Hume, nevertheless, identified a paradox.[11] In precommercial societies, the social group encompassed the full range of social and economic transactions of individuals; calculations of self-interest pervasively shaped friendship, but these bonds drew individuals into transactions that did not have commercial feasibility. Strangers, in contrast, loomed as potential enemies outside the group. The social theorists argued that the advent of commercial society, namely economic exchange in the market across local economies, shifted the calculation of self-interest to the market. This freed friendship from economic self-interest and allowed it to flourish, founded on sympathy and affection. The theorists identified serious problems with embedding economic transactions in bonds of friendship. Their belief that market transactions could stand without friendship implicitly rested on a practical fact. Exchange of capital across local economies cannot rest solely on friendship bonds because intermediaries do not have the time and monetary resources to forge those bonds. Friendship, therefore, fails to fully govern intermediary exchange, but an alternative exists, the building of trust; individuals trust their friends, but people they trust do not have to be friends.

Trust as a bedrock

Everyone participates in forms of community, including family, ethnic group, religion, or common interest group, such as a social club, business association, or professional organization. Persons in a community share beliefs and values, and relations among members are direct, many-sided, and reciprocal.[12] Members trust each other, but they need not be friends. Given that a trustee (actor being trusted) gains by being trusted in the future, then a close community among potential trustors (actors trusting the trustee) leads to greater trustworthiness; this follows for two reasons. First, the trustee expects greater benefits in the future when the relationship with the trustor continues than if it terminates after one exchange. This encourages the trustee to engage in more trustworthiness because the trustee incurs high cost by failure to act that way. Second, extensive communication among the trustor and other actors from whom the trustee might expect to receive trust in the future encourages the

[11] Silver, "Friendship in commercial society."
[12] Taylor, *Community, anarchy and liberty*, pp. 26–33.

trustee to engage in trustworthy behavior because the larger group of trustors will hear about the trustee's behavior. Intermediaries have powerful incentives to participate in these communities of mutual trust as direct adjuncts to their control of exchange of capital because this reduces risks of losses from malfeasant behavior by other intermediaries.[13]

Social norms with sanctions attached reinforce trustworthiness among members of communities of mutual trust. The most common sanction is to restrict exchanges with offending actors; for intermediaries, that is tantamount to terminating their business. Demand for the norm of trustworthiness among intermediaries emerges because each gains a positive externality from trustworthy behavior by other intermediaries, but no individual gains any rights to the control of trust. Intermediaries who gain positive externalities constitute the same group that carries out the behavior and enforces sanctions. Because intermediaries engage in extensive communication among themselves as part of normal business, they can coordinate achievement of the norm of trustworthiness and enforce sanctions against offenders. Members of communities of intermediaries rooted in families, ethnic groups, or religions share beliefs and values and participate in relations that extend beyond intermediary behavior and reach across nations. This supports the norm of trustworthiness among these intermediaries and is a form of social capital that confers advantages over other intermediaries without such roots, especially when international exchange entails extensive political risk. Extra trustworthiness within these communities allows intermediaries to reduce the costs of monitoring malfeasant behavior and enforcing sanctions and allows them to form cooperative ventures more easily. These ventures spread risk, and the pooled financial capital permits intermediaries to engage in larger-scale non-local exchange than anyone could on their own.

Advances in telecommunications that permit almost instantaneous and repeated international exchanges of capital might seem to reduce the problem of trustworthy behavior among intermediary institutions such as banks, stock and bond exchanges, commodity exchanges, and money management firms. This "fallacy of technological intermediation," however, ignores the social-institutional contexts within which these exchanges of capital occur. Trust that exists among these institutions may have impersonal tones, but it remains no less powerful. They create formal bonds through organizations and associations, and members are expected to establish internal controls, supervision, and governance that minimizes errors and monitors the behavior of employees who could place the capital of the firm at risk. This possibility forces these institutions to keep employees within close groups, even though the technol-

[13] The discussion of intermediary trust in this and the next several paragraphs draws especially on Coleman, *Foundations of social theory*, pp. 91–116, 175–96, 241–321. Also see Fukuyama, *Trust*; Shapiro, "The social control of impersonal trust."

ogy of telecommunications might make possible exchanges of capital from dispersed sites of employees.[14]

Intermediaries build communities of trust to reduce the risks of exchanging capital across international boundaries; nevertheless, they still confront a dilemma. If they rely on generalized morality or institutional structures to guarantee trustworthy behavior of partners in exchange, they assume those partners will meet norms of trust expected in intermediary exchange. This oversocialized view of behavior portrays intermediary partners as atomized individuals who follow rules of behavior, but it leaves intermediaries open to opportunistic, malfeasant actions by partners. This dilemma is no less serious than if intermediaries view partners as undersocialized, atomized individuals who exchange opportunistically. Intermediaries, however, may take an alternative tack, the embedding of behavior in actual personal relations and in the structures of those relations, termed networks. This embeds trust in ongoing relations as a way to protect against opportunistic and malfeasant behavior, but it does not guarantee against all such behavior.[15]

Social networks of capital

Because intermediaries face substantial risk when they exchange capital, they look for trustworthy partners to the exchange process. Partners they have dealt with previously rank higher because intermediaries have better information about their trustworthiness. That information costs less than acquiring information about an unknown partner, and it has greater validity and depth because the information was personally obtained. These partners have incentives to remain trustworthy to encourage continued exchange. Social bonds may grow from repeated transactions, thus building trust and discouraging malfeasance. Intermediaries must consider the time and effort devoted to social contacts as scarce resources. If they focus time and energy on building strong ties to a few intermediaries as a means to boost trust and reduce risk from malfeasant behavior, they confront a dilemma. This restricts the range of potential partners, limits the scope of information received about exchange possibilities because they most likely share friends and partners, and reduces the time available to enlarge the number of partners. Paradoxically, intermediaries strengthen their capacity to exchange capital with trustworthy partners and minimize risk if they embed themselves in a maze of non-redundant ties and reduce efforts to maintain a few strong ties. These exchange ties might involve multiple and shifting coalitions of partners in exchange and diverse acquaintances across a range of social and economic groups. Bridges link

[14] Graham, "Cities in the real-time age"; Shapiro, "The social control of impersonal trust"; Thrift, "New urban eras and old technological fears"; Thrift, *Spatial formations*, pp. 213–55.

[15] Granovetter, "Economic action and social structure."

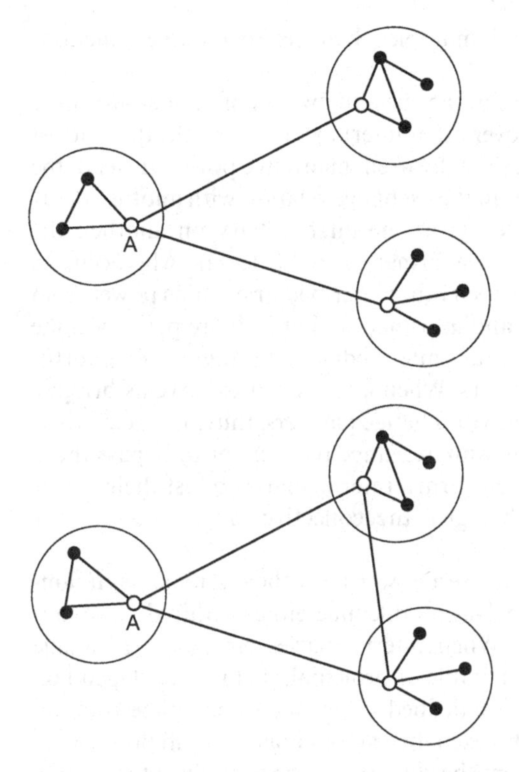

Fig. 2.1. Intermediary A exerts power over other intermediaries in upper portion, but A has less power in lower portion.
Source: author.

different social networks, each of which has some internal cohesion. Intermediaries increase their access to information and their power to control the terms of exchange if they devote time and energy to forging bridges to intermediaries whose social networks do not exchange capital (fig. 2.1). This boosts non-redundant information that the intermediary receives, and the absence of exchange bonds across social networks gives greater power to the intermediary who builds the bridge. That power comes from the inability of intermediaries in other social networks to find substitute exchange partners easily; they must rely on the intermediary that builds the bridge. Intermediaries enhance their effectiveness through constructing bridges that reach pivotal intermediaries in other social networks rather than those less central to networks (fig. 2.1). When intermediaries focus on building bridges, therefore, they augment the flow of information, provide greater diversity of opportunities for exchange, and boost trustworthiness by providing alternative checks on recommendations and multiple channels to enforce sanctions against malfeasant behavior.[16]

The positions of intermediaries in the social networks of capital affect the amount of power they can exert over other intermediaries, but the quantity of capital an intermediary owns does not, by itself, guarantee power. Instead, the power of the intermediary resides in the exchange relation with another intermediary; the greater the dependency of one intermediary on another, the greater the power of the latter over the former. An intermediary who occupies a central position in exchange between other intermediaries gains power from serving as transmitter of capital and information, but as more paths become available, that power declines because intermediaries have greater opportunities to substitute exchange partners. When intermediaries serve as bridges, they gain greater access to alternative exchange partners, thus enhancing their capacity to negotiate better terms with exchange partners or to bypass those partners who do not offer favorable terms. Intermediaries boost their power through cooperative ventures that give the collective greater power over another group.[17]

The capacity of intermediaries to exert power, nevertheless, remains circumscribed; they confront the principal–agent dilemma either within their organization or interorganizationally. When intermediaries engage in exchange relations, they frequently acquire the role of principal, that is, they depend on the actions of other intermediaries, defined as agents, to complete transactions. Intermediary principals who exchange across international boundaries face severe principal–agent problems because they cannot easily enforce sanctions across political borders. Agents have greater access to information about exchange from their end, and intermediary principals cannot perfectly and costlessly monitor actions of agents. This exposes principals to agency losses that increase in severity with greater divergence between the interests of principals and agents and with higher costs of information monitoring. This principal–agent dilemma encourages principals to embed exchange through agents in long-term relations built on trust; that approach lowers monitoring costs and boosts information access.[18]

The embeddedness of intermediaries in social networks of capital provides

[16] Burt, *Structural holes*, pp. 8–49; Granovetter, "Economic action and social structure," p. 490; Granovetter, "The strength of weak ties." Studies that support the arguments in this paragraph include: Erickson, "Culture, class, and connections"; Montgomery, "Weak ties, employment, and inequality"; Uzzi, "The sources and consequences of embeddedness for the economic performance of organizations."

[17] Studies that support the arguments in this paragraph include: Cook, Emerson, Gillmore, and Yamagishi, "The distribution of power in exchange networks"; Emerson, "Power-dependence relations"; Freeman, "Centrality in social networks"; Yamagishi, Gillmore, and Cook, "Network connections and the distribution of power in exchange networks"; Yamaguchi, "Power in networks of substitutable and complementary exchange relations."

[18] Pratt and Zeckhauser, "Principals and agents"; White, "Agency as control."

the glue that bonds flows of commodity and financial capital, but these intermediaries also compete to control exchange of that capital. Their reactions to this competition affect the rise and fall of global metropolises. These intermediaries have at least three alternative ways to react to competition from other intermediaries: alter transaction costs, differentiate or dedifferentiate to control markets, or appeal to force.[19]

Intermediaries alter transaction costs

Intermediaries directly react to competition when they alter the costs of controlling the exchange of commodity and financial capital. Their profit comprises the difference between the price sellers accept in one local economy and the price buyers pay in another. For example, if savers in one nation accept 5 percent interest on deposits whereas borrowers in another nation pay 8 percent, the financial intermediary profits from this spread in interest rates. Similarly, a wholesaler buys toys from a factory in one nation and sells them at a higher price to a retail chain in another nation. The intermediary's profit is the difference between these prices, adjusted for the cost to make the exchange of capital. Intermediaries reduce those costs through: improvements in transportation, communication, and information processing; lowering risks, that is, the probability of losses, from controlling exchange; internalizing intermediary activity within an existing firm or bringing that activity within a firm that previously did not include intermediary activity; and agglomerating.

Improvements in transportation, communication, and information processing

The costs to transport commodities, to travel long-distance to meet buyers, sellers, and other intermediaries, to communicate information, and to process complex information for decision-making constitute highly visible expenditures essential to intermediary activity. When the cost to transport commodities falls, the intermediary may pass along those lower costs to buyers; this stimulates demand and boosts the volume of business of the intermediary. Producers who benefit from lower purchase prices of inputs may reduce their selling prices, thus raising demand for intermediary services. The greater speed of moving and handling goods shortens the time the capital of the intermediary is tied up while in transit, increasing the rate of turnover of capital for new investments. Improvements in passenger travel include lower "ticket" costs

[19] I thank Charles Tilly for the suggestion that intermediary behavior be analyzed as reaction to competition. A preliminary development of this explanation for international intermediaries is in Meyer, "The formation of a global financial center," pp. 98–99; Meyer, "Change in the world system of metropolises," pp. 398–406. The present discussion revises and expands those explanations.

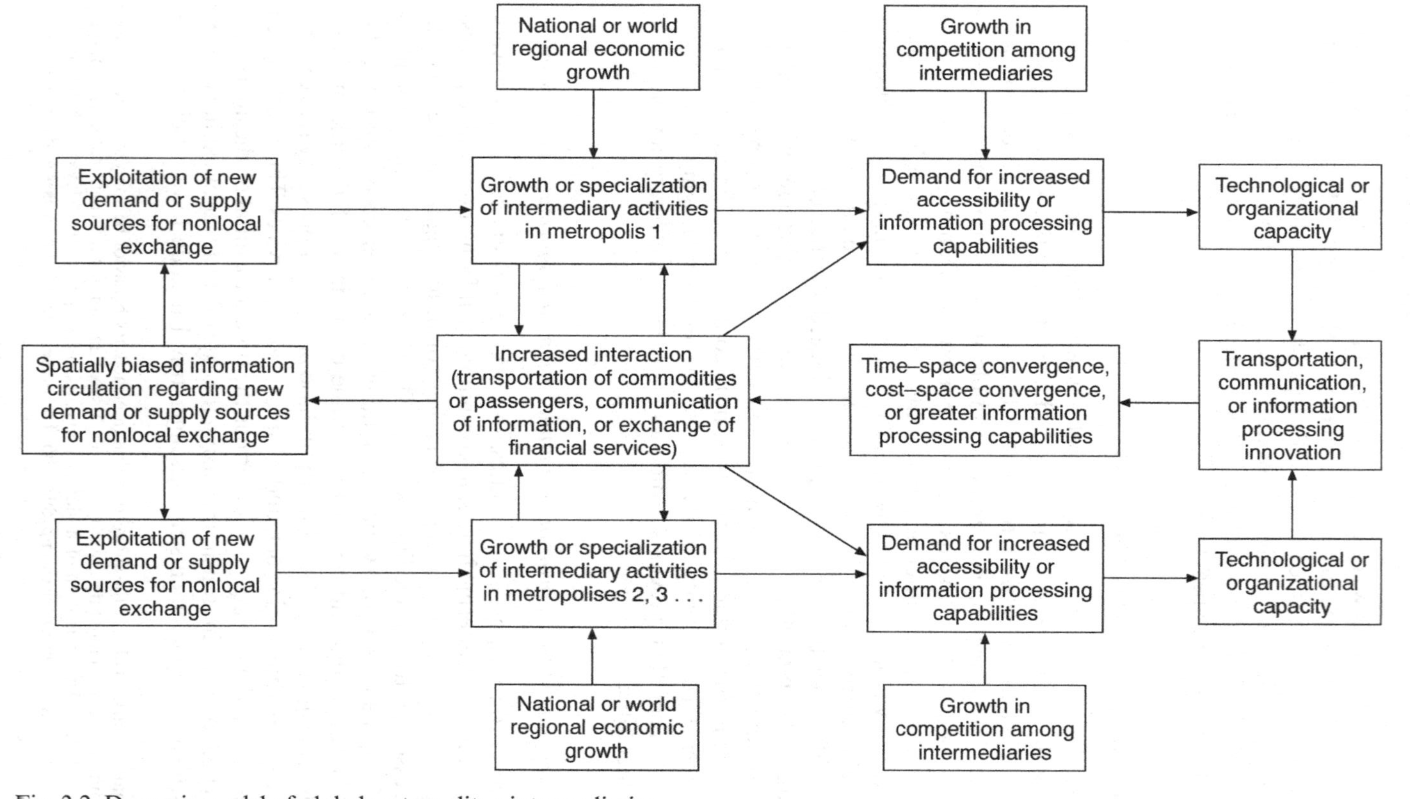

Fig. 2.2. Dynamic model of global metropolitan intermediaries.
Source: author.

and greater frequency and speed of service that reduce time-costs. Advances in communication offer benefits such as transmitting greater volumes of information at faster speeds and at higher quality, and progress in information processing permits intermediaries to improve their capacity to digest and manipulate information for controlling exchange. A heuristic dynamic model specifies the relations between intermediary activity and improvements in transportation, communication, and information processing (fig. 2.2).[20] The model oversimplifies relations among components and only weakly specifies causality; most relations are endogenous. It identifies a circular and cumulative growth process; yet retrogression occurs if some intermediaries fail to compete successfully and the national economy declines relatively or absolutely.

National economic growth, whether in a home nation or elsewhere, generates larger supplies of commodities and financial capital and provides opportunities for intermediaries in a global metropolis to expand their business of exchange. To maintain control of this exchange in competition with local intermediaries or those in other metropolises, intermediaries may raise their degree of specialization, a decision made possible by the larger supplies available for exchange. Demands for better accessibility or for improved information processing capabilities, nevertheless, may not translate directly into a transportation, communication, or information processing innovation. Those innovations depend on supplies of technological factors, such as skilled workers and innovative capital goods firms, and on organizational support, such as cooperative arrangements among intermediary firms and governmental subsidies for infrastructure.[21] Successful introduction of an innovation changes intermediary operations. The faster speed and lower cost of international transportation and communication results in time–space and cost–space convergence (fig. 2.2). The convergence concept, an adaptation of the velocity concept in physics, highlights the principle that intermediaries who control wider territorial exchange with distant places gain larger benefits over time proportional to the greater distance over which they transact exchange than intermediaries who exchange over shorter distances.[22] As faster transportation modes reduce travel time, for example, a destination four times as far as an alternative destination from an origin "converges" on the origin in minutes per year at four times the rate of the closer point. The cost–space convergence principle operates similarly: reductions in the variable costs of transportation and communication confer greater relative benefits on long- than on short-distance exchange. Because transportation and communication carriers

[20] The model blends the insights and diagrammatic models of Janelle, "Spatial reorganization"; Pred, *City-systems in advanced economies*, pp. 173–82.

[21] Lee and Schmidt-Marwede, "Interurban competition?" pp. 501–02.

[22] Janelle, "Central place development in a time–space framework"; Janelle, "Spatial reorganization."

have high fixed infrastructure costs, such as terminals, switching nodes, and lines, they concentrate price reductions per unit distance on longer-distance traffic to stimulate the use of their networks.[23] Innovations that improve information processing capabilities, such as methods of financial analysis, computers, and software, also selectively favor international intermediaries engaged in the widest territorial exchange of capital. They require the largest amounts of information and spread the high fixed costs of information processing innovations over greater volumes, thus lowering the per-unit cost.

Reduced time and cost of interaction across a territory and enhanced information processing capabilities stimulate increased interaction through commodity, passenger, and communication flows and exchanges of financial services, and this translates into heightened interdependence among intermediaries (fig. 2.2).[24] The expansion of intermediary activity in one global metropolis transmits non-local multipliers to intermediaries in another metropolis through backward or forward commodity and financial linkages; these call forth increased intermediary activity. A backward linkage, for example, results when commodity brokers in one metropolis place orders with brokers in another metropolis, and a forward linkage happens if a pension fund in one metropolis forwards money to a merchant bank in another metropolis for investment. These non-local multipliers have feedback effects if the growth of intermediary activity in the metropolis receiving the first round of non-local multipliers, in turn, generates non-local multipliers back in the original metropolis. For example, the merchant bank in the second metropolis might invest capital through an investment bank in the first metropolis. These non-local multipliers need not follow reciprocal paths; complex interdependencies emerge as growth and specialization of intermediary activity unfold in metropolises.

These interdependencies reinforce spatial biases in the circulation of information about demand and supply of capital for non-local exchange (fig. 2.2).[25] Information flows through existing intraorganizational (head office to branches or among branches) and interorganizational linkages among intermediaries. Because those with the greatest control over the exchange of capital generate the most demand for improvements in transportation, communication, and information processing, suppliers of these services eagerly respond; these improvements, therefore, continually favor the most important intermediaries. They are more likely to receive information about sources of new demand and supply faster and with greater redundancy (reinforcing its validity) than intermediaries in other metropolises who participate far less in the global exchange of capital. This permits highly interdependent intermediaries

[23] Taaffe and Gauthier, *Geography of transportation*, pp. 36–45.
[24] For insights on interdependence, see Pred, *City-systems in advanced economies*, pp. 176–82.
[25] This discussion of spatial biases in information circulation rests on Pred, *City-systems in advanced economies*.

to exploit this information earlier and with greater probability of success; thus, they capture control of the exchange, and that, in turn, enhances the growth and specialization of intermediary activity in leading global metropolises.

These improvements in transportation, communication, and information processing impact the non-routine and routine approaches to decision-making about the exchange of capital.[26] Intermediaries with greater control over this exchange have enhanced capacity to centralize non-routine decision-making. Transportation improvements allow them to travel longer distances to meet face to face to negotiate and share complex information; communication innovations allow them to collect and disseminate larger volumes of routine information over wider territories; and information processing innovations enhance their capacity to analyze this information and incorporate it into decisions about the control of the exchange of capital. These same improvements may transform non-routine into routine decision-making and boost the capacity of less-sophisticated intermediaries to collect, manipulate, and disseminate information that previously only the most-sophisticated intermediaries could exploit to control the exchange of capital.

Lower risks in control of exchange

Intermediaries also aim to alter transaction costs by reducing risks in controlling exchange; they construct social networks of capital to achieve that goal, but geopolitical conditions shape those networks.[27] Business ties based on religion, ethnicity, or family provide a means to lower risk in the absence of an empire or of a hegemonic state that exerts indirect control over a multistate territory. Competitive advantages built on religion, ethnicity, or family decline for intermediary firms if international political stability spreads, even though interrupted by war, and if transportation and communication improvements reduce transaction costs. Heightened need for capital beyond what a tightly knit group can raise undermines the advantages of these firms; to compete, they widen their ties.

Internalizing intermediary activities within firms

When firms internalize intermediary activity rather than rely on external firms, they may lower transaction costs. This internalization occurred as firms shifted their organizational structures from single-product, single-function to multinational between the mid nineteenth and early twentieth centuries, and contemporary firms replicate that process as they grow from small to large

[26] This discussion of non-routine and routine decision-making builds on Tornqvist, *Contact systems and regional development*.

[27] All social networks of capital involve risk reduction; this discussion focuses on forms affected by geopolitical conditions.

size; to simplify, only firms that produce goods are considered. Small firms that make one product typically devise a simple organizational structure focused on operations; the president (and owner) administers the firm with a few clerks, and most workers engage in production.[28] Limited time and financial resources of the president restrict the firm's access to information about non-local sources of inputs and markets for outputs. Its scale of production remains too small to gain discounts from sellers or transportation agents; therefore, the firm relies on wholesalers to purchase inputs and distribute outputs.

If the scale of production grows, the firm may internalize purchasing and/or sales, and improvements in transportation and communication encourage that reorganization because they permit a firm to replicate wholesale services at a low cost and thus maintain better resource management. Complex products such as sophisticated machinery require a trained sales force to sell, install, and maintain equipment, and timely control of the distribution of perishable goods may require continuous monitoring by one organizational unit.[29] Internalization of purchasing and sales typically coincides with a restructuring of the firm into departments because the president cannot manage all these functions. Greater scale of production and the location of several plants at different sites also contribute to an organizational form that includes departments of finance (intermediary activity) and manufacturing, as well as purchasing and sales. If the firm has multiple plant sites, the manufacturing department controls the allocation of commodities among plants, also a form of intermediary activity. The corporate headquarters, consisting of the president and the department heads, acquires the intermediary role within the firm and with other firms; it sets policies and controls the exchange of information and allocation of capital among departments.[30] In a similar way to intermediary firms such as banks and wholesalers, the corporate headquarters must access specialized information about supply sources, markets, and financial capital.

The multiproduct, multifunctional organization, also known as the multidivisional firm, adds another intermediary hierarchical level, and the multinational extends this structure to the global scale by adding an international division; the largest multinationals may organize the firm along product lines or world regions. The corporate headquarters of the multinational appraises, coordinates, and determines long-term goals and policies for the enterprise. Senior corporate officials, along with the head of the financial unit, operate a

[28] Chandler and Redlich, "Recent developments in American business administration and their conceptualization."

[29] Porter and Livesay, *Merchants and manufacturers*. Transaction-cost economics offers a more detailed rationale for the internalization of intermediary activity; see Williamson, *Markets and hierarchies*; Williamson, *The economic institutions of capitalism*.

[30] Chandler and Redlich, "Recent developments in American business administration and their conceptualization."

miniature capital market, allocating capital to divisions, and they maintain regular contact with leading participants in global financial intermediation such as commercial and investment banks. The corporate headquarters invests surplus funds, acquires capital for allocation among divisions, and buys and sells divisions. Beneath this level, the divisional headquarters includes intermediary activities similar to those of the single-product, multifunctional organization.[31]

Multinational corporations base their headquarters in global metropolises to optimally access financial and service firms that assist the strategic planning and management of the corporation and to utilize air passenger and telecommunication services for their global control operations. They cluster their corporate headquarters in the leading global metropolises (New York, London, and Tokyo), whereas national metropolises house a few multinational corporate headquarters of indigenous firms. Because highly developed nations spawned most multinationals, their corporate headquarters remain concentrated there. These firms, especially in manufacturing and resource extraction (mining and oil), establish world-regional or divisional headquarters closer to operations to maintain better control over them. They place these headquarters in major metropolises to access finance, service firms, and air passenger and telecommunication services. If multinationals establish major facilities in a nation, they may institute a hierarchical structure of control with high-level administrative units at pivotal metropolises to access other intermediaries and routine management of operations at mines or factories. Their upper administrative level within the nation reports either to a world-regional headquarters or directly to corporate headquarters.[32]

Agglomeration

Intermediaries face powerful incentives to agglomerate to reduce transaction costs. They use face-to-face contact to coordinate, to negotiate, to share complex, specialized, and confidential information, and to build trust among themselves and with sophisticated service firms. Intermediaries need not avoid competitors to control international capital flows successfully, and they may cooperate on some exchanges of capital to share risk. Access to information and capital and acquiring skill in using those assets influence their degree of success; a competitor in the same agglomeration does not affect that negatively. Instead, intermediaries in other agglomerations may pose competitive

[31] Chandler, *The visible hand*, pp. 455–83; Chandler and Redlich, "Recent developments in American business administration and their conceptualization"; Lorsch and Allen, *Managing diversity and interdependence*; Stopford and Wells, *Managing the multinational enterprise*; Williamson, "The modern corporation," pp. 1555–56.

[32] Cohen, "The new international division of labor, multinational corporations and urban hierarchy"; Hymer, "The multinational corporation and the law of uneven development"; Sassen, *The global city*, pp. 168–91.

threats if they have better access to information and capital to control exchange. Critical information for exchange comes embedded in local and non-local social networks of capital rooted in intermediary exchange partners; these may exist as friends, families, and religious and ethnic groups. Social networks often interweave and support cooperative, as well as competitive, ventures. The "hub and spoke" structure and high fixed costs of telecommunication and passenger infrastructure reinforce the agglomeration of intermediaries. They locate at these hubs to optimize access to telecommunication services and to use passenger services for distant face-to-face communication. Once an agglomeration of intermediaries becomes a better operational base than alternative sites, in-migration of intermediary firms reinforces its attractiveness.[33]

As intermediary firms accelerate international expansion, their confrontations with different laws about finance, taxes, and trade generate sizable demands for specialized advisory services, such as law, accounting, and management consulting, to handle complex transactions. These producer services need proximity because they communicate face to face among themselves and with clients in financial firms, trading houses, and corporate and divisional headquarters of non-financial firms, and international firms among these producer services build a global branch network to meet their clients' demands. Leading world cities, such as New York, London, and Tokyo, develop the greatest concentrations of producer services, and significant clusters appear in other world cities. These agglomerations attract offices of major controllers of capital exchange, such as commercial and investment banks and upper-level corporate administrative offices.[34]

Intermediaries differentiate or dedifferentiate to control markets

Intermediaries may also employ differentiation or dedifferentiation of their organizational structure as a competitive weapon. They dedifferentiate, that is, fuse functions within the organization that previously were housed in distinct organizations, as a strategy to offer clients diverse intermediary services that generate synergy from the package, to spread risk across more activities, or to reduce transaction costs.[35] If the intermediary firm retains specialization in the newly internalized activity at the same level as existed in the outside firm and does not shift the operational base, then dedifferentiation does not fun-

[33] Meyer, "A dynamic model of the integration of frontier urban places into the United States system of cities"; Pred, *City-systems in advanced economies*; Sassen, *The global city*, pp. 126–67; Thrift, "The fixers."

[34] Cohen, "The new international division of labor, multinational corporations and urban hierarchy"; Daniels, *Service industries in the world economy*; Daniels and Moulaert, *The changing geography of advanced producer services*; Sassen, *The global city*, pp. 90–191.

[35] Rueschemeyer, *Power and the division of labor*, pp. 141–69; Tilly, *Big structures, large processes, huge comparisons*, pp. 1–13, 43–53.

damentally alter the activities of intermediaries; transactions are within the firm rather than among firms. Specialization, either dropping a less-specialized role for a more-specialized one or adding the latter while retaining the former, is a competitive weapon. Growth of international exchange that stimulates national and global economic growth and development encourages the intermediary to take advantage of superior access to information and of greater capital and to shift to higher levels of specialization (fig. 2.2). This requires the exchange of a larger volume of commodity and financial capital in a particular line to compensate for the elimination of other lines. Commodity trading illustrates the dynamics of specialization in reaction to competition; the same rationale fits other forms of intermediation, such as commercial and investment banking.[36]

Economic growth and development in one or more nations encourages growing numbers of unspecialized wholesale firms that trade a basket of goods, such as grain, textiles, hardware, and metals, to emerge in various metropolises to control this international exchange. Firms compete, but growth in total business also boosts opportunities for firms with the greatest capital and access to information to specialize in selected items, such as grain or textiles. Specialization becomes attractive if the wholesaler can achieve economies of scale in the control of international exchange; those economies rest on large-volume purchases and sales on each transaction. The focus on fewer lines of goods permits the wholesaler to concentrate fixed costs of information access for each good on a larger volume, thus reducing the unit cost of information. The wholesaler also may invest capital in more sophisticated information gathering and processing and in improving the quality of information.

Because the specialized wholesale firm engages in large-scale wholesaling of a commodity, it must exchange over a wide territory. If the original, less-specialized wholesalers also control exchange over a large territory to gain enough business to make normal profits, the specialized wholesaler will pose a competitive threat in that larger territory because it will gain economies of scale that enable it to undercut less-specialized wholesalers. If those firms have sufficient capital and access to information, they also can react to this competition by specializing; alternatively, they can retreat from global competition and focus less-specialized wholesaling on a region of the globe. That possibility arises if economic growth and development boosts market size sufficiently that additional less-specialized wholesalers can operate in the region; otherwise, some terminate business. Because specialized wholesalers offer the best prices on the buy and sell sides, less-specialized ones have an incentive to

[36] This originally drew on the wholesaling case formulated by Vance, *The merchant's world.* For greater specification of the argument and an extension to all forms of non-local exchange, see Meyer, "A dynamic model of the integration of frontier urban places into the United States system of cities." A simpler version of the argument for the international scale is in Meyer, "Change in the world system of metropolises," pp. 403–04.

obtain commodities from specialized wholesalers, thus creating a hierarchical structure of wholesaling. Less-specialized wholesalers collect goods within regions and funnel them to specialized wholesalers; in turn, they transfer commodities globally to those less specialized. Specialized wholesalers bypass lower levels of the hierarchy whenever they find buyers or sellers for large-volume trades, such as a government or another specialized wholesaler. Similar rationales extend both to increased specialization in wholesaling, such as dealing only in individual metals (copper, iron), and to other intermediary activities, such as banking.

Intermediaries appeal to force

Because international intermediaries exchange capital across political boundaries, they confront the need or opportunity to "appeal to force." If foreign competitors utilize the power of their nation-state to disrupt the capital exchange of others, intermediaries might convince their nation-state to retaliate with sanctions against competitors outside the boundaries of the home state. By forcing competitors to pay higher protection costs, intermediaries gain protection rent, the profit or difference between the protection costs intermediaries face and those of their competitors.[37] Intermediaries also may take a benign approach and convince their political leaders to negotiate with leaders of other nation-states, such as over taxes or tariffs, to mute hostile interference in international exchange. Intermediaries confront a dilemma if they request that their government enforce rules of behavior for foreign competitors on domestic soil. If they lobby their government to reduce barriers to local operations of foreign intermediaries, competition for domestic and international business intensifies. Alternatively, if domestic intermediaries lobby their government to restrict the local operations of foreign intermediaries, the domestic intermediaries will lessen their possibilities to specialize, face reduced capital liquidity, and hinder their access to information about the exchange of capital because critical information comes from other international intermediaries. Although the nation-state is an ally of intermediaries in battles with foreign ones, coercive means may gain only temporary advantages and disrupt international exchange; that reduces opportunities to control exchange. When international intermediaries accumulate large amounts of capital, they become vulnerable to coercion from their nation-state to support foreign policy goals.[38] Opportunity to profit from supporting those efforts, such as supplying capital or controlling the exchange of military goods, might turn into a loss if the nation's efforts fail; thus, benign forms of "appeal to force," offer attractions. These include direct subsidies for international

[37] Coleman, *Foundations of social theory*, pp. 260–63; Lane, *Profits from power*, pp. 12–13.
[38] Tilly, *Coercion, capital, and European states, AD 990–1990*.

exchange, such as payments to export commodities produced in the home nation, or indirect subsidies from global municipalities and the nation to supply economic data and to build and maintain infrastructure, such as ports, airports, and telecommunications.

Dynamics of global metropolises

This theory implies a sequence of changes in intermediaries and their global metropolitan bases (fig. 2.3), but it abstracts from other economic, cultural, and political activities in those cities; the synthesis, nevertheless, offers guide-posts to interpret changes in global metropolises.[39] An individual or organization identifies an opportunity to operate as an unspecialized intermediary in the international exchange of commodity or financial capital. That opportunity can have diverse roots, such as colonial expansion or national economic growth and development. This intermediary chooses a base with optimal access to specialized information about the demand and supply of capital. That base incorporates high-quality telecommunications to collect and disseminate information and high-quality passenger service to engage in face-to-face contact with intermediaries at other locations. Subsequent intermediaries at the same level of specialization are lured to that site for similar reasons. The initial intermediary also attracts them because they can exchange information face to face and cooperate on some exchanges in order to reduce risk. This growing agglomeration demands improvements in transportation, communication, and information processing that, when implemented, boost the appeal of the metropolis as an intermediary base. Growing numbers of ancillary business services emerge, such as law and accounting, that support the exchange activities of intermediaries.

At some point, differential world regional growth and development may open opportunities for less-specialized intermediaries to locate outside the original agglomeration at sites with better access to information about buyers and sellers of capital. A nation may offer protection rent to attract these intermediaries or support indigenous ones. In the original metropolis, some intermediaries who have greater capital and better access to information respond to this competition by specializing; they retain control of exchange at the larger territorial scale. Other intermediaries who do not specialize must contract their territory of operation; some go out of business, whereas others remain less specialized if nearby territory experiences economic growth and development. Firms that specialize exchange capital with less-specialized intermediaries, forging hierarchical linkages, and sometimes they internalize these bonds within family firms or ethnic groups, especially under international political uncertainty.

[39] This synthesis draws on Meyer, "Change in the world system of metropolises," pp. 406–09.

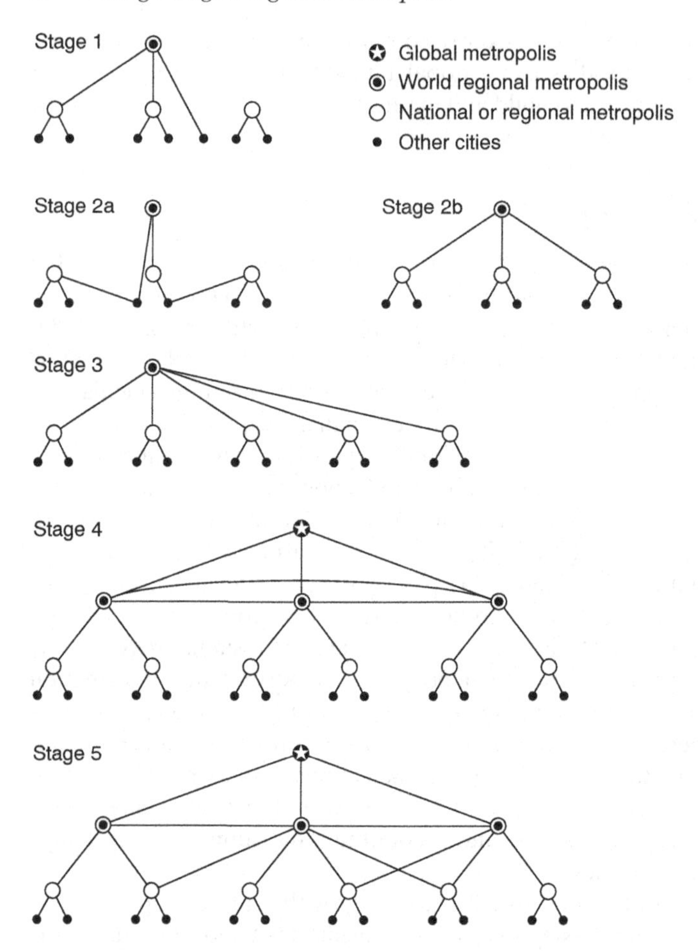

Fig. 2.3. Long-term dynamics of the world system of metropolises.
Source: Meyer, "Change in the world system of metropolises," fig. 1, p. 408.

This sequence of specialization and competition continues if the world economy grows. At each point, the most-specialized, highly capitalized intermediaries with the best access to specialized information about the global economy have optimal capacity to shift to higher levels of specialization; their global metropolitan bases stay at the top of the hierarchy of metropolises. Nations may offer unregulated business environments or try regulation to enhance competitive advantages to support this dominance of their intermediaries. The effectiveness of either strategy partly depends on the strategies of other nations; benefits, therefore, remain problematic. World regional growth and development may allow intermediaries in formerly unimportant global metropolises to rise rapidly in specialization. They may gain initial advantages from a protective nation-state or from cultural or ethnic ties that permit them to compete against foreign intermediaries that are more specialized and capitalized.

Because specialization of intermediaries corresponds to the value of capital for exchange, less-developed nations will not have specialized intermediaries. Those in global metropolises of highly developed nations have greater access to information and lower cost; they always control the exchange of capital of less-developed nations. Intermediaries in metropolises of less-developed nations, therefore, have few interlinkages, whereas those in global metropolises of highly developed nations forge a dense web of intermetropolitan linkages based on a finely honed set of specializations. These formulations provide a framework to interpret changes in Hong Kong as the meeting-place of the Chinese and foreign social networks of capital from the Canton days before 1840 to the present. These intermediaries of capital make Hong Kong the global metropolis of Asia.

3

From Canton to Hong Kong

> The Opium Trade continues to be conducted on board of the Ships at Lintin with no material interference or interruption from the Chinese Government . . . As the traffic at Lintin however is not now confined to Opium alone but is extended to transhipment of goods of every description by which means all Port charges are evaded it is probable that this illegal trade which is annually increasing must soon attract the more serious attention of the Canton Govt.[1]

Two protagonists

The transfer of control over the island of Hong Kong from China to Britain in 1842 under the Treaty of Nanking set the official start date for Hong Kong, but that date fell well along the path of the transformation of Asian trade and finance that intermediaries in Hong Kong helped shape. The late eighteenth century offers a better perspective on this path. At that point the two main protagonists, China and Britain, stood at dramatically different junctures. The Manchu conquerors, in the midst of their long reign as the Qing dynasty from 1644 to 1912, devised a sophisticated two-prong strategy to retain power as a tiny minority amidst the Chinese. First, they preserved the social and political order of imperial Confucianism and integrated their rule with Chinese culture, and they encouraged mutual dependence, backed by moral approval, leading from rural peasants through the governing elite to the emperor. Second, to preserve power as an alien dynasty, the Manchus supported scholarship to maintain the loyalty of the Chinese gentry-elite, whose numbers always exceeded opportunities for government office. The Manchus, nevertheless, still battled eruptions of sedition, and their military and bureaucratic control from Beijing down to the village remained weak. The territorial and population size of China and complex regional economies prevented tight reins in an era of

[1] Canton Committee, end of February 1827, Records of the East India Company; quoted in Morse, *The chronicles of the East India Company trading to China, 1635–1834*, vol. IV, pp. 134–35.

poor transportation and communication. The Qing court retained the allegiance of lower-level officials, by allowing them to operate with discretionary power, and of landlords and merchants, by not siphoning off their wealth. China experienced vigorous territorial, economic, and demographic expansion during the eighteenth century. Interregional trade in bulk commodities (grain, cotton) and high-value goods (tea, silk, fine porcelain, precious metals) increased, and foreign trade surged eightfold. Merchants and bankers elaborated new trading and financial instruments and institutions to lubricate the economy. These economic changes supported a doubling of the population from about 143 million in 1741 to about 295 million in 1800. As frontier expansion added populations with fragile allegiance to the Qing state and the heart of China gained massive numbers of people, the creaky administrative machinery strained to retain control on a small revenue base that remained below 5 percent of gross national product. Into this breach stepped governors of provinces and local leaders, eager to assert power.[2]

The failure of per capita income to rise much pointed to looming problems in the nineteenth century. Inequality in the distribution of farmland and exploitation of landless peasants by landlords can be dismissed as explanations for this failure because inequality levels did not differ much from other countries and owner cultivators were prominent; the problems were rooted elsewhere. Owner and tenant farmers increased the productivity of their land with greater applications of labor on smaller parcels; the per capita acreage of cultivated land fell about 25 percent to 0.5 acres over the course of the eighteenth century. Output per unit of land rose to extraordinary levels, but farmers gained smaller increments of output for additional inputs of labor. Rural wages fell, encouraging households to internalize handicraft production or to trade their tiny farm surpluses for local handicrafts made by low-wage workers. The growth of interregional and international trade during the eighteenth century mostly tracked the doubling of the population. Excluding rent payments to landlords, farmers exported no more than 4 percent of their production outside the region. This low share, spread across the vast rural population, consituted a large volume; doubling that over the century boosted trade enormously. The volume of low-value, bulk commodities, nevertheless, masked the high share that luxury items (tea, silk, fine porcelain, precious metals) for the upper class formed of the value of interregional and foreign trade. Assuming the upper class represented only 2.5 percent of the population, their numbers reached about 7.5 million by 1800. This vast market could not power economic development because an army of low-wage, highly skilled craftworkers supplied their needs; no factories of that era could compete with them. In contrast, the United States and Britain built industrial powerhouses

[2] Fairbank, *China*, pp. 143–91; Ho, *Studies on the population of China, 1368–1953*, appendix I, pp. 281–82; Moulder, *Japan, China, and the modern world economy*, pp. 50–60; Naquin and Rawski, *Chinese society in the eighteenth century*, pp. 97–137, 219–26.

on production for the mass market that totaled about the same numbers as the Chinese upper-class market.[3]

In the late eighteenth century the British invested substantial resources in constructing canals and roads that lowered the costs of communicating and of shipping bulk commodities. Reorganization of land-holdings and capital investments in land raised land and labor productivity in agriculture, and swelling urban populations whose per capita incomes climbed increased the markets for agricultural and manufactured products; investments in industrial technology lowered the prices of goods. Factories and workshops shifted production towards standardized, moderately priced goods that the majority of the population demanded. The best export markets for British goods outside Europe also consumed those commodities, reinforcing the shift of manufacturing to mass market production. Its economy sucked in surging volumes of consumer goods, such as tea, and of raw materials, such as cotton, that it could not supply, and investors, merchants, and industrialists searched for global markets to keep industrial engines running. During the last two decades of the century, industrial production doubled and foreign trade soared threefold in value. In contrast to the limited reach of China outside its borders, the political and financial elite of Britain backed a naval policy that protected far-flung trade and finance. Profits from those economic activities, directed from London, financed the government and boosted the wealth of the landed aristocracy.[4] The positions of China and Britain on opposite sides of the world symbolized their separation on most measures, including population size, economy, and political organization. They remained woefully ignorant of each other's political economy, a condition that continually haunted their intercourse; that contact solidified first at Canton.

The Canton nexus of trade

A seat at the table

By the start of the eighteenth century, rivalry over trade in Asia had crystallized as a broad division of trade empires: the British, under the East India Company, dominated India, and the Dutch, under their East India Company, controlled trade in the Indonesian archipelago. Spain commanded the Philippines, and Manila ships brought silver to Asia from Spanish America. The rest of Asian trade remained open to competitive battles, and China

[3] Chao, *Man and land in Chinese history*, table 5.1, p. 89, and pp. 221–28; Elvin, "The high-level equilibrium trap"; Ho, *Studies on the population of China, 1368–1953*, table 13, p. 56; Perkins, *Agricultural development in China, 1368–1968*, pp. 116–24, table G.1, p. 314. The United States and British populations totaled, respectively, about 5 million and 9 million around 1800; Weber, *The growth of cities in the nineteenth century*, tables 2, 18, pp. 22, 46.

[4] Cain and Hopkins, *British imperialism*, pp. 53–104; Floud and McCloskey, *The economic history of Britain since 1700*; Landes, *The Unbound Prometheus*, pp. 41–77.

loomed as one prize; yet, the Qing dynasty had little interest in foreign trade and tried to restrict it. The vast Chinese economy produced most essential goods; foreign trade accounted for a trivial share of the economy, and customs revenue contributed little to state coffers. The Beijing government let local officials and a Hoppo, or revenue commissioner, appointed by Beijing, handle most trade decisions at ports such as Ningpo, Amoy, Canton, and Macao. The British made feeble efforts to trade with China through Canton during the 1630s, and they nibbled at the China trade through Tongking (Vietnam), Taiwan, and Amoy by the 1670s. Nevertheless, "the English had now, at the opening of the eighteenth century, thrust their feet over the threshold of the China trade, but had not yet obtained a seat at the table." By the 1740s, however, from three to nine British ships called at Canton annually; because the British government had granted the East India Company a monopoly over the trade of its subjects in Asia, these ships had Company charters. Private traders, nevertheless, continually thwarted that monopoly at Canton, just as they had in India, because the Company used them as an adjunct to its trade. Swelling numbers of British and other foreign traders posed problems for the Chinese because they did not care to stimulate foreign trade; by 1760, they constructed a system that brought revenue to the government and controlled foreigners. The Chinese confined most foreign trade to Canton, located at the southern limits of the empire and far from Beijing.[5]

Conflict at Canton did not loom immediately because the East India Company dominated trade at the end of the eighteenth century. Its operations compared favorably to efficient private traders, and its London bureaucracy effectively managed the supply of tea to sell at auctions for the British market. London headquarters, the principal in the business transactions, employed sophisticated financial accounting to monitor expenditures and revenue; profits funded overall expenses, including the empire in India. The headquarters sent detailed instructions to its supercargoes, the agents on the ships who purchased tea. Time-lags to receive instructions typically reached six months, and the directors of the Company had to wait another year to learn results. This raised the specter of malfeasance because the Company's agents in Canton had more information about supplies and prices than the headquarters in London, and they might grasp opportunities for private gains at the Company's expense. It minimized this principal–agent dilemma through careful bureaucratic control and generous remuneration to its agents, supercargoes, and other staff in Canton. As long as the tea trade stayed within orderly bounds, this efficient bureaucracy could manage the growth of trade.

[5] Fairbank, *Trade and diplomacy on the China Coast*, pp. 23–53; Furber, *John Company at work*; Furber, *Rival empires of trade in the Orient, 1600–1800*; Greenberg, *British trade and the opening of China, 1800–1842*, pp. 41–48; Morse, *The chronicles of the East India Company trading to China, 1635–1834*, vol. I, pp. 1–99 (quote, p. 99), table, pp. 307–13; Sargent, *Anglo-Chinese commerce and diplomacy*, pp. 5–12.

Between 1775 and 1804, the trade of the Company with Canton surged. The number of ships calling more than tripled from the level of the 1740s to a range between fourteen and twenty-nine annually, and the size of ships jumped from 800 tons before 1796 to 1,200 tons after that date as the Company leveraged its large capital position to achieve economies of scale in shipping. After Britain reduced tea duties, annual legal imports soared quickly from 6 to 15 million pounds; they doubled to 30 million pounds by 1830. The tea trade with China became one pillar of the British political economy; the government relied on tea duties for about one-tenth of its revenue, and the Company gained its entire profit from the tea trade. By 1800, growing imports of tea and, to a lesser extent, silk, posed a dilemma; the Company had exported gold, silver, bullion, and coin to China to pay for tea because few British manufactures or other goods fetched profitable prices in China. This pressured the Company to acquire precious metals, forcing British merchants to engage in costly, far-flung trade. But acquisition of precious metals did not draw on the productive capacities of the British economy. The solution to the problem of the China trade lay in India.[6]

The East India Company held the monopoly of British trade in both India and China, but it could not dominate all facets of trade. Because the Company controlled imports of tea, it focused on buying tea in Canton, and from 1800 to 1820, the majority of the revenue to purchase it came from exports of raw cotton from India to China. The Company manipulated purchases of cotton in India through its monopoly, but it could not control sales in China because the Nanking region of the lower Yangtze River valley produced vast quantities of cotton; Canton prices reflected the success or failure of the China crop, not India prices or growing Chinese demand. As a large, bureaucratic firm with high fixed costs, the Company could not readily adjust purchases of cotton and shift to other commodities when the cotton trade remained unprofitable. China's limited demand for other goods kept those trades small, and these were precisely the type that the Company could not operate at low cost. Even if it tried to operate flexibly, the Company faced principal–agent problems; the London headquarters would have to give discretion to its agents in India and Canton, opening opportunities for malfeasance. The Company gradually licensed private traders to sell cotton on commission at Canton for merchants in India. It also monopolized the production, manufacture, and sale of opium at auctions in India, and private traders carried opium to Canton for sale. By keeping supply fixed and relatively small, "the company's revenues were secure, and the Chinese were debauched no more than was necessary." The Chinese government prohibited its merchants in Canton from

[6] Greenberg, *British trade and the opening of China, 1800–1842*, pp. 3–9; Morse, *The chronicles of the East India Company trading to China, 1635–1834*, vol. II, table, pp. 436–51; Mui and Mui, *The management of monopoly*; Pritchard, *The crucial years of early Anglo-Chinese relations, 1750–1800*, appendix V, p. 395.

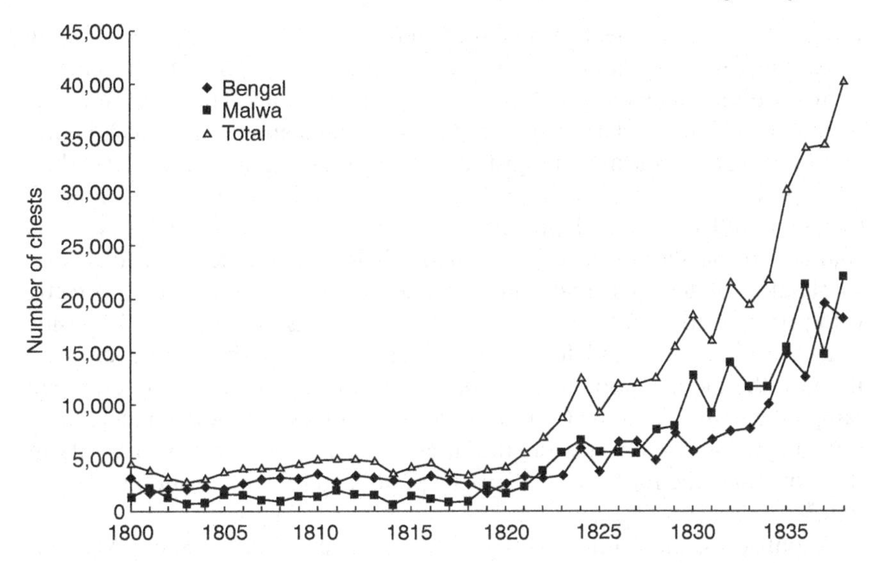

Fig. 3.1. Opium shipments to China, 1800–1838.
Source: Greenberg, *British trade and the opening of China, 1800–1842*, appendix 1D2, p. 221.

trading in opium or dealing with foreigners who traded in it. Because the Company held the lucrative monopoly to export tea from China, it stayed on safe legal grounds with the imperial government and avoided direct contact with opium in China.[7] The comfortable monopoly functioned well prior to 1820; the volume of opium sent to China stayed between 3,000 and 5,000 chests annually (fig. 3.1). This exchange of opium and other commodities in the China trade at Canton made it the meeting-place of the social networks of capital of the Chinese and foreign merchants.

Canton mercantile system

The political economy of the Canton mercantile system rested on the weak capacity of the Chinese state to exert power. The Beijing government appointed a Viceroy and Hoppo (revenue commissioner) to enforce regulations and collect fees and customs duties, while the Manchu commander provided military muscle. Local authorities exerted substantial control because the Beijing-appointed officials and the national military could not police the vast expanses of the Pearl River delta and other coastal areas of Kwangtung

[7] Greenberg, *British trade and the opening of China, 1800–1842*, pp. 79–92; Morse, *The chronicles of the East India Company trading to China, 1635–1834*, vol. I, pp. 215–16, vol. II, pp. 344–46, vol. III, pp. 328–83, vol. IV, pp. 20–195; Owen, *British opium policy in China and India*, pp. 1–48, 80 (quote).

Province. The government granted the Cohong, a weak association of Chinese merchants (hongists), a monopoly over the sale of staple goods (tea, silk); in return, Cohong merchants enforced security regulations. These intermediaries in international exchange, paradoxically, dreaded appointment as hongists and even worse, as senior hongist, and they often faced bankruptcy. Their monopoly came riddled with holes; government officials exacted "squeeze" (payoffs), and these merchants could not trade freely. Other Chinese merchants and the shopmen, who operated officially as retailers but illicitly as wholesalers of tea, silk, and opium, evaded regulations and traded directly with private British and American merchants. Hongists engaged in illegal side deals with shopmen and foreign merchants to lessen their disadvantages. Because the lucrative opium trade operated outside the law, it underpinned rampant corruption in the Canton trade. The Chinese and foreign merchants bribed government officials, and the latter exacted squeeze from merchants; at the same time, the East India Company piously stood aside from the corruption.[8]

To outsiders, the Canton mercantile system appeared chaotic, a view reinforced by bribery and the armed smuggling vessels that plied the waters of the Pearl delta. Behind the scenes, social networks of capital bound powerful mercantile firms in London, Calcutta, Bombay, Boston, and New York that underwrote a global trading enterprise which imported and exported at Canton to a value exceeding $20 million annually in each direction by 1820. East India Agency Houses headed this hierarchy of traders; as many as thirty of these firms existed at any time, but about six large firms dominated commerce. London merchants and politicians founded most firms and organized them as partnerships with a headquarters in London and branch offices in India, principally in Calcutta and Bombay; sometimes branches formally organized themselves as separate partnerships, which reduced the principal–agent dilemma by making merchants principals. They sometimes started business at Canton and moved to Calcutta or Bombay to exploit those contacts; others relocated from the Indian metropolises to Canton. They utilized these contacts upon their return to London where they founded an East India House or joined another as a partner. The leading Canton firms, Jardine, Matheson & Company and Dent & Company, had roots in these early East India Houses, and the venerable Baring Brothers, one of the greatest London firms, had family representatives in Canton. East India Houses in London were the global hub of the social network bridges to pivotal mercantile centers, and the experience of many partners at different metropolises in Asia gave them multiple bridges to local networks of merchant firms. Because partners often followed different paths among metropolises, redundant network

[8] Cheong, *Mandarins and merchants*, pp. 14–18; Greenberg, *British trade and the opening of China, 1800–1842*, pp. 41–74.

ties were minimized and this maximized information flow to the London firm and enlarged its range of business opportunities. Family members, often with common Scottish ancestry, comprised the core of many firms; this reinforced the mutual trust required to engage in long-distance trade in an era of poor communication, and it provided a means to enforce sanctions for malfeasant behavior. Merchants often participated in illegal activities, such as the opium trade; thus, they needed confidence in associates. Because no single activity reached a scale sufficient to support specialization, even the largest firms operated diverse mercantile activities, including banking, bill-brokerage, shipowning, freighting, and insurance, and most firms were commission agents.[9]

The complicated nature of the commission business, combined with great distances in Asia and between Asia and Europe in an era of poor communications, required sophisticated partners in Canton. The firms bought and sold commodities for principals in India or Britain and charged them a commission based on the price of the commodity, and they also levied commissions on other services, including currency exchange (cash, bills of exchange), warehousing, freight, and insurance; total charges could climb to 8 percent of the value of the product.[10] To boost profits, firms took on more risk and traded on their own account, but to acquire ownership of commodities required extensive capital; losses quickly eroded the capital of small firms. Large firms that invested capital in owning goods specialized in a few commodities, allowing them to focus their collection of information. They handled greater volumes of commodities and invested in larger, faster ships and bigger warehouses; these strategies cut the cost per unit to trade each commodity and increased the return on capital. This shift to occasional ownership of commodities, specialization, and a larger scale of operations, however, did not gain momentum until after the mid 1820s; prior to 1815, few firms had a permanent base in Canton. Their diverse trade was costly because firms peddled multiple commodities among numerous ports, and merchants had difficulty acquiring sufficient information to trade effectively; profits suffered as losses on some commodities wiped out gains on others. The opium trade remained small, and private firms could not handle tea, the largest Chinese export, because the East India Company monopolized it.[11]

[9] Greenberg, *British trade and the opening of China, 1800–1842*, pp. 22–40, 75–84, 144–52; Morse, *The chronicles of the East India Company trading to China, 1635–1834*, vol. III. The official firm of Jardine, Matheson & Company was founded in 1832, but the principals had entered the India–China trade in 1802 (William Jardine) and 1815 (James Matheson) and participated under various partnerships. Rather than follow those different firm names through various changes, the Jardine Matheson name is used as a label. See Keswick, *The thistle and the jade*, pp. 256–57.

[10] American firms operated similarly; see Lockwood, *Augustine Heard and Company, 1858–1862*, p. 7.

[11] Cheong, *Mandarins and merchants*, pp. 62–78; Greenberg, *British trade and the opening of China, 1800–1842*, pp. 22–40, 75–84, 144–52.

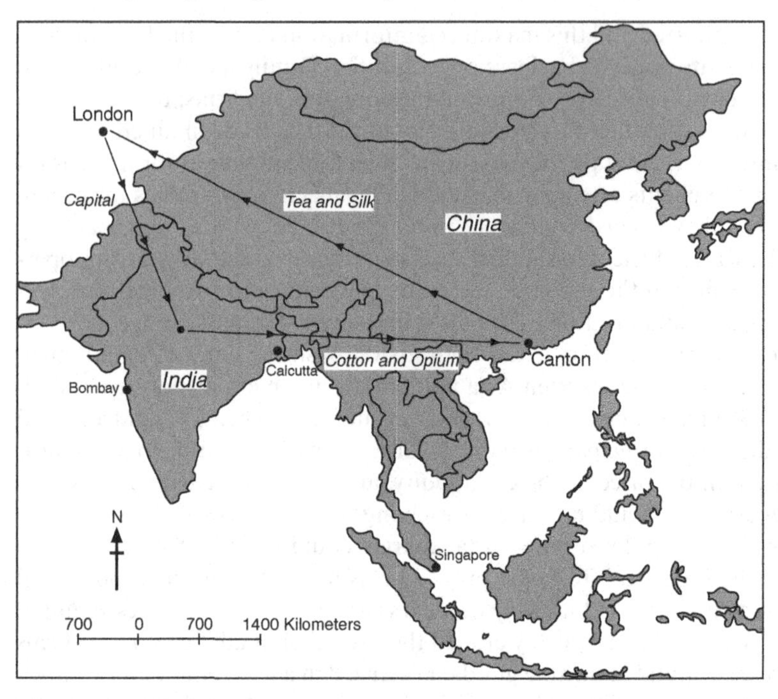

Map 1. The global triangle of trade, *circa* 1820.
Source: author.

The Canton mercantile system becomes fully explicable when viewed as part of a global triangle of trade that linked Britain (London), India (Calcutta, Bombay), and China (Canton) (map 1). The Britain–India side arose from investment of financial capital by the East India Company and East India Agency Houses in cotton, opium, and indigo in India. Cotton and opium had no markets in Britain, but they had markets in China; thus, the India–China side of the triangle was formed. Private British traders bartered cotton and opium for silk and other goods and for silver, specie, and bills of exchange. The Company purchased tea and silk for sale in Britain, translating commodities into financial capital in London and completing the triangle. American merchants from New York and Boston brought furs, silver, specie, and later, opium, to Canton; they purchased tea and silk goods for export to the United States; and they received bills of exchange on London from hongists to pay their London creditors. American merchants ranked second to the British as the largest exporters of tea at Canton, and their long-standing trade with Mexico and South America gave them greater access than the British to Spanish silver production. To protect its opium monopoly, the Company required British traders to carry only Indian opium in its territory of control; this gave Americans, led by Perkins & Company in Boston, free rein to acquire

Turkish opium and export it to Canton, and they garnered huge profits.[12] Between 1800 and 1820, therefore, Canton stood enmeshed within a global metropolitan system of cities; London headed the hierarchy. Its merchant banks and trading firms, including the East India Agency Houses, had the greatest capitalization and specialization, and they supplied credit that lubricated global commodity flows. Merchant houses in New York and Boston and branches of East India Agency Houses in Calcutta and Bombay occupied the second tier; they relied on London credit to complete their trade through Canton. Small-capitalized, unspecialized commission houses in Canton relied on credit from firms based in other metropolitan centers: British firms looked to Calcutta and Bombay, and American firms looked to New York and Boston. Direct credit links from Canton to London did not prevail at this time, except through the East India Company, because the small-scale private Canton firms did not have the capacity for direct exchanges with well-capitalized, highly specialized London firms.

Trade crescendo

The East India Agency Houses and large American firms increasingly challenged the staid monopoly of the East India Company in China. Just as the solution to the China trade problem was found in India, the challenge of private merchants to the Company also arose there. The British parliament terminated the monopoly of the Company over the India trade in 1813, partly under pressure from Lancashire textile centers. The Company had benefited from exports of Indian textiles, but private British traders had no loyalty to that trade; instead, they exported huge quantities of low-cost Lancashire textiles to India after 1815. This created pressure to boost the India–China trade because China offered the best opportunities to acquire remittances to return to England. Competition in the India–China trade intensified between 1815 and 1820 as the number of British merchant firms jumped from eighteen to thirty-two in Calcutta and from eleven to nineteen in Bombay. This spilled over to Canton where heightened competition to sell cotton and opium reduced prices and growing demand for Chinese exports of tea, silk, and silk goods raised prices, thus undermining profits. The under-capitalized hong merchants, however, could not adequately finance their side of the surging trade, and this pressure intensified when silk exports from Canton spiked upwards after Britain cut tariffs on silk imports in 1825. Because the raw cotton trade from India to China stagnated during the 1820s, the opium trade was left as the best means to fuel trade expansion.[13]

[12] Downs, "American merchants and the China opium trade, 1800–1840"; Greenberg, *British trade and the opening of China, 1800–1842*, pp. 18–85, 144–60; Hao, "Chinese teas to America – a synopsis," pp. 12–16; Wakeman, "The Canton trade and the Opium War," pp. 166–69.

[13] Greenberg, *British trade and the opening of China, 1800–1842*, pp. 16–17, 85–94, 180.

Swelling numbers of private merchants engaged in the India–China trade challenged the monopoly of the East India Company over the production, manufacture, and sale of Bengal opium after 1815. The fixed supply the Company sold and greater competition among private merchants to purchase it raised the price of Bengal opium in India. The Company gained monopoly profits, but private merchants found their profit margins crimped at Canton; they sought an alternative: Malwa opium from the west coast of India. Malwa had entered international trade before 1800, but its poor quality kept demand low. Around 1819, major British firms, such as the predecessor of Jardine, Matheson & Company, began experimenting with the sale of Malwa at Canton. Their business networks that linked Canton and Calcutta with Bombay, the trading capital of western India, gave them the capacity to exploit this source of opium. Improvements in the quality of Malwa and its lower price caused imports to China to surge after 1819 (fig. 3.1), and leading merchants, such as James Matheson, formed specialized agencies for opium trading to react to this new competitive regime. The East India Company employed political pressure to stop Malwa exports, but termination of its monopoly of Indian trade in 1813 left the Company with only weak leverage over Bombay merchants, native states, and Portuguese enclaves on the west coast. The Company drifted from its "principled" stance of restricting opium production, which generated handsome profits, towards boosting supplies of Bengal opium. Growing numbers of private traders at Canton who flouted prohibitions on opium imports undermined the authority of the Chinese government; in 1821 it pressured local officials to issue stronger edicts against the opium trade. In response to greater enforcement, private merchants shifted operations from Whampoa, near Canton, about 100 kilometers south to the "outer anchorages" near Hong Kong island. They established floating warehouses to exchange opium for silver and to exchange other import and export commodities; the weak Chinese navy could not threaten heavily armed merchant vessels and smugglers. British and American trade had largely separated from the Canton mercantile system: the Chinese government lost import duties, opportunities for local officials to "squeeze" traders declined, and the Company lost much of its control over private traders.[14] Leading merchants still exchanged information and made deals at their offices in Canton, but most smaller traders did not have offices there. The efficiency of the Canton mercantile system as an information exchange had declined; this cumbersome system could not survive in an era of enlarged global trade.

The East India Company faced the worst fears of a monopolist, a competitor with lower prices and a growing supply. Swelling shipments of Malwa opium after 1819 propelled the number of chests sold in China to a level that

[14] Greenberg, *British trade and the opening of China, 1800–1842*, pp. 33–50, 118–31; Owen, *British opium policy in China and India*, pp. 80–115.

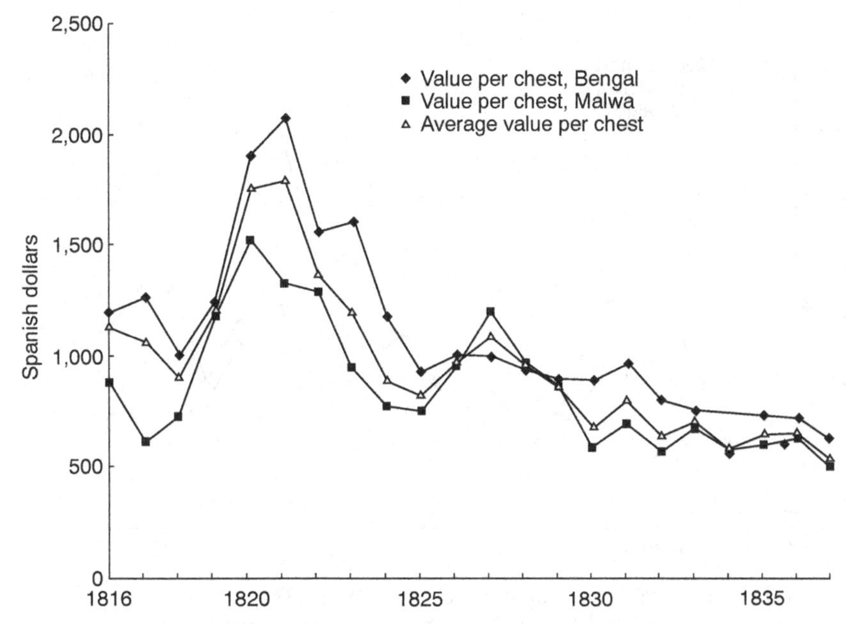

Fig. 3.2. Value per chest of opium imports at Canton, 1816–1837.
Source: Chang, *Commissioner Lin and the Opium War*, appendix B, p. 223.

consistently surpassed the Company's Bengal opium by 1828 (fig. 3.1). Bengal's price plunged 50 percent from 1821 to 1830, but this only kept pace with a similar fall in Malwa's price (fig. 3.2). Total sales value of Bengal in 1830 had merely returned to the level it reached in 1821, despite a doubling of production, whereas the rising tide of Malwa brought to China by 1830 boosted its total sales value significantly above the level of the early 1820s (figs. 3.1 and 3.3). The opium trade had entered a new phase; the drug flooded into China, exploding fourfold in volume, and its payments jumped 50 percent. Dramatic outward shifts in the supply curve of opium caused prices to decline, but total revenue rose, implying that the Chinese increased their consumption (figs. 3.1–3.3). The Company faced a dilemma: it had to react to competition, but it could not control the sale of opium at Canton because private merchants dominated that trade and if the Company tried to sell opium, China would revoke its monopoly of trade there.

Balance of trade in 1828

The triangular trade that had underpinned the Canton mercantile system since 1800 remained vividly outlined in the balance of trade in 1828, but its structure had shifted ominously (table 3.1). Direct imports of commodities

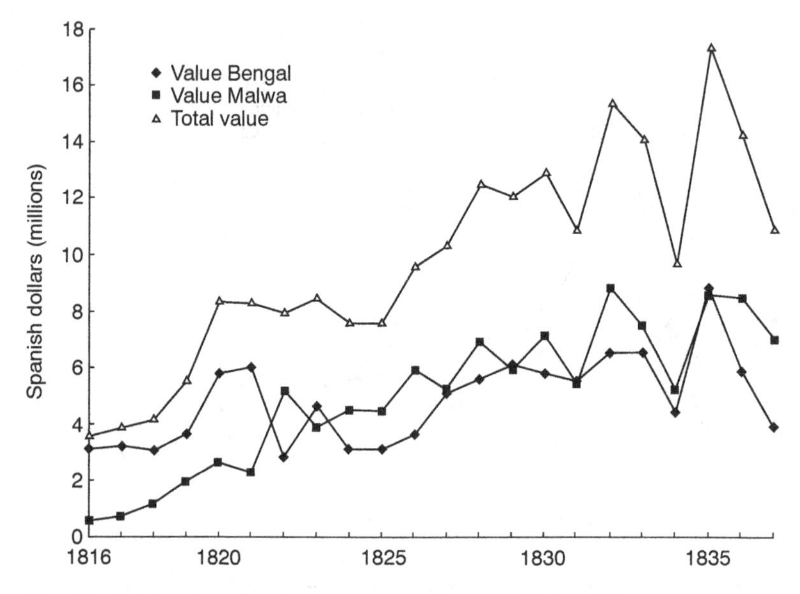

Fig. 3.3. Total value of opium imports at Canton, 1816–1837.
Source: Chang, *Commissioner Lin and the Opium War*, appendix B, p. 223.

from Europe and America continued to form a small share (19 percent) of total imports because those regions produced little that China demanded. The value of British woolens brought to Canton in 1828 stood in the same range as a decade earlier; private British traders had no incentive to bypass the monopoly of the East India Company because woolens never generated profits. Textile imports, nevertheless, had shifted subtly; British and American firms brought cotton goods to Canton, whereas a decade earlier none had arrived, indicating that cotton textile manufacturers from Lancashire, England, had commenced their intrusion. The structure of intra-Asian trade at Canton had been transformed as Asia's share of total imports of commodities rose from an annual average of 69 percent around 1818 to 80 percent in 1828. The India–China trade shifted decisively to opium, the value of which was double that of raw cotton, whereas a decade earlier raw cotton's value had been triple opium's value. At 46 percent of total imports, the opium trade fueled the business of British and American merchants who accounted for 80 percent of all imports, but the Company could not react to their competitive threat without incurring the wrath of Chinese officials. Tea exports remained the reliable, dominant trade, and the Company captured the bulk of it on the British side based on their monopoly, while Americans ranked second, unfettered by legal restraints. Exports of Chinese textiles (raw silk, silk goods, nankeens) had shifted; private British traders exported textiles to the value of $3.6 million by 1828, more than double the amount a decade earlier, whereas the

Company had virtually abandoned that trade.[15] As a specialized monopoly firm with high fixed costs, the Company could not compete against unspecialized private trading firms with lower fixed costs and wider social networks of capital that bridged to mercantile centers in Asia, Europe, and North America. Firms required those networks in order to reach fragmented, small-scale markets for Chinese textiles.

This restructured balance of trade at Canton had another critical component. From 1801 to 1825, China had a total net inflow of $74.7 million of silver, but in 1828, net outflow reached $4.0 million, a sensational turnaround; silver accounted for 20 percent of the value of total exports (table 3.1). The backing for China's currency declined as silver exports increasingly paid for opium imports, and private merchants handled this exchange of opium for silver outside the Canton mercantile system, signifying deeper erosion of the East India Company's monopoly of the China trade. Leading American merchants, chiefly from Boston, contributed to that erosion during the 1820s. Perkins & Company was a hub with social network bridges to prominent individuals and firms bound by kinship, friendship, business, and politics that spanned the globe from its Boston base, including representatives in Canton and in London; the latter was a partner of Baring Brothers, the premier British merchant firm. During the first four decades of the nineteenth century, members of the Perkins extended family and associates of the firm spawned other prominent American firms in the China trade. They had a superb information network for trade, access to immense capital resources, and experienced merchants. At Canton, they imported huge amounts of specie, including silver, that totaled $37 million from 1820 to 1830 and became major suppliers of bills of exchange drawn on English merchant houses. The entry of American bills, with Baring Brothers as chief guarantor, allied powerful English mercantile firms with East India Agency Houses and premier American merchant firms as threats to the Company's monopoly over the China trade.[16] Hub firms in each group forged non-redundant network bridges to other groups, thus generating wide-ranging information flows among firms and numerous opportunities to cooperate in competition with the Company. Its prominence at Canton had declined considerably by 1828, and it accounted for only 19 percent of imports and 34 percent of exports (table 3.1). External, highly capitalized and specialized financial intermediaries in London, and secondarily in

[15] Greenberg, *British trade and the opening of China, 1800–1842*, pp. 7–32; Morse, *The chronicles of the East India Company trading to China, 1635–1834*, vol. III, pp. 328–66, vol. IV, pp. 196–370.

[16] Downs, "American merchants and the China opium trade, 1800–1840"; Greenberg, *British trade and the opening of China, 1800–1842*, pp. 67–165; Hao, *The commercial revolution in nineteenth-century China*, p. 122; Johnson and Supple, *Boston capitalists and Western railroads*, pp. 21–24; Morse, *The chronicles of the East India Company trading to China, 1635–1834*, vols. III, IV; Pitkin, *A statistical view of the commerce of the United States of America*, ch. 6, table 18, p. 303.

Table 3.1. *Balance of trade at Canton, 1828 ($ '000s)*

	British				
	East India Company	Private	Total	American	Grand total
			Imports		
European and American goods					
Woolens	2,702	103	2,805	324	3,129
Cotton goods	70	185	255	174	429
Furs	0	0	0	269	269
Metals	241	18	259	644	903
Other	0	0	0	147	147
Subtotal	3,013	306	3,319	1,558	4,877
Asian goods					
Raw cotton	1,837	3,767	5,604	0	5,604
Opium	0	10,909	10,909	817	11,726
Sandalwood	92	198	290	127	417
Tin	0	115	115	13	128
Other	0	1,078	1,078	126	1,204
Subtotal	1,929	16,067	17,996	1,083	19,079
Total goods	4,942	16,373	21,315	2,641	23,956
Silver	0	0	0	732	732
Bills of exchange	0	0	0	657	657
Total imports	4,942	16,373	21,315	4,030	25,345
			Exports		
Goods					
Tea	7,670	871	8,541	2,777	11,318
Raw silk	0	2,529	2,529	144	2,673
Silk goods	0	461	461	1,053	1,514
Nankeens	3	649	652	325	977
Other	3	1,745	1,748	252	2,000
Total goods	7,676	6,255	13,931	4,551	18,482
Silver	0	4,703	4,703	0	4,703
Port expenses	432	295	727	98	825
Total exports and expenses	8,108	11,253	19,361	4,649	24,010

Source: Morse, *The chronicles of the East India Company trading to China, 1635–1834*, vol. IV, pp. 181–82.

Boston and New York, therefore, underwrote enhanced liquidity and specialization of credit markets at Canton. Rather than the Canton agglomeration acting as a competitor to London intermediaries, it operated as an extension of them.

When Thomas Raffles, an employee of the East India Company, maneuvered the establishment of Singapore in 1819 as a "free port" (free of customs and local trade charges) to attract trade with the East Indies archipelago and directly challenge Dutch dominance of the region, he unwittingly provided another opening for private British traders to challenge the Company's monopoly over the China trade. After the Dutch officially agreed to its status in 1825, Singapore flourished; private British and Chinese merchant firms flocked to the port to establish branch houses, instantly creating an intermediary agglomeration of unspecialized firms. British firms, including Jardine, Matheson & Company, used Singapore as a transshipment point for trade between Britain and China. Ships paid a commission to a merchant house and received new bills of lading, thus circumventing the prohibition on direct trade by British firms between Britain and China. This trade created a politically powerful constituency of British firms with headquarters or patrons in London who cooperated to repel attempts of the Company to change the status of Singapore.[17] These battles undergirded its emergence as an intermediary agglomeration, not immediate trade prospects within the Indonesian archipelago.

The opium trade ratchets higher

The challenge of private merchants to the East India Company intensified when exports of Malwa opium from the west coast of India to China turned sharply upwards again following 1827, but shipments of the Company's Bengal opium stagnated from 1826 to 1830 and prices slipped (figs. 3.1–3.2). Bengal administrators, under prodding from Company directors in London, momentously decided to abandon moral positions against producing too much opium for the China market and to plunge ahead with a plan to enlarge output (fig. 3.1). The Company stopped hindering exports of Malwa and opened the port of Bombay to exports by payment of a license fee.[18] The Company's decision to boost supplies of Bengal opium as its reaction to the competitive threat of Malwa forced private opium merchants to restructure and achieve economies of scale to cope with the inevitable fall in price. That concentrated the opium trade in fewer, larger firms with global-spanning network bridges that bound China, India, and Britain, and within Britain their ties reached to the heart of the industrial, financial, and political elite.

[17] Greenberg, *British trade and the opening of China, 1800–1842*, pp. 87–99; SarDesai, *British trade and expansion in Southeast Asia, 1830–1914*, pp. 32–40.

[18] Owen, *British opium policy in China and India*, pp. 98–109.

They would pose formidable opposition to the monopoly of the Company in China.

Developments in Britain gave the final blow to this monopoly. Starting in the mid 1820s, Lancashire cotton textiles offered competition to Chinese cotton goods. By 1831, swelling sales encouraged Jardine, Matheson & Company to establish a department in the firm to handle cotton goods, and they appointed a special acquisition agent in the city of Manchester. Its textile leaders grasped the lead to terminate the East India Company's monopoly over the China trade on its renewal date in 1834. The campaign intensified by 1829, and the textile industry joined forces with private merchants to form a juggernaut that made the result preordained: the textile industry could count on representatives from numerous towns and cities with factories and commerce that depended on the industry, and the greatest mercantile houses in London and Liverpool backed the Canton traders. These allies had broader goals than simply to terminate the monopoly in 1834; they aimed to completely reorganize trade with China, but their successful battle raised a new political element. Merchant firms in London and Canton henceforth blatantly appealed to force to defend and expand their trade; starting in 1834, the British government put its full weight behind the China trade. Prescient opium traders recognized that larger shipments and the expected termination of the Company's monopoly would tempt merchants to enter the China trade. Jardine Matheson, which accounted for one-third of the opium imports at Canton in 1829, led the reaction to looming competition. Between 1831 and 1833, it implemented a strategic plan and deployed extensive capital resources to build specialized opium ships and achieve economies of scale. The firm invested in a fleet of clipper ships that could complete up to three roundtrips a season between Calcutta and Canton (Lintin Island) and in a fleet of brigs and schooners to redistribute opium along the China Coast.[19] It could pay higher prices for opium in India, transport the high-value commodity rapidly to Canton in technologically advanced ships, and sell opium at lower prices than competitors through an efficient distribution system. The fast ships, large fleet, and wide market area also gave the firm better access to information about prices and markets, and thus an even greater edge against competitors.

The decision of the East India Company to engage in unbridled opium production around 1830, coupled with the strategic plan of Jardine Matheson, suggest that opium traders envisioned outward shifts in the supply curve to capture greater sales, and they guessed correctly: opium shipments to China soared (fig. 3.1). But their expectations about outward shifts in the demand curve failed to materialize. Prices of opium imports fell slightly, yet total revenue did not rise much after 1830 even as its volatility increased (figs. 3.2–3.3); this discrepancy implies that traders kept opium off the market to

[19] Greenberg, *British trade and the opening of China, 1800–1842*, pp. 101–95.

sustain prices. Opium traders faced a slippery slope because they could not continue to import large supplies of opium without bankrupting themselves; but an alternative loomed: expand distribution channels in China outside the Canton system, an approach that required the power of the British state. On the Chinese side, higher levels of opium imports at Canton during the 1830s, compared with the 1820s, translated into a torrent of silver flowing from China. Between 1827 and 1834, the net annual drain reached over $6 million four times, and from 1827 to 1849 that level became the annual average. The total exodus of $134 million of silver during that period represented a 40 percent reduction in the money supply, and that collapse in the backing of the currency, in the absence of governmental currency reforms, translated into deflation. From 1820 to 1850, the price level fell almost 10 percent; China had slipped into prolonged economic difficulty.[20] The afflictions of China, however, did not reduce to a simple equation of opium imports, silver exports, deflation, and economic destitution.

Protagonists on a collision course

Down the spiral inside China

By 1800, population pressure, a decline in the marginal productivity of rural labor, and emergent problems of political control eroded revenue that the Qing state required to cope with challenges. Yet, the educational system produced a swelling supply of candidates with little hope of gaining political offices, and government corruption rose as office-seekers offered bribes. Administrative stress penetrated the three specialized superintendencies that transferred revenue to support the bureaucracy in Beijing and the idle Manchu nobility nearby. The spiraling costs of the bloated bureaucracy and workforce of the grain-tribute superintendency that forwarded the food supply tax to Beijing required higher land taxes, and this encouraged corruption and bribery to avoid taxes; the gentry shifted the tax burden towards poorer households. Corruption and incompetence at the Yellow River Conservancy, the superintendency that managed and maintained the river channel and the intersecting Grand Canal that brought grain to Beijing, imperiled its capacity to handle growing ecological challenges and threatened the entire logistical transport system for moving grain. Revenue from the salt monopoly tax fell as smugglers increasingly avoided taxes. Inadequate government revenues and

[20] Chang, *Commissioner Lin and the Opium War*, p. 23; Hao, *The commercial revolution in nineteenth-century China*, p. 122; Wang, "Secular trends of rice prices in the Yangzi Delta, 1638–1935," table 1.1, fig. 1.1, pp. 40–48. The large amount of opium imports around 1838 need some adjustment downwards to correspond to actual sales. A mad rush to buy opium in India led to oversupply in China; thus, the stocks that remained unsold rose. See Cheong, *Mandarins and merchants*, pp. 134–37.

the excess supply of individuals without clear channels of upward mobility within the bureaucracy created a vigorous market for the purchase of degrees and offices. Revenue from these purchases, excluding the grain-tribute, soared from under 17 percent of central government revenue during the eighteenth century to 36 percent from 1821 to 1850. Expanded patronage ranks of office-holders required increased local and provincial tax revenue; thus, surcharges were added to land taxes. Because those taxes were denominated in silver, which climbed in value as it drained out of China, farmers' real tax burden rose dramatically, intensifying rural misery. During the early 1830s the silver drain and deflation contributed to the focus on opium as an evil to suppress, and heavy addiction among army troops raised opium addiction to a crisis level because their reduced effectiveness struck at the Manchus' ability to retain control; nevertheless, easy solutions were not apparent. The elite consumed much of the imported drug, and an eradication campaign would fail without their support. Coolie laborers, chair-bearers, and boatmen, the heaviest smokers outside the elite, consumed lower-quality domestic opium; campaigns against them would embroil the government in suppressing the domestic economy. Because criminal gangs controlled most of the opium distribution channels that connected wholesalers, retailers, and consumers, a campaign against these groups and opium farmers would strain governmental resources.[21] The scholar-elite led the campaign to stop the opium import trade, even though that group included substantial numbers of addicts, but they aimed to regain political power rather than to eliminate the evils of opium. Because economic restrictions that the Qing imposed on foreign merchants at Canton kept them locked in the status of principals, they could not bypass their Cohong agents for information about internal Chinese politics. Foreign merchants, therefore, remained ignorant of literati motivations and machinations over the opium trade, the merchants' most lucrative business.

Resurgent literati

The weakened Qing state opened a window of opportunity for the literati to regain their position as political power-brokers that they held under the Ming dynasty. The literati included those who had passed a rigorous examination system for degrees, but the scholar-elite with sufficient resources to live in Beijing and a somewhat larger group who lived in major cities formed those with the greatest influence. Their control of the examination system conferred a potent means of exerting power through patronage that eased the path of other literati up the examination hierarchy. The literati combined a focus on local social ties with aggressive efforts to forge bridges to other social networks

[21] Jones and Kuhn, "Dynastic decline and the roots of rebellion," pp. 107–29; Spence, "Opium smoking in Ch'ing China"; Wang, *Land taxation in imperial China, 1750–1911*, pp. 9–18, table 3.4, p. 61, and p. 114.

of literati throughout China, making the scholar-elite a potent information and influence network that could challenge the power of the Qing state. The Manchus aggressively resisted the formation of cliques or factions within literati ranks to prevent them from pursuing political influence, but this resistance lapsed during the early 1800s under monumental problems of falling marginal productivity of farm labor, economic stagnation, ecological decline, and bureaucratic paralysis besetting the Qing state. As the crisis of the 1830s deepened, the Manchu emperor turned to members of the Spring Purification Circle, a literati faction that built its position on moral censure of bureaucratic officials, for advice and as counter-weights to other factions in the government that fought over measures to resolve the crisis. Many government officials opposed a trade embargo against the British to stop opium imports because the Chinese navy could not stop heavily armed British opium vessels; the Napier affair of 1834, when the British navy pounded Canton batteries, demonstrated that the Chinese military could not defeat the British. In 1836, therefore, leaders within the bureaucracy promoted legalization of opium to lower its price and make it unprofitable for the British, but that approach was naïve because foreign opium merchants, led by Jardine, Matheson & Company, had already devised sophisticated strategies to achieve economies of scale. The Spring Purification Circle saw the legalization initiative as an opportunity to exert moral censure; they mobilized their network of members of the scholar-elite to successfully discredit proponents of legalization, thus gaining favor with the emperor. That victory placed them on a confrontation path with foreign merchants at Canton.[22]

Flush from their victory, the Spring Purification Circle sought another coup to solidify political gains in Beijing without provoking a trade war with Britain, an outcome that would have little support in the Qing court. In 1838, they proposed an anti-opium campaign among consumers to shutter the opium trade indirectly. This would also serve as a ploy to demonstrate their superior policy skills. Kwangtung Province and its port of Canton had two advantages for the experiment: the weak, resident Manchu bureaucracy would exert little interference on this periphery of China; and the Circle's local base of wealthy landed members of the scholar-elite could fund enforcement and potent scholar academies in Canton could lead the eradication effort. They convinced the emperor to appoint Lin Tse-hsu, a politically powerful member of the Circle, to lead the campaign at Canton. Initial success against consumers and the falling price of opium, partly caused by trader speculation, emboldened Lin in 1839 to pressure the British to surrender 20,000 chests of opium. The spectacle of the burning chests encouraged Lin to press Captain Elliot, British superintendent of trade at Canton, to get opium traders to post bonds that guaranteed their exit from the opium trade. Elliot refused that

[22] Polachek, *The inner Opium War*, pp. 17–124.

request, and Lin then pressured the Qing court to end trade with the British on January 5, 1840.[23] The Circle suffered a fatal flaw; they saw the power of the British state filtered through the presence of merchants at Canton, but the Circle did not know that those merchants had network bridges that reached to the mercantile and industrial heart of Britain and to the seat of power in government. These tragi-comic events at Canton pitted two protagonists with little insight into the underpinnings of each other's stance.

Trade vortex at Canton

As the literati strove for domestic political power during the 1830s, they maintained unwavering self-confidence that their policies and actions would cause the British to capitulate. On the other hand, the British were full of visions of expansive global trade; partnership networks and branches of their commission firms, accepting houses, agency houses, and import/export firms spanned the Americas, Europe, the Middle East, and the Far East. Paradoxically, restrictions of the Chinese government that concentrated foreign trade at Canton focused the business of the richest, most sophisticated Chinese merchants on contact with foreign firms, primarily British, that had backing from powerful firms in London, thus creating bonds of trust among these Chinese and foreign merchants that placed them as a group in opposition to policies of the Chinese state. Along with the successful campaign to end the monopoly of the East India Company over the China trade during the early 1830s, the same merchants and industrialists mobilized to change the policy of the British government towards China, and Jardine Matheson, greatest of the opium merchants, was the pivot of this effort. Their extensive investments in specialized opium ships and larger scale of operations poised them to grasp opportunities from expanded trade with China, and their partners, agents, and relatives bridged to each of the social networks of capital in Canton, Calcutta, Bombay, London, and Manchester, giving them unsurpassed capacity to lead the political effort. The firm and its allies increased efforts over the decade and blatantly appealed to force as they pressured the British government to take an aggressive posture to get the Chinese to remove trade restrictions embodied in the Canton mercantile system. By 1834, the British government had acceded to the broader aims of this potent mercantile network, but it remained reticent to challenge China forcefully. Political maneuvers to influence British policy continued as letters, private petitions, pamphlets, and resolutions bombarded Lord Palmerston, the Foreign Secretary, and these pressures worked; by 1836, the Foreign Office recognized the need to station a warship on China's coast.[24]

[23] Polachek, *The inner Opium War*, pp. 125–52.

[24] Chang, *Commissioner Lin and the Opium War*, pp. 51–69; Chapman, *Merchant enterprise in Britain*; Fairbank, *Trade and diplomacy on the China Coast*, pp. 23–38; Graham, *The China station*, pp. 48–71; Greenberg, *British trade and the opening of China, 1800–1842*, pp. 178–95; Polachek, *The inner Opium War*, pp. 107–12.

British pressure on China intensified and collided with the anti-opium campaign of the literati at Canton. In 1839, tensions reached feverish levels when Lin Tse-hsu demanded the surrender of opium, detained British traders, and burned 20,000 opium chests; traders departed for Macao in June. When news of the Canton confinement of British merchants reached London, William Jardine, who had arrived to influence the China policy of the British government, mustered almost 300 firms connected to the cotton industry to ask Foreign Secretary Palmerston to intervene to protect the China trade. Jardine met with Palmerston and later provided a document that proposed a blockade of principal ports along the China coast to enforce four demands: an apology for the insult to the British at Canton; payment for opium burned there; the signing of a commercial treaty fair to both parties; and the opening of additional ports, such as Foochow, Ningpo, and Shanghai, to trade. Occupation of several islands, such as Hong Kong, Jardine argued, might help enforce those demands. Jardine also sent Palmerston a memorandum that detailed the full structure of armed forces required and the military strategy to follow; Palmerston utilized Jardine's strategy in instructions sent to Canton. The British aimed to demonstrate naval might to the Chinese to coerce them into agreeing to their demands, but they had no intention of engaging in a mainland war of attrition. The British government placed its political and military might at the disposal of the merchants at Canton and their allies in the mercantile houses and factories of Britain.[25]

The Opium War and the Treaty of Nanking

American merchants eagerly filled the breach opened by the departure of the British from Canton and exported tea to Britain and America, while British firms focused on smuggling opium along the Kwangtung and Fukien coasts. British merchants, such as Jardine, Matheson & Company, worked surreptitiously through American firms, underscoring the bonds of trust among key intermediaries. Lin convinced the emperor officially to terminate trade with the British in January 1840, but he underestimated the capabilities of the enemy. Because Lin believed the British were mere sea marauders who would attack Canton to acquire booty to support their war, he concentrated defense preparations around Canton and assured the emperor of British intentions. The chasm that separated the Chinese and British economies became grimly evident during the Opium War. China massed troops for battle, deployed a feudal military with simple weapons, and relied on primitive, slow logistical support for the troops, whereas the British deployed heavily armed troops with advanced weapons and operated swift, mobile warships and logistical support vessels. The first phase of the war commenced when British forces arrived in

[25] Chang, *Commissioner Lin and the Opium War*, pp. 71–195; Fairbank, *Trade and diplomacy on the China Coast*, pp. 82–83; Graham, *The China station*, pp. 73–108.

June 1840; they blockaded Canton with a few ships, but the bulk of the force headed to the lower Yangtze valley. After several minor defeats, the Qing court agreed to negotiate, but talks at Canton failed. The war entered a new phase in August 1841, with the arrival of a powerful British industrial force, and it came to a brutal end during the Yangtze valley campaign in 1842. The Chinese belief in their capacity successfully to throw mass armies at the enemy in land warfare fell in ruins before modern military technology. Just before the British attack on Nanking, emissaries from the emperor declared that China would negotiate British terms. The purge of literati from positions of influence at the Qing court and in the provinces paved the way for negotiations, and the Manchus chose appeasement to save the dynasty. So long as the British kept to their trade demands and treaty ports, the Manchu and British empires occupied common ground for negotiations; thus, the Manchus could continue the Chinese strategy to keep foreigners at the periphery of the empire. The Treaty of Nanking was signed in August 1842, and its key provisions met the aims of British merchants at Canton, led by Jardine and Matheson, and of Foreign Secretary Palmerston: reparations; the opening of five ports, at Canton, Amoy, Foochow, Ningpo, and Shanghai, to trade; diplomatic exchange; British consuls based in each port; termination of the Cohong monopoly; uniform modest tariffs; and the award of Hong Kong as British territory. A prominent member of the Canton Cohong, on the Chinese side, and Alexander Matheson, on the British side, led the tariff negotiations. The documents maintained silence on the opium trade, the root of the collision between the two empires, because neither side found a solution acceptable to the other; thus, future trade moved in a legal channel through treaty ports and an illegal channel offshore. The British gained their treaty ports and Hong Kong, but they started twenty frustrating years dealing with the Qing government and finally resolved trade issues only after fighting another war.[26]

These agreements demonstrated unequivocally that Chinese and British merchants at Canton drew on their long-standing bonds of trust to cooperate in the creation of a mercantile system that benefited both sides. They envisioned each treaty port as a hinge: Chinese merchants at the port brought interior products for export and redistributed imports, and British merchants based there controlled imports and exports. This conception, nevertheless, mistakenly assumed that the capitalization and specialization of Chinese and British merchants would stay fixed. The freedom to trade at treaty ports without interference from the Chinese government dramatically reduced the political costs of operating, thus opening doors for hordes of unspecialized Chinese, British, and other foreign merchants to enter the competitive fray.

[26] Chang, *Commissioner Lin and the Opium War*, pp. 195–208; Fairbank, *Trade and diplomacy on the China Coast*, pp. 84–151; Fairbank, "The creation of the treaty system," pp. 216–23; Graham, *The China station*, pp. 108–39; Polachek, *The inner Opium War*, pp. 17–18, 152–53, 209–12; Wakeman, "The Canton trade and the Opium War," pp. 195–212.

That competition forced larger-capitalized firms to reduce costs through investments in technology, organizational changes, and specialization to achieve economies of scale. Yet, the agreements prevented British merchants from trading outside the treaty ports; as some firms raised their capitalization and specialization they were tempted to push into China to react to competition. Chinese merchants had no restrictions on their use of capital to specialize and expand into foreign trade, even as they retained control of interior China trade. This embedded instability in the new mercantile regime of treaty port trade.

4

Hub of the China trade

The new tariff and port regulations are really very moderate and favourable, and if strictly adhered to by the Chinese, our trade with England is sure to increase very much. The drug trade continues to prosper.[1]

Trade dilemma

In their home nations, British merchants observed bustling urban centers and American merchants saw a prosperous agricultural economy with rising farm incomes. The merchants viewed these as lucrative markets for commodities and manufactures, but they remained woefully ignorant about the Chinese market.[2] Under the new trade regime, foreign merchants needed market information; yet, prohibitions on travel in China prevented them from gaining personal acquaintance with its internal economy. The lower Yangtze valley was one of the few areas the foreign merchants knew about, based on the later military campaigns of the Opium War; nevertheless, outsiders readily misinterpreted it. Extraordinarily high productivity per acre disguised low marginal productivity of labor and limited peasant purchasing power. Rising opium consumption by the gentry and urban professional classes hid the deflation that wracked China. Captain Charles Elliot, the British Chief Superintendent of Trade at Canton during the late 1830s, never believed that a vast Chinese market was waiting to be plucked by the merchants – but that remained a minority view. Alexander Matheson, a principal in the greatest China house, Jardine, Matheson & Company, expressed the majority view: trade would grow at the treaty ports and opium would remain lucrative. His enthusiasm could not disguise the fact that foreign merchants had entered uncharted

[1] Letter of Alexander Matheson, July 31, 1843, from the Private Letter Books, Jardine Matheson Archives, Cambridge University Library; quoted in Greenberg, *British trade and the opening of China, 1800–1842*, p. 214.

[2] Allen, "Agriculture during the industrial revolution"; Landes, *The Unbound Prometheus*, pp. 68–77; Rothenberg, *From market-places to a market economy*.

waters; they confronted a bewildering variety of regional economies without the mediating services of the Cohong at Canton. Social systems, clan organizations, currencies, standards of weights and measures, and dialects varied across these regions, and the Chinese language formed an immense barrier. Their frustrating experience with the Cohong also taught them that few Chinese merchants had adequate capital to finance trade. They borrowed at high interest rates from the British, an indicator of inadequate capital accumulation and capital markets in China, and the government often "squeezed" the few rich merchants at Canton, such as Howqua, one of the world's wealthiest, to underwrite projects. For all of the optimism of the British and American merchants, therefore, their rapidly expanding economies threw off far more surplus than the Chinese economy could absorb.[3]

Large, highly capitalized trading firms, such as Jardine Matheson, must have recognized that competition for business would intensify. The influx of traders at Canton after the termination of the East India Company monopoly in 1834 anticipated the events which would occur when four more ports opened to trade after the Treaty of Nanking. Numerous unspecialized traders would shuttle among ports acquiring small quantities of tea, silk, and other items and peddling diverse imports. These firms offered stiff competition: they avoided fixed investments in offices and warehouses, management costs stayed low because the ship owner/captain or supercargo handled trade, and only a few people shared the profits. The Chinese trade policy of "most favored nation" exacerbated this competition; whenever any Western nation extracted a trade concession from China, it immediately extended that concession to other nations. This intensified competition among firms because they traded on an equal basis. To counteract this, large trading firms needed to invest capital in a new organizational structure, including branch offices, and in an expanded infrastructure of offices, warehouses, and wharves in multiple ports. The comprador, an organizational innovation, became a key component of their strategic plan and constituted a packaged solution to increased risk and greater competition in the China trade.[4]

Compradors

The comprador brought talents to bridge the chasm between the Chinese and Western economies and operated simultaneously as a hired strategist and manager for the foreign firm and as an independent merchant. Fluency both

[3] Graham, *The China station*, p. 171; Greenberg, *British trade and the opening of China, 1800–1842*, pp. 50–74, 214; Hao, *The comprador in nineteenth-century China*, pp. 24–25.

[4] Fairbank, *Trade and diplomacy on the China Coast*, pp. 195–99. Prior to 1843, the foreign merchants at Canton had worked with the *ya-hang*. These licensed brokers acted as independent commission agents, but they differed from compradors who were employed by the foreign firms and had more complex relations with them; see Hao, *The comprador in nineteenth-century China*, p. 2.

in Chinese and in pidgin English, the lingua franca of the treaty ports, placed the comprador in a powerful position to deal with both sides of trades. As a strategist, the comprador brought expertise and information about Chinese products, markets, mercantile channels, and finance. The foreign firm paid a modest salary and office expenses, and the comprador hired, supervised, and guaranteed the Chinese staff who included a purser, a shroff (expert in metal currencies), a bookkeeper, godownmen (warehouse workers), office boys, and coolies. Commissions on trading and banking for the house and on his own account generated income for a comprador, and sometimes he traded on joint account with the firm. The comprador also operated as an independent merchant either on commission or on speculation through taking ownership; he kept separate books and correspondence for those transactions. Foreign firms benefited because the comprador opened additional business opportunities, and firms even sent compradors abroad to search out markets for the firms and themselves. The comprador also operated as treasurer of the firm and maintained an account with a native bank, an essential relation to disburse and collect payments; the "comprador orders to pay" served as checks written on the accounts. He retained the interest earned on the account, which represented payment for his expertise, effort, and guarantees for the account. Earnings from these activities enabled the successful comprador to accumulate substantial capital; some ceased operating in that capacity and acted as independent merchants. Cantonese merchants met little competition for comprador positions: they had "large face," an appellation signifying high standing in the Chinese mercantile community; they spoke pidgin English and had extensive experience of dealing with British and American merchants in the export and import trades; they had the largest number of bridges to mercantile networks in interior China; and they had the greatest wealth and largest capital to finance trade. For the remainder of the nineteenth century, natives of Canton dominated the comprador ranks, filling as many as 90 percent of the positions within the largest British and American firms. A referral system whereby Cantonese merchants recommended peers and family members as compradors and guaranteed their trustworthiness contributed to their long-term dominance. In Jardine, Matheson & Company, which attracted the most skilled, richest merchants as compradors, one family kept the position for decades. The powerful position of a comprador made him a target for accusations that he "squeezed" the employer or engaged in surreptitious graft, but lengthy family dominance of comprador positions refutes that interpretation. No evidence exists that a larger proportion of compradors were dishonest relative to other individuals in comparable positions of authority; in fact, leading compradors were paragons of integrity.[5] The squeeze that compradors

[5] Fairbank, *Trade and diplomacy on the China Coast*, p. 219; Hao, *The comprador in nineteenth-century China*, pp. 2–177, appendices A–D, pp. 227–34; Lockwood, *Augustine Heard and Company, 1858–1862*, pp. 40–46; Viraphol, *Tribute and profit*, pp. 223–31.

exerted, instead, represented charges for sophisticated services and guarantees that they provided for their employers.

Foreign firms grasped the comprador institution to bridge to the China trade, but paradoxically that institution raised the principal–agent dilemma anew. At Canton, foreign firms served as agents for principals based in India, North America, and Europe; only large firms such as Jardine Matheson were principals during the decade preceding the Opium War, but following that war every firm of consequence that traded with China hired a comprador. That made these firms principals, regardless of their status as agents *vis-à-vis* firms headquartered elsewhere. The continual frustration that foreign firms expressed with the comprador institution derived mostly from the principal–agent dilemma, not any deficiency or dishonesty on the part of compradors. Foreign firms depended on compradors for information and trade opportunities, and this pivotal network position gave them strength in dealings with their employers. Because the power of foreign firms ended at the borders of the treaty ports, they could not enforce sanctions against compradors outside those sites; therefore, foreign firms had strong incentives to build bonds of trust with compradors to protect against malfeasance. The comprador institution, and talented individuals attracted to serve in the positions, provided a smooth mechanism for the transfer of capital and mercantile expertise from Canton to other treaty ports. Compradors solved the problem of institutional access to the China market, but trading firms also had to devise an organizational structure to manage operations in one or more ports; those decisions depended on the scale and scope of their business.

Alternatives for a mercantile headquarters

Small-scale, unspecialized trading firms that sent vessels from port to port used their ships as headquarters, and they rented warehouse space or purchased ship-repair services from other firms whenever they required these. If a trading firm operated at a somewhat larger scale, but only sent a few ships each season back-and-forth between Asia and Europe or North America, a small office in one port sufficed, perhaps with a part-time comprador. Trading firms that operated at a still larger scale and dealt with several ports needed to access partners and correspondents in London and New York (or Boston) and to access office managers and compradors in treaty ports and in other ports of Asia. The vast distances and long time lapse for communications between the West and Asia placed a premium on basing sophisticated management at one port in Asia, designated the headquarters, to access information about markets (supply, demand, currencies, politics) and to implement overall policy decisions of the senior partners in Europe or North America. Other ports housed branch offices with their Western manager and staff and the comprador and his department.

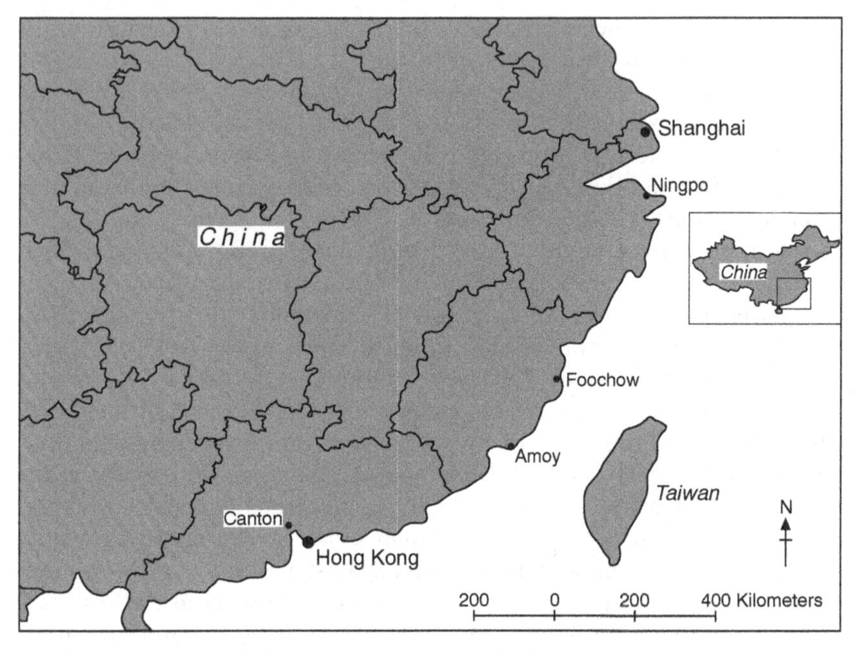

Map 2. Hong Kong and the treaty ports, *circa* 1845.
Source: author.

Singapore, whose merchants controlled the trade of the Malayan and Indonesian archipelago, straddled the Straits, one of the world's busiest shipping channels that linked Bombay and Calcutta, headquarters of the huge Indian trade, and East Asian ports; India, in turn, occupied an intermediate point on routes to and from Europe. The 1,600 miles and more to the China coast, however, made Singapore an impractical site for a headquarters of a firm with heavy investment in the China trade. Delays in accessing information, inability to participate in social networks of merchants in Chinese ports to hear information and to cooperate on ventures, and time-lags in responding to events, such as supply gluts, currency fluctuations, and political crises, would make the firm uncompetitive. Canton, the largest foreign port in China, remained attractive because trading firms had long-standing ties to its merchants, but this hotbed of literati-led antagonism towards the British, though not the Americans, after the Opium War made it a difficult place to maintain a headquarters (map 2). Amoy fronted a poor hinterland that offered limited trade opportunities, and merchant channels to interior China did not radiate from it; therefore, Amoy did not attract many headquarters, although its prominence in the junk trade to the Straits (Malaya) encouraged some British firms to move in. Foochow offered little initially because Canton merchants successfully lobbied the Chinese government to prohibit tea exports from

Foochow, and that policy remained in force for a decade after the Treaty of Nanking. Ningpo dominated the coastal junk trade, but its merchants had poor access to interior China. Shanghai stood out because its merchants controlled access to the trade of the Yangtze valley, the richest, most commercialized area of China. Leading firms in the China trade had to base an office, either the headquarters or a major branch, in Shanghai if they hoped to retain their position in trade.[6]

The trade giants choose Hong Kong

Management center and information node

Hong Kong, nevertheless, had the critical feature: status as a British colony backed by the full legal and military authority of the home nation – and the governor of Hong Kong also served as Her Majesty's Plenipotentiary and Superintendent of Trade for China. The British navy based a squadron there to protect British interests in the Far East, and by 1844, it became the headquarters of the China station of the navy. Until the next outbreak of hostilities, in the late 1850s, however, British naval forces remained a shell, capable mainly of suppressing pirates and intimidating a weak Chinese navy. For the great China houses, Jardine, Matheson & Company and Dent & Company, Hong Kong had the administrative and military features that they preferred for a headquarters; both immediately established themselves there. They gained most of their profits from their duopoly over the opium trade. It remained illegal in China and subject to suppression and harassment at any time; thus, treaty ports, though under British consular protection, did not offer security for illegal activities. A headquarters at Hong Kong provided political and social access to the governor to influence British policies regarding the China trade. British naval protection added a bonus because it reduced the need to defend their headquarters and opium godowns (warehouses) from the Chinese navy and pirates; their opium clippers at sea and ships at receiving stations along the China coast, however, remained heavily armed.[7]

The organizational structure and operations of Jardine, Matheson & Company provide clues to the formation of Hong Kong as the hub of Asian trade and finance. The firm commenced building construction in the midst of the Opium War in 1841, testimony to its confidence that Hong Kong offered the best location for a headquarters, and by March 1844, the headquarters staff, numbering about twenty, relocated from Macao to Hong Kong. Donald Matheson managed the headquarters, while David Jardine managed the

[6] Fairbank, *Trade and diplomacy on the China Coast*, pp. 193–95, 291–331; Graham, *The China station*, pp. 230–53; Polachek, *The inner Opium War*, pp. 237–57; SarDesai, *British trade and expansion in Southeast Asia, 1830–1914*, p. 50.

[7] Graham, *The China station*, pp. 230–75.

Canton branch; Alexander Matheson supervised them, but by 1847 he returned to London and started service as a director of the Bank of England, solidifying the firm's access to the best information about global trade and finance. Jardine's sent Alexander Dallas, a junior partner, to open a branch office in Shanghai in 1844, the same year the headquarters shifted to Hong Kong, signifying the importance the firm attached to a presence at the gateway to the Yangtze valley. Within two years, the firm added a shore branch at Amoy, giving it branches in that port, Canton, Shanghai, and Macao. By 1846, Jardine Matheson had refined a hierarchical organizational structure and a well-oiled, routine operation of the opium trade, with Hong Kong as corporate headquarters. The head office contained a large Chinese staff under the supervision of a comprador to support opium activities, including accounting, ship repairs, and warehouse operations. Opium clippers brought Malwa and Bengal opium from Bombay and Calcutta to Hong Kong for storage in a warehouse, and on a biweekly schedule, one clipper headed to southern stations and another to eastern stations to deliver opium to receiving ships anchored offshore along the China coast; on the return voyage, these clippers collected proceeds from each station. The receiving ships numbered around ten by the 1850s, and each housed an opium comprador or an interpreter and a shroff (assessor of metal currencies). The Hong Kong headquarters sent monthly instructions to commanders of the receiving ships and the staff settled financial accounts and sent gold or silver to India for payments, keeping the balance to finance additional trade. Whenever necessary, Jardine Matheson cooperated with Dent & Company and engaged in selective price wars at receiving stations along the China coast to eliminate competitors.[8]

The operations of Jardine Matheson and other firms made Hong Kong a pivot of the information network in Asia. Jardine's bridges to firms in Bombay and Calcutta kept it appraised of market information regarding opium crop reports, auctions, market trends, and government reports; Jardine's incorporated this information in instructions to commanders of its receiving ships along the China coast. Major trading firms in Hong Kong also published formal business circulars and "prices current" for distribution in Asia, Europe, and North America, but the most important business news, because it was confidential, circulated in letters that head offices of large trading firms sent customers. Commercial and political news from Hong Kong also circulated through newspapers underwritten by leading British merchant houses: the *Canton Register*, funded by the Matheson family, which moved from Macao to Hong Kong in 1843; and the *China Mail*, partially supported by Dent & Company, which started publication in 1845. Dense social, economic, and political exchanges among merchants and government officials in venues such

[8] Fairbank, *Trade and diplomacy on the China Coast*, p. 233; Le Fevour, *Western enterprise in late Ch'ing China*, pp. 16–22; Keswick, *The thistle and the jade*, p. 258; Reid, "The steel frame," pp. 26–29.

as the Hong Kong Club (1846), the branch of the Royal Asiatic Society (1847), and the Hong Kong Chamber of Commerce (1861) generated commercial and political information that entered circulars, newspapers, and private correspondence.[9]

Seeds of the Hong Kong agglomeration

When the premier firms in the China trade, Jardine Matheson and Dent, ensconced their headquarters in Hong Kong, it became the hub of Asian trade and finance. Their trade, families, friends, and partners constructed social network bridges to London, the global pivot of trade and capital, to the leading mercantile centers in India (Bombay, Calcutta), to Southeast Asia (Singapore), and to China, including the treaty ports and offshore opium receiving stations. These bridges formed a relatively non-redundant set of contacts across the channels of trade and finance, providing these firms with extraordinary access to commercial and political intelligence. Firms that hoped to compete needed access to information that circulated in the Hong Kong business and political community, and the organization of cooperative ventures with these leading firms and others also occurred more readily in Hong Kong than in other ports. Both firms served as bankers during the 1840s and 1850s, before exchange banks became securely established, and they owned insurance companies; thus, traders could readily arrange loans and insurance in Hong Kong. Jardine Matheson and Dent, therefore, became the seed for the agglomeration of headquarters and branches of other trading firms. That seed sprouted quickly as firms headquartered in Bombay, Calcutta, and Canton either moved their headquarters or founded a branch in Hong Kong in a pell-mell rush that started with land purchases as early as 1841. A substantial number of prominent firms, such as Gilman & Company and Gibb, Livingston & Company, as well as smaller ones, joined the trade giants by 1845, and this continued through the 1840s and 1850s. British firms dominated, but an international contingent with representatives from India (Parsee and other Indian firms), Germany, France, Portugal, the Netherlands, and the United States joined them. These firms increased the array of bridges from Hong Kong to the social networks of capital in major global metropolises. Starting in the 1840s, Chinese merchants from Canton and other ports in China also slipped into Hong Kong or sent family members to operate as merchants. They circumvented the requirement that they should acquire customs passes to trade at Hong Kong, and the refusal of British officials to enforce those rules gave Chinese merchants leeway to join international intermediaries as full partners in trade with China and the rest of Asia. The main contin-

[9] Endacott, *A history of Hong Kong*, pp. 70, 119; Kwan, *The making of Hong Kong society*, pp. 28–40; Le Fevour, *Western enterprise in late Ch'ing China*, pp. 18–21; Lockwood, *Augustine Heard and Company, 1858–1862*, pp. 11–12; Welsh, *A borrowed place*, p. 140.

gent of Chinese merchant firms, however, arrived in the 1850s; the number of brokers/traders, compradors, and merchant houses reached 63 by 1855 and 128 by 1861. American firms always enjoyed better relations than the British with China's government, and they did not suffer from the taint of the Opium War. Most American traders stayed in Canton during the 1840s, but as turmoil intensified there, they relocated to Hong Kong. The top firms, Augustine Heard & Company and Russell & Company, transferred their headquarters from Canton to Hong Kong by the mid 1850s and established organizational structures that mirrored the British giants.[10] During the 1840s and 1850s, therefore, Hong Kong had become the premier meeting-place of the foreign and Chinese social networks of capital in Asia. Firms headquartered at ports elsewhere inside and outside Asia henceforth depended on Hong Kong firms for access to the most sophisticated global information as it affected Asian trade and finance. Physical movements of ships and commodities comprised one part of those operations, but they were not critical features; other ports, especially Shanghai, built huge shipping businesses based on commodity flows. Firms in Hong Kong, however, dominated decision-making about the control of the exchange of commodity and financial capital.

Their agglomeration inspired Henry Pottinger, the first governor of Hong Kong, to proclaim it "the grand emporium of Eastern Asia."[11] The crush of Opium War commerce, warships, and troops quickly gave way to peacetime pursuits as the arrival of trading firms, government officials, and naval facilities set off a construction boom to develop offices, godowns, wharves, retail stores, houses, streets, and other facilities of a new port. As of 1851, the 1,520 non-Chinese (British, Americans, Parsees, Germans) formed a tiny superstructure of government administration and business headquarters in a total population of 32,983, underscoring that Hong Kong was a Chinese city. Comprador departments in trading firms accounted for a small number of Chinese, but the vast majority worked as laborers in construction, shipping and ship repairs, retailing, and the host of other occupations that keep a small mercantile city functioning. During the 1840s, many Chinese emigrated from nearby Kwangtung Province, poor refugees from economic decline and turmoil, and as unrest intensified in Canton during the 1850s, wealthy Chinese families from there also joined the flight to Hong Kong. Throughout these unsettled political conditions, Hong Kong's trade swelled; between 1844 and 1861, the number of ship arrivals rose almost fivefold from 538 to 2,545 and

[10] Banister, *A history of the external trade of China, 1834–1881*, pp. 31–35; Bard, *Traders of Hong Kong*, pp. 51–108; Hao, *The commercial revolution in nineteenth-century China*, pp. 75, 252; Kwan, *The making of Hong Kong society*, table 3.8, p. 77; Lockwood, *Augustine Heard and Company, 1858–1862*, pp. 6, 56, 81–90; Morse, *The international relations of the Chinese empire*, vol. I, p. 562; Tsai, *Hong Kong in Chinese history*, pp. 36–64; Wright, *Twentieth-century impressions of Hongkong, Shanghai, and other treaty ports of China*, pp. 210–34.

[11] Foreign Office Records 17/183, China correspondence, Public Record Office, London; quoted in Fairbank, *Trade and diplomacy on the China Coast*, p. 124, note 53, p. 499.

Table 4.1. *Number of non-Chinese adult male civilians in ports, 1850–1859*

	1850	*1855*	*1859*
Hong Kong	404	377	1,462
Canton	362	334	127
Amoy	29	34	45
Foochow	10	28	57
Ningpo	19	22	49
Shanghai	141	243	408
Total	965	1,038	2,148

Source: Morse, *The international relations of the Chinese empire*, vol. I, pp. 346, 359, 362.

total tonnage climbed almost sevenfold from 189,257 tons to 1,310,388. The quadrupling of the number of non-Chinese adult male civilians in Hong Kong from 377 in 1855 to 1,462 in 1859 implies that the headquarters and branch offices of merchant trading firms substantially enlarged their activities (table 4.1). Because trade with Britain languished until after 1857, the expansion of trading firms must have signaled the growth of Hong Kong's trade within Asia. Although Hong Kong was a mighty center of trade in Asia as of 1861, its population size of 119,321, of whom 97 percent were Chinese, compared to London's total of 2,803,989, made Hong Kong appear as a tiny outpost of the Empire to British visitors.[12]

The China trade to the early 1860s

Under the leadership of the duopoly of Jardine, Matheson & Company and Dent & Company, Hong Kong operated as the administrative and financial center of the opium trade; this powered its early rise as the hub of Asian trade and finance. The compound annual growth rate of the value of opium imports to China was 3.3 percent between the 1830s and 1840s, and this rate increased to 5.1 percent between the 1840s and 1850s. The dislocations of the Opium War and reorganization of trade under the treaty port structure delayed the expansion of the opium trade, but after 1846, imports decisively exceeded the 30,000-chest level annually and more than doubled by 1854 (fig. 4.1). Turmoil

[12] Endacott, *A history of Hong Kong*, pp. 66–70; Morse, *The international relations of the Chinese empire*, vol. I, pp. 342–46; Tom, *The entrepot trade and the monetary standards of Hong Kong, 1842–1941*, appendices 2, 4, 18, pp. 105, 107, 157–58; Tsai, *Hong Kong in Chinese history*, pp. 19–23; Weber, *The growth of cities in the nineteenth century*, table 18, p. 46.

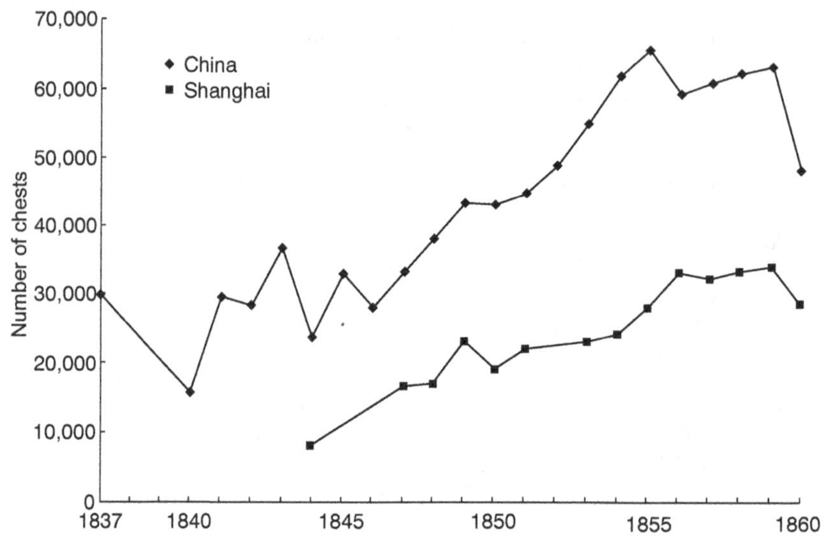

Fig. 4.1. Opium imports to China and Shanghai, 1837–1860.
Sources: Fairbank, *Trade and diplomacy on the China Coast*, p. 229; Hao, *The commercial revolution in nineteenth-century China*, table 9, p. 130; Morse, *The international relations of the Chinese empire*, vol. I, pp. 358, 465, table G, p. 556.

around Shanghai, the leading opium import center, caused the drop in imports in 1860 as the Taiping rebels threatened and Western allies battled with China. Surging revenues from opium imports after 1846 enabled the giant merchant firms to finance their trade expansion in other lines during the 1850s. Opium increasingly served as a medium of exchange in the treaty ports, and by the late 1840s and early 1850s, compradors of the large trading firms began to take opium on "upcountry" trips as "currency" to purchase tea and silk. Collaboration between British and Chinese officials expedited the stunning growth of opium imports. At each port along the coast, Chinese officials blatantly extracted bribes for allowing safe transfer of opium to shore, and this "squeeze" continued in the interior. The opium trade remained public knowledge; city directories of ports such as Canton and Shanghai listed opium receiving ships based offshore, owners of opium firms, and officers in charge, much like they might report advertisements for mercantile firms dealing in tea or silk. This smooth operation made opium legalization in China under the Treaty of Tientsin in 1858 a moot point.[13] Because the British continued to dominate China's trade, this exchange offers a window on the prospects of the merchant firms.

Alexander Matheson's enthusiastic projection that British exports to China would prosper fell wide of the mark; instead, Britain incurred a substantial

[13] Hao, *The commercial revolution in nineteenth-century China*, pp. 55–60, table 6, p. 69; Morse, *The international relations of the Chinese empire*, vol. I, pp. 540–617.

Fig. 4.2. Total British exports to, and imports from, China and Hong Kong, 1834–1904.
Source: Sargent, *Anglo-Chinese commerce and diplomacy*, chart A.

trade deficit with China in direct legal trades (fig. 4.2). Total exports to China (including Hong Kong) did not surpass the level of 1844 until after 1857, whereas imports to Britain rose from a bottom in 1840, during the Opium War, to a peak in 1857, followed by a dramatic plunge during the Taiping rebellion. The direct export trade of Britain with China maintained a narrow path after 1843, just as earlier; textiles accounted for most exports (fig. 4.3). The woolen textile trade languished until 1859, when it first surpassed the level of 1844, and cotton goods, the export hope of the Lancashire manufacturers, never decisively exceeded £2 million sterling between 1844 and 1858; then, cotton goods surged to a sharp peak between 1859 and 1861, before collapsing. The mass of poor rural peasants continued to elude the market grasp of British textile factories. Raw silk imports followed a volatile path during the 1840s and 1850s, with a spike between 1854 and 1857, whereas tea imports climbed steadily as per capita consumption rose, despite prohibitive tea duties in Britain (fig. 4.4). Following the plunge in 1857, under pressure from Taiping rebels in the Fukien tea district, imports surged to a new high in the early 1860s. The coincidence of the rise in tea imports to Britain after 1850 and the formation of a plateau in opium imports to China starting around 1853 (fig. 4.1) ended the large net outflow of silver (and gold) from China. This shift to relative balance in the inflow and outflow of these precious metals closed the deflationary chapter for the Chinese economy.[14]

[14] Gardella, "The boom years of the Fukien tea trade, 1842–1888," p. 42; Sargent, *Anglo-Chinese commerce and diplomacy*, pp. 131–40.

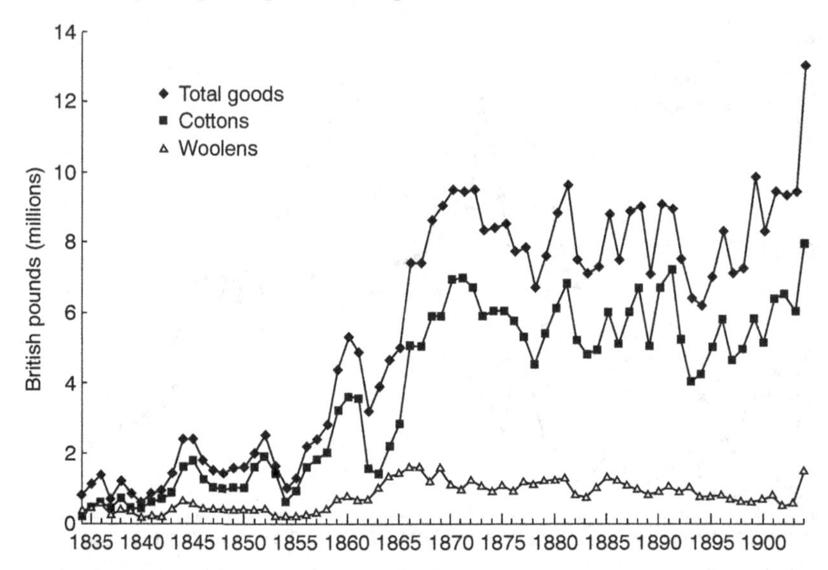

Fig. 4.3. British exports to China and Hong Kong for total goods and textiles, 1834–1904.
Source: Sargent, *Anglo-Chinese commerce and diplomacy*, charts B, D, F, pp. 329–31.

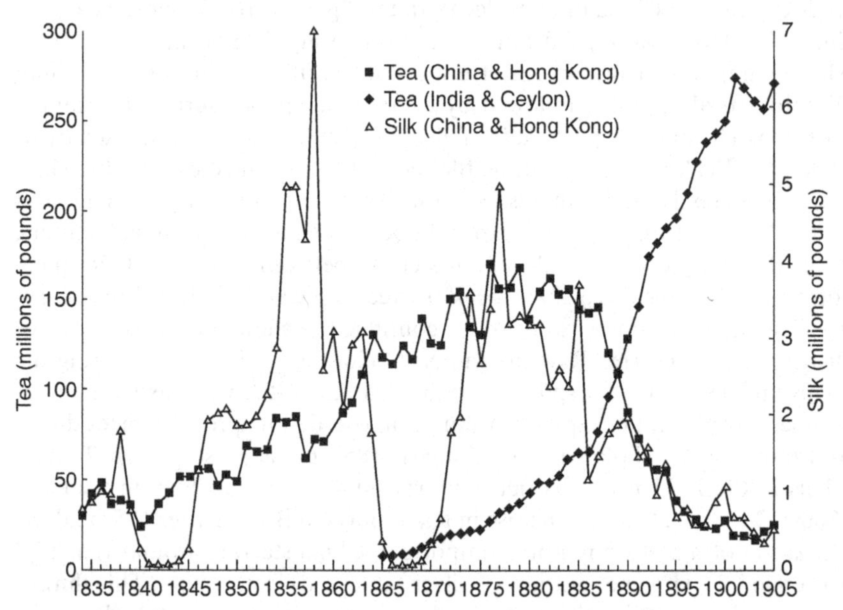

Fig. 4.4. British imports of tea and raw silk from China and Hong Kong, 1834–1904, and of tea from India and Ceylon, 1864–1904.
Source: Sargent, *Anglo-Chinese commerce and diplomacy*, charts C, E, G, pp. 330–32.

China trade through the treaty ports

Overall supervision of the Britain–China trade emanated from the Foreign Secretary in London, but the governor of Hong Kong, who also served as Chief Superintendent of Trade, exercised the main oversight. British consuls at each port supervised trade and British citizens; they received directions from, and reported to, the Chief Superintendent at Hong Kong, thus making it the administrative hub of the treaty ports from the British perspective. Because they dominated the China trade, their merchants exerted the greatest influence on the growth of treaty ports and on the emergence of trade management, finance, transportation, and communication bonds between the treaty ports and Hong Kong. Some British and American firms placed headquarters in the treaty ports, and their shipping bypassed Hong Kong when large-scale trades filled ships and there was thus no requirement for transshipment at Hong Kong to collect or redistribute full cargo loads. Changes in the import trade of opium and export trades of tea and raw silk accounted for most of the differential commercial expansion of the treaty ports. Foreign merchants could not participate in China's interior trade, but their incursions into the coastal trade began to alter inter-port relations. The British, drawing on the advice of their leading merchants, astutely chose treaty ports that housed Chinese merchants with links to the existing foreign and coastal trade of China and with superior bridges to interior merchant networks.[15]

Canton: decline of the national monopoly

The Cohong merchants of Canton learned ruefully that what a government gives, it can take away. Their tight control of trade in the regional hinterland of Kwangtung and Kwangsi Provinces rested on long-standing commercial ties, but the Qing government artificially enlarged this hinterland to Central China, including Hunan and Hupeh Provinces, and even beyond, by granting Canton merchants the national monopoly over trade with foreigners covering imports (opium, raw cotton, cotton goods) and exports (tea, raw silk). This prevented Hankow merchants, who dominated the interregional trade in tea of the middle Yangtze valley for over a century before the Treaty of Nanking (1842), from participating much in international trade through Shanghai. British and American traders recognized that Shanghai, gateway metropolis to the Yangtze valley and the commercialized heart of China, would become a major base for foreign merchants, but they did not intend to abandon Canton. They believed trade would prosper there and tried to make it work efficiently. During the 1845–46 trading season, shortly after the treaty ports

[15] Coates, *The China consuls*, pp. 7, 101; Fairbank, *Trade and diplomacy on the China Coast*, p. 155.

opened, the Canton customs office collected 73 percent of the total customs receipts of the five treaty ports, or four times the receipts collected at Shanghai. Canton and Shanghai continued to collect the overwhelming share of customs receipts, but Canton's lead over Shanghai fell to only 1.3 times by the 1850–51 season, as Canton's trade quickly intertwined with that of Hong Kong.[16]

The diminished stature of Canton extended beyond its falling share of customs receipts; from 1837 to 1852, the annual value of British imports to Canton fell 40 percent and British exports from Canton plunged 58 percent. Chinese merchants elsewhere threatened the hegemony of Canton merchants over foreign trade, but they refused to meekly relinquish control of the export trade. They appealed to force to gain the support of the Qing court to hinder exports from Shanghai and to prohibit tea exports from Foochow following the Treaty of Nanking, even though Foochow merchants enjoyed better access to the Fukien tea districts. Tax rebellions by rural residents in Fukien Province, among others, finally undermined this power play of Canton merchants; by 1854, the Fukien governor, with the acquiescence of the Qing court, began to allow tea exports through Foochow as a way to raise tax revenues. The jump in the export of tea from Shanghai after 1845 testified to severe erosion of government efforts to hinder its exports, and between then and the early 1850s, the tea export trade split (fig. 4.5). Production areas in Fukien, Kiangsi, Chekiang, and Anhwei Provinces started to export through Shanghai, and after 1853–54, Fukien teas exited at Foochow; Canton merchants captured the southern edges of Central China's production areas, such as Hunan Province. The decline in tea exports from Canton after 1845 eventually threw out of work up to 100,000 tea porters and 10,000 boatmen on the borders of Kwangtung Province, the core hinterland of Canton; they became ready recruits for the brewing Taiping rebellion. The erosion of tea exports from Canton accelerated during 1852 and 1853 as Taiping rebels disrupted tea distribution channels on their push north from Kwangsi through the middle Yangtze valley to Nanking, and Shanghai merchants capitalized on this disruption. Except for a brief recovery in 1854, Canton's exports plunged as Shanghai merchants maintained a substantial share, and Foochow merchants commenced their reach into the Fukien tea districts. The decline of British and American trade at Canton, nevertheless, did not motivate firms to abandon their headquarters or branch offices immediately. The number of adult male non-Chinese civilians in treaty ports provides a surrogate for the number of merchants and their staff; a small number of missionaries and consular staff formed the rest of the males. Their numbers rose 41 percent from 1845 to 1850, implying that foreign trading firms still hoped that Canton would regain luster

[16] Endacott, *A history of Hong Kong*, pp. 74–75; Fairbank, *Trade and diplomacy on the China Coast*, table 10, p. 262, and pp. 272–79; Rowe, *Hankow*, pp. 52–89, 122–25; Skinner, "Regional urbanization in nineteenth-century China," pp. 212–15.

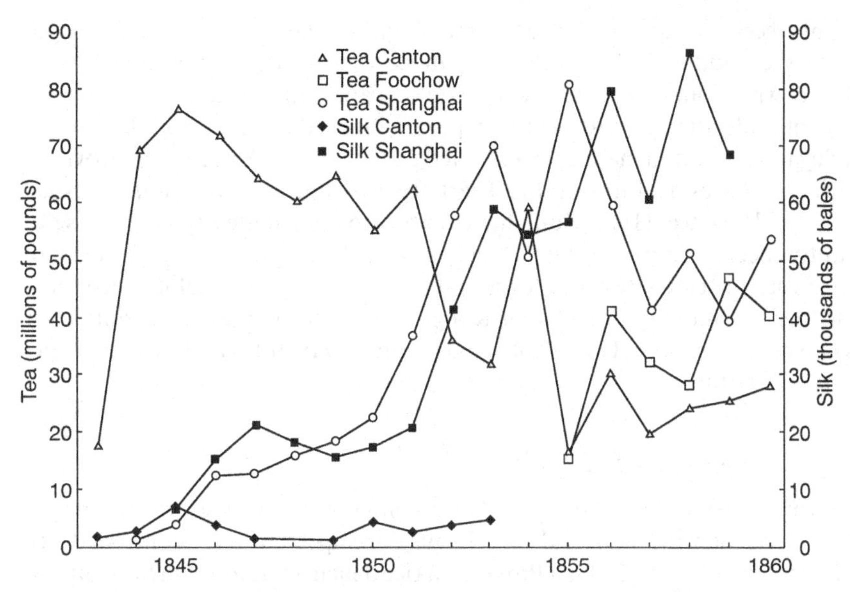

Fig. 4.5. Exports of tea and silk from Canton, Foochow, and Shanghai, 1843–1860. *Source:* Morse, *The international relations of the Chinese empire*, vol. I, table F, p. 366.

even in the face of declining trade, but they must have reassessed trade prospects between 1850 and 1855 because the number of males fell 10 percent (table 4.1). In 1856, literati-led turmoil culminated in the burning of factories (trading establishments) where foreign merchants lived and transacted their business. That event prompted the prominent American trading firms, Augustine Heard & Company and Russell & Company, to relocate their headquarters to Hong Kong, and numerous other firms did the same or shuttered branch offices; the number of males at Canton plummeted 62 percent between 1855 and 1859.[17]

The Taiping rebellion and turmoil in Canton hastened its descent from national monopoly metropolis for the foreign trade of China to regional metropolis for South China, but those events did not cause its decline. The opening of other treaty ports, principally Shanghai, but also Foochow, undermined the hegemony of Cantonese merchants over foreign trade before 1850. As merchants in those ports asserted control over the foreign trade of their domestic regional hinterlands, Canton's decline became irreversible. Its mer-

[17] Fairbank, *Trade and diplomacy on the China Coast*, pp. 102, 288–308; Gardella, "The boom years of the Fukien tea trade, 1842–1888," pp. 34–36; Graham, *The China station*, pp. 230–53, 299–329; Kuhn, "The Taiping rebellion," pp. 273–75; Lockwood, *Augustine Heard and Company, 1858–1862*, pp. 6, 81–90; Morse, *The international relations of the Chinese empire*, vol. I, p. 365; Wakeman, *Strangers at the gate*, pp. 98–100; Wang, *Land taxation in imperial China, 1750–1911*, pp. 114–15.

chants received another blow when merchants in Hankow, regional metropolis of the middle Yangtze valley, expanded their trade with Shanghai merchants who dominated the lower Yangtze valley and asserted control over foreign trade after it became a treaty port.[18] Blocked from the middle Yangtze valley trade, Canton merchants retreated to control over Kwangtung Province, with its rich agriculture of the Pearl River delta, and over impoverished Kwangsi Province. Hong Kong merchants also contributed to the demise of Canton because most of the largest-capitalized, most-specialized firms with the greatest social network bridges within Asia and with the global economy had their headquarters in Hong Kong. Canton firms needed to shift their Asian management to Hong Kong and retain only their South China management in Canton.

Amoy: invasion of the junk trade

Fukienese merchants at Amoy ranked second, after Cantonese merchants, in the junk trade with Siam and the Malayan archipelago. They controlled rice imports from Siam to Fukien Province, a rice deficit area, through familial and ethnic bonds with Fukienese emigrant merchants based in Siam's ports who handled the export side of the trade, but the rice trade peaked a decade prior to the opening of Amoy as a treaty port in 1843. The export trade in silks, fine china, nankeens, and other high-value Chinese goods from Amoy to Siam and the Malayan archipelago remained minor because Amoy merchants had no special control over those goods. To British merchants, with their larger, efficient ships, Amoy's trade seemed ripe for grasping, but they miscalculated the effectiveness of long-standing familial and ethnic trade bonds between Fukienese merchants in Amoy and Siam. Those social network bridges gave Fukienese lower costs through enhanced trustworthiness and superior access to information. Firms such as Jardine Matheson imported Chinese from Singapore and Penang, who had British citizenship, to work for them, and by the late 1840s, Anglo-Chinese in Amoy even outnumbered British natives, but those individuals were not central participants in the social networks of the Fukienese in Amoy and Siam. Before 1850, British merchants at Amoy attained some success in the trade with the Straits (Malaya, Singapore), where they held a competitive advantage through British firms based there, yet they made few inroads into the junk trade between Amoy and Siam until the mid 1850s, when British merchants from Singapore and Malayan ports started to capture the trade of Siam. This gave the British power at both ends of the trade link, that is, they bridged the networks of merchants. American traders, likewise, fared poorly at Amoy until the mid 1850s, although their trade expanded somewhat after that. These frustrations of British and American

[18] Rowe, *Hankow*.

merchants at Amoy, however, exclude two trades: opium stations of Jardine Matheson and other firms plied a lucrative business offshore, and foreign merchants entered the coolie trade that had existed at Amoy for several centuries. The British started with a shipment to Havana, Cuba, in 1847, and exports of coolie labor reached substantial numbers by 1852. Growth in branch offices and staff of foreign merchant firms at Amoy (table 4.1) track their inroads into the coolie trade and the gains of British firms in the trade of Amoy with Siam and the Malayan archipelago beginning in the mid 1850s. Amoy, however, was only one of several collection points for coolie labor, most of which came from Kwangtung and Fukien Provinces. Hong Kong shipping firms and brokers managed this lucrative trade, and Chinese compradors and other middlemen participated. Their firms occupied the hub position with the widest array of bridges to social networks of merchants in all regions that demanded coolie labor.[19]

Foochow: instant tea port

Chinese merchants at Foochow, similarly to those at other treaty ports, participated in the thriving junk trade of the China coast, but this diverse, small-scale shipping did not offer lucrative opportunities for big British and American trading firms. Their high fixed costs of headquarters, branches, ships, and warehouses required them to specialize in large-scale trade. As of 1850, Foochow housed only three foreign merchants among the ten males; missionaries accounted for the rest (table 4.1). Foreign firms did not ignore Foochow; opium receiving stations of the great trading houses floated offshore and handled about $2 million of business annually by the mid 1840s. Chinese merchants from Shanghai set the base for changes in Foochow through earlier purchases of tea in the Fukien districts for their thriving coastal trade to North China. When Shanghai opened as a treaty port, the muddled distinction between domestic coastal trade and foreign exports offered profitable opportunities for its Chinese and foreign merchants; they built local bonds of trust and conspired to evade the efforts of Canton merchants, supported by the Qing court, to thwart foreign tea exports at Shanghai. During the late 1840s and early 1850s, foreign merchants such as Russell & Company, the largest American house, followed the example of Shanghai Chinese merchants. Russell's partner, John Griswold, started sending compradors and staff to the Fukien districts to purchase tea. Because tax rebellions during the deflation and depression of the 1840s had undermined revenues in Fukien Province, the tea districts offered tempting possibilities to garner transit and customs taxes

[19] Fairbank, *Trade and diplomacy on the China Coast*, pp. 215–34, 313; Gardella, "The boom years of the Fukien tea trade, 1842–1888," table 9, p. 45; Griffin, *Clippers and consuls*, appendix 4C, p. 402; Morse, *The international relations of the Chinese empire*, vol. I, pp. 363, 401–02; Tsai, *Hong Kong in Chinese history*, pp. 24–26; Viraphol, *Tribute and profit*.

if tea exited through Foochow, instead of by the circuitous trip to Shanghai or even longer trip south to Canton. With the implicit approval of Beijing, the Fukien governor changed course and opened the floodgates for tea exports from Foochow. In 1853, a high Chinese government official in Shanghai, Wu Chien-chang, who knew Griswold, must have received word about the policy change instituted by the Fukien governor. Wu urged Griswold to send Russell's comprador to the Fukien districts and arrange to ship tea through Foochow to avoid disruptions caused by Taiping rebels and Triad uprisings west and southwest of Shanghai; Russell & Company took the advice.[20] This collusion between foreign and Chinese merchants and government officials in Shanghai exemplifies the extraordinary local social network bonds within that city and the wide-ranging bridges to other social networks in Central and North China that participants forged.

Large British and American trading firms in Shanghai monitored the experiment of Russell & Company as tea exports from Shanghai more than tripled between 1850 and 1853 (fig. 4.5), but turmoil along routes from the Fukien tea districts to Shanghai threatened this lucrative trade. Foreign firms faced a dilemma as soaring trade taxed the capacity of Chinese merchants ("teamen") to fund purchases of raw tea and forced the foreign firms to lend larger amounts of capital to teamen. If big foreign firms hoped to benefit from the swelling export trade from the Fukien tea districts, they must rapidly build a large branch-office base at Foochow and vertically integrate their operations backwards into the tea districts. Their swift reaction to this new form of competition, similar decisions, and mutual support suggest that these firms operated within a local social network based on mutual trust. Augustine Heard & Company, Jardine, Matheson & Company, and Dent & Company opened branches at Foochow in 1854 and 1855, and they rapidly expanded their tea exports there (table 4.1 and fig. 4.5). They quickly institutionalized the tea trade at Foochow; by 1856, they had a well-honed tripartite system in place that ensured a high flow of tea from districts of Fukien for sale at competitive prices in Foochow. Foreign firms sent silver and opium with their compradors "upcountry" to purchase tea; they contracted with Chinese teamen, often giving them credit, to purchase tea; and they bought tea on the open market in Foochow. The Hong Kong headquarters of these firms occupied the pivot of the tea trade: they monitored the entire sequence of activities through detailed communications with their branch offices, and they kept themselves informed of wholesale prices in London and New York. Although the volume of tea exported from Foochow fluctuated between 1856 and 1860 as political turmoil in the interior affected supply and financial crises in Europe and North America affected demand (fig. 4.5), Foochow maintained its secure,

[20] Fairbank, *Trade and diplomacy on the China Coast*, table 10, p. 262, and pp. 288–321; Gardella, "The boom years of the Fukien tea trade, 1842–1888," pp. 34–36; Griffin, *Clippers and consuls*, pp. 295–96; Hao, *The comprador in nineteenth-century China*, p. 76; Morse, *The international relations of the Chinese empire*, vol. I, pp. 360–61.

prosperous status as a specialized center of tea exports for almost three more decades.[21] Yet, Foochow remained a branch-office metropolis; its firms depended on information and capital from firms at the hub of Asian trade and finance, Hong Kong, and secondarily, from Shanghai.

Ningpo: underwriters for Shanghai

The British believed that Ningpo, a center of the junk trade, would offer lucrative opportunities because its merchants dominated a rich farm hinterland in northeastern Chekiang Province, and their wholesale trade connections extended from Hangchow to Nanking and Beijing along the route of the Grand Canal. These merchants, joined by Fukienese, also traded with Japan, and engaged in an active import trade with the North China coastal ports (fruits, bean cake) and with the Fukien ports and Taiwan (wood, fruits, rice); export trade to the south consisted of raw cotton, cotton cloth, and silks. Ningpo's merchants were small-scale, unspecialized traders and wholesaled little of the high-value commodities of tea and raw silk, specialties of big British and American merchant firms. After their arrival in Shanghai in 1843, British and American merchants quickly learned that the merchant agglomeration at Shanghai dominated the coastal junk trade and the lower Yangtze valley trade. Compared to Ningpo-based Chinese merchants in 1844, Shanghai traders imported five times (by value) as much from North China and almost seven times as much from the south. Few foreign merchants settled in Ningpo; as of 1855, only five of the twenty-two males held that occupation (table 4.1). Foreign trade remained tiny and primarily comprised local traffic between Ningpo, acting as a satellite, and Shanghai. Ningpo merchants had shrewdly identified trends decades earlier; by 1796, so many merchants had relocated to Shanghai that they financed construction of their own guildhall, and some accumulated enough capital to expand into banking before 1821. By the time foreigners arrived in 1843, Ningpo merchant families in Shanghai had founded "native banks" and laid their base for dominating Chinese banking there for the rest of the century.[22]

Shanghai: Chinese and foreigners cooperate

Chinese merchants at Shanghai warmly welcomed foreign merchants, whereas their reception ranged from hostile at Canton to lukewarm at other treaty

[21] Gardella, "The boom years of the Fukien tea trade, 1842–1888," pp. 42–44; Griffin, *Clippers and consuls*, pp. 294–96, appendix 4C, p. 403; Hao, *The comprador in nineteenth-century China*, pp. 77–82; Hao, *The commercial revolution in nineteenth-century China*, pp. 146–62; Lockwood, *Augustine Heard and Company, 1858–1862*, pp. 43–46; Morse, *The international relations of the Chinese empire*, vol. I, p. 360.

[22] Coates, *The China consuls*, p. 22; Fairbank, *Trade and diplomacy on the China Coast*, p. 155, table 10, p. 262, and p. 312; Griffin, *Clippers and consuls*, appendix 4C, p. 403; Johnson, *Shanghai*, pp. 155–75; Jones, "Finance in Ningpo"; Morse, *The international relations of the Chinese empire*, vol. I, p. 359; Shiba, "Ningpo and its hinterland."

ports. Shanghai's merchants did not fear that foreigners would challenge their hegemony over the lower Yangtze valley trade. That business required a rich social network of mercantile relations to provide intelligence about supply and demand, trustworthy relations to keep costs low, and cheap labor to handle trade. High-cost foreign firms needed compradors to trade; therefore, those firms could only enter large-scale export trades such as tea and silk and had to turn domestic distribution of textiles and opium imports over to Shanghai Chinese. The literati failed to mount successful attacks against foreign merchants, probably because Shanghai Chinese merchants fiercely opposed those actions as detrimental to their new-found opportunities to engage in foreign trade. The initial appearance of Canton merchants in Shanghai raised fears among foreign firms that the Cantonese would try to extend their monopoly control, but those concerns proved groundless because Shanghai Chinese merchants refused to support them. Foreign trading firms stampeded to Shanghai to gain access to the richest market in China, immediately transforming it into the treaty port, after Canton, that housed the largest-scale, most highly specialized mercantile firms, along with smaller-scale, less-specialized firms. The largest American house, Russell & Company, acted with unseemly haste to have a partner on the scene in 1843 with an appointment as *de facto* American consul. By 1844, only a year after opening as a treaty port, Shanghai housed eleven British and American merchant firms with a total of twenty-three merchants, and these included the giants, Jardine, Matheson & Company and Dent & Company. Merchants typically formed about 80 percent of the foreign males; between 1850 and 1859, their number soared from 141 to 408 (table 4.1). At that point, the Shanghai merchant agglomeration ranked second to Hong Kong and was over three times larger than Canton's. As entrepot for the richest, commercialized part of China, Shanghai offered the most profitable base for importing opium, the "mother" of all trades. The major opium firms, Jardine Matheson and Dent, located receiving ships near Shanghai by early 1844, and that year, trading firms sold about 8,000 chests, comprising one-third of total opium imports to China. The volume rose to over 30,000 chests by the late 1850s, when Shanghai's share consistently exceeded 50 percent of total opium imports (fig. 4.1). In both 1850 and 1860, opium imports represented about 50 percent of total commodity imports at Shanghai, and cotton textiles ranked a close second. Opium imports held up at Shanghai throughout the Taiping disturbances, implying that drug addiction overwhelmed other considerations.[23]

The capacity of Shanghai Chinese merchants to handle an instantaneous

[23] Fairbank, *Trade and diplomacy on the China Coast*, pp. 135, 155, 229, 303–12; Gardella, "The boom years of the Fukien tea trade, 1842–1888," p. 36; Griffin, *Clippers and consuls*, pp. 295–96, appendix 4C, p. 400; Hao, *The comprador in nineteenth-century China*, pp. 52, 76–77; Hao, *The commercial revolution in nineteenth-century China*, pp. 129–32; Hinton, *The grain tribute system of China, 1845–1911*, pp. 1–15; Huang, *The peasant family and rural development in the Yangzi Delta, 1350–1988*, pp. 44–48; Johnson, *Shanghai*, pp. 155–75; Kuhn, "The Taiping

jump in foreign trade from nothing to $6 million in 1844 and a doubling of that total trade to $12 million within three years offers conclusive testimony to the financial capital and liquidity concentrated in Chinese firms. The Ningpo *pang* (clique) of Shanghai, pivotal mobilizers of this capital through their guild and the movement of its members into other organizations, relied on kinship and locality (Ningpo) ties to add recruits, implying that they combined redundant social network ties with non-redundant ties as each clique member bridged to their kinship networks. This approach gave members access to a wide net of information and cooperative relations to mobilize capital, and the structure of the clique reinforced trust among members. The Ningpo *pang* successfully inserted members as bridges in the exchange of capital between foreign traders and bankers and their Chinese counterparts in Shanghai; between hinterland producers, merchants, and native bankers and Shanghai (foreign and Chinese) merchants and bankers; and between native bankers and the government. The native banks of the Ningpo *pang* operated like other banks and accepted deposits, remitted funds intra- and interregionally, and issued and discounted commercial paper. The Ningpo *pang*, though a leader among the guilds based on native-place and common trades, was just one among many guilds; most of them shared the Ningpo *pang*'s structure. Their merchants held membership in several guilds, and they focused on services, insurance, standards for trade, and charitable and benevolent activities, rather than on monopolizing or otherwise limiting trade. The Ningpo *pang* and other guilds looked familiar to the largest British and American firms. Jardine, Matheson & Company and Dent & Company participated in a "Scottish clique" that rooted in Scotland and London and was bound by family, ethnic, and school ties. Similarly, Russell & Company and Augustine Heard & Company were active in a "Boston clique" with extreme locality ties in that city and nearby towns; family, friendship, and school ties also undergirded this group. The entry of the Taipings into the middle Yangtze valley in 1853 accelerated capital accumulation in Shanghai. Shansi banks, who ordered their Hankow branches to shift capital and operations to Shanghai, commanded greater capital than native banks, controlled more of the interregional transfer of capital, and provided financing for the national and provincial governments; they also served as "bankers" for native banks, a form of wholesale banking. Numerous wealthy merchants and landlords in the Yangtze valley also fled to Shanghai during the Taiping rebellion, contributing their capital to city firms. By 1858, Shanghai's International Settlement housed about seventy native banks, and by 1864, it had twenty-four Shansi banks. This swelling volume of Chinese capital in Shanghai merchant firms

rebellion"; Le Fevour, *Western enterprise in late Ch'ing China*, pp. 15–19; Lockwood, *Augustine Heard and Company, 1858–1862*, p. 5; Lubbock, *The opium clippers*, pp. 276–77; Morse, *The international relations of the Chinese empire*, vol. I, pp. 346–57, 464–68; Murphey, *Shanghai*, table 2, p. 120; Polachek, *The inner Opium War*; Reid, "The steel frame," pp. 28–29; Rowe, *Hankow*, pp. 52–62, 90–121.

and banks propelled the city to the apex of the financial center hierarchy in China.[24]

Shanghai capital underwrote hinterland purchases that Chinese merchants made of raw silk and tea, the leading commodities sold to foreigners. The decline in tea exports from Canton after 1845 mirrored in reverse surging tea exports from Shanghai as its merchants and their foreign allies redirected the interior tea trade (fig. 4.5). Raw silk exports also soared as Shanghai's Chinese merchants diverted from the domestic market a small share of the huge silk production near the city; Canton merchants lost all competitive relevance by 1846 (fig. 4.5). Foreign demand for raw silk, as for tea, was such a small share of Chinese production that domestic demand and supply set silk prices for foreign markets. The financial underpinnings of the tea and silk trades altered during the two decades after Shanghai opened as a treaty port. Chinese merchants and bankers initially relied on their capital, supplemented by credit from large foreign firms such as Jardine Matheson, to purchase tea and silk in production districts, but explosive export growth from 1849 to 1853 temporarily overwhelmed Chinese firms' capacity to finance purchases. At that time, foreign firms in Shanghai experimented with sending compradors to purchase tea "upcountry" in the Fukien districts, and they employed the "Soochow system"; compradors were sent into the silk district of Soochow with opium to exchange it for raw silk. After 1853, foreign firms in Shanghai reduced direct interior purchases through their compradors and purchased commodities mostly in the Shanghai market.[25] This coincided with continued capital accumulation by local merchants and banks and new capital that wealthy gentry, merchants, and Shansi banks brought to Shanghai when the Taipings arrived in the Yangtze valley. Foreigners and Chinese, therefore, exchanged commodities and capital more efficiently at Shanghai than through *ad hoc* incursions of foreign firms into the hinterland. They could not compete with the superior social networks of capital that Chinese merchants and bankers forged.

Crisis in treaty port trade

The robust trade at Shanghai, disrupted during the Taiping rebellion, and the decline of trade at Canton, punctuated by Sino-British friction, pointed to the same brewing crisis that the Treaty of Nanking had not resolved. The expand-

[24] Greenberg, *British trade and the opening of China, 1800–1842*, pp. 22–40; Johnson and Supple, *Boston capitalists and Western railroads*, pp. 21–24; Johnson, *Shanghai*, pp. 122–54, 202–06; Jones, "The Ningpo pang and financial power at Shanghai"; Lockwood, *Augustine Heard and Company, 1858–1862*, p. 55; McElderry, *Shanghai old-style banks (ch'ien-chuang), 1800–1935*, pp. 11–14, 62–71.

[25] Hao, *The commercial revolution in nineteenth-century China*, pp. 60–64; Johnson, *Shanghai*, pp. 216–21; Le Fevour, *Western enterprise in late Ch'ing China*, pp. 148–52; Li, *China's silk trade*, p. 70; Lockwood, *Augustine Heard and Company, 1858–1862*, pp. 46–47; Morse, *The international relations of the Chinese empire*, vol. I, pp. 358–59, 466–67; Murphey, *Shanghai*, table 1, p. 119.

ing network of Chinese merchants in treaty ports north of Canton, especially in Shanghai, grasped new opportunities to prosper in trade with foreigners, but the scholar-elite vehemently opposed the treaty port settlement and aimed to regain political power. Looming in the background stood the Manchu government, weakened by defeat in the Opium War and its signature attached to an odious treaty. Resolution of this crisis by the early 1860s set the course for Sino-foreign trade and finance into the early twentieth century.

In 1842, the Manchus purged members of the scholar-elite from the central government and provincial bureaucracies, allowing them to implement a policy that avoided military confrontation with the British. Within two years, nevertheless, scholar-elite opposition crystallized, and shrewdly spread a net to rally factions throughout China, regardless of their public policy views. They proposed to revive scholarship-based patronage built on the examination system and mobilized their social networks to implement a two-prong strategy to return to power: absolve leaders of the literati for the military defeat of the Opium War and the humiliation of the Treaty of Nanking; and spread the views that a cowardly Manchu military lost to the British and the Manchus caved in to the demands of foreigners in the treaty negotiations. The literati fomented a confrontation at Canton because the Manchus continued to see it as part of China's periphery and they had few resources to station major military units there. The Manchus relied on Chinese gentry and wealthy Cantonese to lead and fund paramilitary units to suppress lawless elements, and they maintained scholar-elite in the provincial administration under the governor. When the literati instigated unrest in Canton directed against foreign merchants, the British chose to avoid confrontation because its relevance to trade had declined and Hong Kong served as the management center of Asian trade; Shanghai had become the entrepot to Central China for foreign firms. The British failure vigorously to pursue with the emperor in 1850 the issue of residence at Canton convinced him that the British were weak and unwilling to fight another war. The entire premise of Manchu diplomacy to appease treaty demands so as to avoid war collapsed; the scholar-elite regained power when the emperor opened his administration to them.[26]

The crisis in Kwangsi during the 1840s contributed to this breakthrough for the literati. Deflation, crop failures, and redirection of the tea trade to Shanghai that threw numerous porters and boatmen out of work, wracked this poor province, and in-migration of bandits from southeastern coastal provinces, who fled British pirate suppression, added professional lawbreakers. The weak Qing government could not organize a costly campaign to confront rampant civil disorder, and Kwangsi, which contributed minimally to central government revenues, remained peripheral to Manchu concerns; but failure to suppress disorder proved disastrous. The Taipings, a revolutionary

[26] Graham, *The China station*, pp. 230–53; Polachek, *The inner Opium War*, pp. 137–257; Wakeman, *Strangers at the gate*, pp. 90–105.

movement in Kwangsi and Hunan Provinces, emerged in 1850 to threaten the foundations of the Qing state, swept through the Yangtze valley, and captured Nanking by 1853. The literati gentry led the effort to defeat the Taipings and built military command structures cemented by kinship and pervasive loyalty networks rooted in the examination system and associated patronage. Local militarization gave lower-level gentry opportunities to entrench themselves in local government, and upper-level gentry formed regional armies; these organizations gained the capacity to siphon off tax revenues. By 1864, the Manchu and literati armies, with assistance from foreign allies, destroyed the Taipings. The devolution of authority to gentry at the local and regional level created powerful alternatives to the Manchu bureaucracy and military during the turmoil of the 1850s; the weakened Qing never recouped that authority during the post-war years. While members of the scholar-elite mobilized militia and armies to fight the Taipings, literati gentry in Kwangtung organized militia to battle growing disorder, and they deployed supporters in Canton to harass foreign merchants. Conflict heated to a fever pitch by 1856, when British naval forces attacked Canton and the factories of merchants were burned. In 1858, an Anglo-French force captured Canton; then, the allies transferred pressure to the north where the Taiping rebellion made the Qing vulnerable. Britain, France, Russia, and the United States extracted the Treaty of Tientsin in 1858 and applied additional pressure on a resistant government in the settlement of 1860.[27]

The treaty settlements of 1858 and 1860 and negotiations over the next several years imposed on a weak Qing state a greater insertion of foreigners into China's trade. The treaties legalized the opium trade, tripled the number of treaty ports, and opened the Yangtze River to trade as far as Hankow; under the terms of a perpetual lease, the British acquired Kowloon, the peninsula across the harbor from Hong Kong. Foreigners could participate in the coastal and river trade of China, a ratification of practices never granted under the Treaty of Nanking; that trade had grown rapidly during the 1850s. The Chinese Maritime Customs, the most extraordinary outcome of the settlements, became the pivotal institution of the foreign treaty rights. A British citizen served as Inspector-General and reported to the Chinese government; Robert Hart held the office from 1863 to 1908. The Maritime Customs set incorruptible standards and generated large revenues for the government that served as collateral for foreign loans, and commissioners of the Maritime Customs also served as financial counselors to the government.[28]

[27] Fairbank, *Trade and diplomacy on the China Coast*, pp. 243–61; Kuhn, *Rebellion and its enemies in Late Imperial China*, pp. 135–225; Kuhn, "The Taiping rebellion"; Michael, *The Taiping rebellion*, pp. 3–71, 198–99; Polachek, *The inner Opium War*, pp. 258–71; Wakeman, *The fall of imperial China*, pp. 163–74; Wakeman, *Strangers at the gate*, pp. 132–56.

[28] Dean, *China and Great Britain*, pp. 42–79; Fairbank, *Trade and diplomacy on the China Coast*, pp. 462–63; Fairbank, "The creation of the treaty system," pp. 249–63; Graham, *The China station*, pp. 385–86.

Unimportance of the China trade

The attention of Britain and its allies to the China trade, nevertheless, overstates its importance; it probably never exceeded 2 percent of total international trade. British leaders in the China trade gained little from exchange; as of 1827, before efforts to gain greater entry to the China trade, British exports to China constituted under 1 percent of total British exports. By 1854, a decade into the expansion of treaty port trade, British exports to China still fell well below 3 percent of total exports, and imports from China stayed far below 8 percent of total imports. India remained the crown jewel of British trade in Asia. Similarly, exports of the United States to China from 1831 to 1860 never exceeded 2 percent of its total exports, and imports from China remained under 6 percent of total imports. The unimportance of the China trade to Britain comports with the restrained approach its government took towards China. Viewed from China, the Opium War and battles over the next two decades that culminated in the settlements of 1858 and 1860 loom large, but in Britain's expansive geopolitical arena, Europe dominated. The permanent China squadron of the British navy based at Hong Kong never amounted to more than about six warships to police pirates and defend treaty rights. Britain did not covet imperial control of China, meaning that it did not want a second India, and its Foreign Office resolutely opposed a land war in China because it knew that naval force alone would never defeat the Chinese military. If China had maintained a powerful navy equivalent to what it possessed at the start of the fifteenth century, no treaty port settlements would have existed or their achievement would have come at enormous cost.[29]

The rise of Hong Kong and the elaboration of treaty port trade constituted a small political-economic process on the global stage from 1800 to 1860, but Chinese and foreign merchants and financiers forged intermediary bonds that served as the foundation for continued growth of Hong Kong as the hub of Asian trade and finance into the next century. Just when the British and other foreign trading firms believed that they had finally solidified the political status of the China trade with the settlements of 1858 and 1860, the underpinnings of trade and finance shifted. Chinese merchants and bankers accumulated capital and transportation and communication technologies changed. Chinese and foreign intermediaries, therefore, had to react to new conditions of competition and alter their behavior.

[29] Graham, *The China station*, pp. 407–21; Remer, *The foreign trade of China*, pp. 231–33; Schlote, *British overseas trade from 1700 to the 1930s*, appendix tables 4, 5, 18, 19, 24, 25, pp. 121–26, 156–60, 170–74; Schran, "The minor significance of commercial relations between the United States and China, 1850–1931," table 37, pp. 239–40.

5

Chinese and foreign social networks of capital

> In all the old staple branches of China commerce, the broker is taking the place of the merchant, or the merchant is becoming a virtual broker. When the period of transition is over, no doubt most of the larger firms with capital and credit will have transferred themselves from produce to industrial and financial enterprises – and will become, in fact, private bankers. Those of them, who, like ourselves have the management of steam lines and insurance offices are the luckiest, and it is significant that Jardines, Heards and others are all struggling to develop this branch of business.[1]

Transformation of transportation and communication

Technological innovations after 1860 that sharply raised the quality and speed of transportation and communication and lowered their cost enhanced the exchange of commodity and financial capital over the vast distances that separated Europe and North America from the Far East. Fixed dates mark these innovations: the opening of the Suez Canal in 1869 and the immediate rush to build steamships for Europe–Asia trade, and the start of telegraphic communication between Europe and the metropolises of Singapore, Hong Kong, and Shanghai by 1871. Their timing implies that they reshaped Asian trade and finance and powered economic growth and development of the region into the early twentieth century.[2] The innovations produced time–space and cost–space convergence: distant places drew closer together, and the most widely separated, which included Asia, Europe, and North America, captured greater relative benefits (fig. 2.2). This stimulated larger flows of commodities, passengers, and information, but the impact of transportation and communication improvements on trading and financial firms in Hong Kong and else-

[1] F. B. Forbes to M. Cordier, August 17, 1872, Frank Blackwell Forbes's Letter Books; quoted in Liu, *Anglo-American steamship rivalry in China, 1862–1874*, p. 139.

[2] Latham, *The international economy and the undeveloped world, 1865–1914*, pp. 17–39.

where needs scrutiny. Some innovations had immediate repercussions, whereas others generated complex feedbacks or did not reach fruition until decades later.

The Suez Canal and the steamship

Since the early 1840s, the Peninsular and Oriental Steam Navigation Company had provided steamship service between England and Calcutta, including transshipment overland across the isthmus of Suez, and it extended service to Hong Kong by 1845. These ships carried passengers and mail, but few cargoes could withstand the expense of overland transshipment. Fast clipper ships carried high-value cargoes of textiles from Britain to Asia and tea and silk on return voyages, whereas low-value commodities such as jute traveled on slower freighters. Steamships that circumnavigated Africa via the Cape of Good Hope gave clippers little competition because the cost of coal was prohibitive, and there was little opportunity to acquire cargoes along the African coast. From 1845 to 1855, economic growth in Europe and Asia stimulated greater flows of passengers and commodities between them. This sparked demands for improved accessibility, exhibited as renewed interest in a canal across the isthmus of Suez, but the business elite in Britain and France underwrote little of its cost. Their skepticism cautions against overstating the Suez Canal's impact on trade between Europe and Asia. When the canal opened in 1869, it instantly halved the 11,560 miles from Liverpool to Bombay and cut the distance to China by up to one-third for direct shipments. Trading firms quickly transferred high-value commodities to steamships because those goods benefited the most from regular service, the smaller insurance premiums that the improved quality of service made possible, and the lower interest costs on commodities with shorter transit time. By the end of the 1870s, steamships had driven clippers from the outbound Britain–China trade in textiles and the inbound trade in tea, raw silk, and silk goods, but lower freight rates contributed little to this rapid switch because freight costs constitute only small shares of the selling prices of high-value commodities. The competitive benefits for steamships over sailing vessels fell with distance; substantial freight cost advantages of steamships across the vast distances that separated Europe and North America from the Far East did not arise until after 1890. Improvements in steamships came incrementally from various sources, not from revolutionary changes induced by the Suez Canal and technological changes in steamships.[3]

[3] Farnie, *East and west of Suez*, pp. 7–54; Fletcher, "The Suez Canal and world shipping, 1869–1914," pp. 558–61; Harley, "The shift from sailing ships to steamships, 1850–1890," tables 2, 3, pp. 227–28, appendix 1, pp. 232–33.

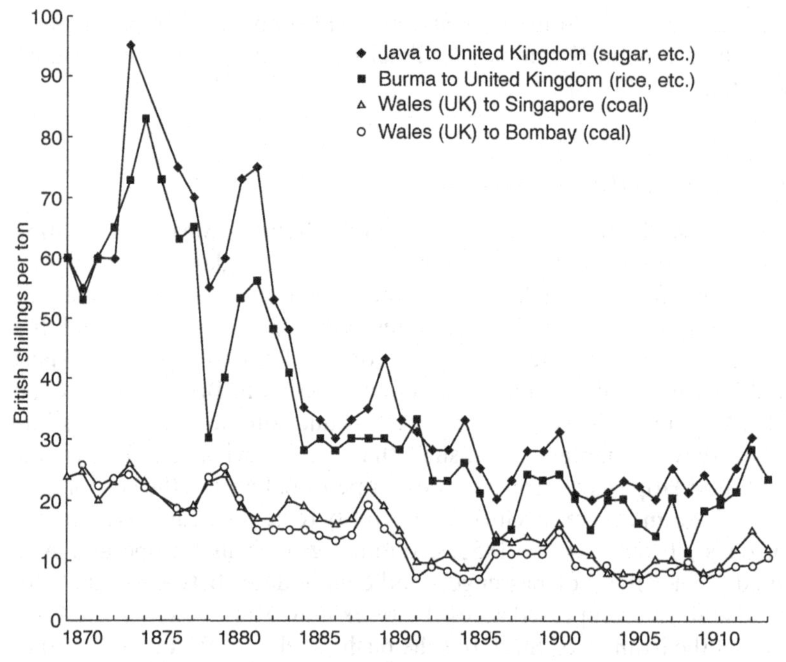

Fig. 5.1. Freight rates between Britain and Asia, 1869–1913.
Source: Latham, *The international economy and the undeveloped world, 1865–1914*, appendix 1, p. 193.

From 1869 to 1913, freight rates for a mix of high- and low-value commodities fell about two-thirds, implying a long-term stimulus to greater commodity shipments, but the decline proceeded sporadically (fig. 5.1). Rates did not fall until the early 1880s, over a decade after the Suez Canal opened, and the decline had run its course by the early 1890s; rates stayed within a band for the subsequent two decades. Steamships gradually displaced sailing vessels in the low-value, bulk commodity trades, first for shorter voyages and then for longer ones, shifts that lasted into the early twentieth century; yet, changes in freight rates and in the trade of the Far East did not move synchronously. China's exports were stagnant until the late 1880s, in a period which covered the opening of the Suez Canal, the steamship takeover of China's high-value commodity exports, and much of the fall in freight rates during the 1880s; Chinese exports then rose sharply to 1914 while freight rates remained in a band. Exports from the plantation and mining economies of Southeast Asia doubled in value from £19 million in 1870 to £38 million in 1890, implying some impact from the Suez Canal, the introduction of steamships, and the decline in freight rates; nevertheless, this export growth paled compared to the almost threefold rise to £106 million between 1890 and 1913 while freight rates stagnated.[4] Although the Suez Canal and the steamship benefited Asian trade, economic stagnation and growth in China and the intensification of European and American imperial exploitation of Southeast Asian colonies had greater impacts on the volume of trade. Trading firms utilized steamships, which had greater flexibility than sailing vessels, as finely honed organizational tools. The telegraph also offered such a tool to those firms, as well as to financial firms, but its impact also should not be overstated.

The telegraph alters intermediary activity

In 1870 and 1871, private companies laid telegraph cables across Siberia to Vladivostok and in the Sea of Japan and extended submarine cables from India to most major trading and financial centers of Asia, including Penang, Singapore, Batavia, Saigon, Hong Kong, Shanghai, Nagasaki, and Vladivostok. This gave Southeast and East Asia telegraphic connections to Europe via both India and Russia, and London was the global hub of the network that also included an Atlantic link to New York. Time-lags in information about the prices of commodities and currencies collapsed to about one day for the metropolises of Asia, Europe, and North America that were nodes on the network, this representing the culmination of a progressive reduction of time-lags during the 1860s as telegraphic short-cuts entered service. For traders and financiers, this eliminated risks of knowing spot (current) prices in widely separated markets and focused risk on speculation about future prices. Reduced risk from shorter time-lags between orders and actual purchases or sales, however, eroded the competitive advantages of large trading firms. Fast clippers and steamships that cost more than slow sailing vessels had conferred disproportionate information advantages on large trading firms in Asia because they had greater capital to purchase and charter those ships. They leveraged their capital to enter riskier transactions, whereas small firms could accept only limited risk; one miscalculation would decimate their capital. Because the telegraph company paid the fixed costs of construction and spread them over numerous customers, small firms could afford to send short telegraphic messages such as prices and orders to buy and sell commodities and currencies. The time between an order from London to buy in China and delivery in London for sale fell from about four months to only two, thereby reducing the amount of credit required; thus, small firms could now obtain the necessary credit. At each metropolis, all firms could obtain quotes of current and future prices, whereas previously only large trading firms had the sophisticated knowledge and judgment to identify those prices. In stan-

[4] Harley, "The shift from sailing ships to steamships, 1850–1890," pp. 222–25; Hsiao, *China's foreign trade statistics, 1864–1949*, table 1, pp. 22–25; Simkin, *The traditional trade of Asia*, table 5, pp. 308–09.

dardized trades, such as the export of tea and silk from China to Britain, small brokers executed orders from British buyers and competed with commission houses. The brokers' low overhead allowed them to operate with the slim profit margins that characterized these competitive trades, something large firms, such as Jardine, Matheson & Company, with their extensive support staffs and large infrastructures of docks, warehouses, and offices, could not do. These organizational and operational changes encouraged businesses to devise standardized indices of commodities, such as grades of tea. Brokerage firms, therefore, did not need to hire experts in commodity characteristics as the large trading firms had done, another erosion of their competitive advantage; this standardization contributed to advances in forming futures markets.[5]

Completion of the telegraph to Asian metropolises by 1871 reinforced London's dominance as the global trading and financial center, but the telegraph did not create that status. By the second half of the eighteenth century, London merchants had replaced Amsterdam merchants as the leading traders and suppliers of credit. During the first half of the nineteenth century, numerous international merchants moved their headquarters to London or established major branch offices there, and some leading London firms, such as Barings and the Rothschilds, shifted into merchant banking. After about 1830, British multinational banks with headquarters in London started branch-office networks in colonial possessions and in other countries, and by the time the telegraph reached the Far East, they already had branches to offer telegraphic transfer services to trading houses and brokers for shifting credit between London and Asian metropolises. The agglomeration of international trading and financial firms in the City, the area of London that housed them, created an enormous pool of capital and information; most new firms that aimed for global operations or those shifting to that enlarged scale of business had to base their senior officers there to participate in London's social networks of capital.[6] Because its firms operated at the hub of the global telegraphic network, they exploited business opportunities that required large amounts of capital before firms in other metropolises mobilized. On a smaller scale, trading and financial firms in Hong Kong, Singapore, and Shanghai reinforced their dominance in Asia through telegraphic connections to lower-level metropolises where many of them housed branches. The extension of the telegraph to Asia, the opening of the Suez Canal, and the switch to steamships

[5] Farnie, *East and west of Suez*, pp. 185–86; Lockwood, *Augustine Heard and Company, 1858–1862*, pp. 104–06; Marriner, *Rathbones of Liverpool, 1845–73*, pp. 112–18.

[6] Cain and Hopkins, *British imperialism*, pp. 141–60; Chapman, *The rise of merchant banking*; Chapman, *Merchant enterprise in Britain*; Jones, *British multinational banking, 1830–1990*, pp. 13–30; Marriner, *Rathbones of Liverpool, 1845–73*, pp. 99–111; Neal, *The rise of financial capitalism*; Steele, *The English Atlantic, 1675–1740*; Szostak, *The role of transportation in the Industrial Revolution*, pp. 49–80.

altered transaction costs for firms based in the metropolises. They reacted to new forms of competition and changed their specialization, organizational structure, and modes of operation, but pressures to change did not come only from technological innovations, most of which took decades to reach fruition. Trading and financial firms also contended with country-specific and region-wide changes in the economy, society, and polity, as well as with Chinese competitors.

Chinese intermediaries in China

The treaty settlements of 1858 and 1860 gave foreign merchants wide latitude to capture control of the trade of China through new treaty ports on the Yangtze and along the North China coast, the right to travel in China on passports, low tariffs, an honest customs service, and the right to participate in shipping on the Yangtze and along the China coast. Foreign merchants also had extensive capital to fund large-scale trades and sophisticated organizations with social network bridges to merchants in Europe who purchased Chinese exports and supplied imports; these advantages buttressed their confidence about profiting in the China trade. Treaties, nevertheless, could not resolve the dilemma that foreign merchants faced previously: lack of knowledge about the internal complexity of the Chinese market, the formidable language barrier, differences in currencies and weights and measures, a legal system without clear protections for property rights, and reliance on personal ties to deal with local bureaucracies. They still required compradors as inside contractors to bridge the gap between their firms and the Chinese market, continuing their principal–agent dilemma. As principals they needed a management layer of high-cost foreign employees to monitor their agents, the compradors. Even with trustworthy compradors, and most were, foreign firms had greater administrative costs than Chinese firms, thus placing them at a competitive disadvantage.[7]

The Chinese mercantile hierarchy

The foreign merchants' delusions about their capacity to wrest trade from Chinese merchants quickly crashed following the end of the Taiping rebellion. Between 1865 and 1870, foreign observers at most treaty ports witnessed Chinese merchants briskly reassert control over trade in domestic commodities and in import distribution and export collection. They had low overhead, excellent access to market information, low-cost credit from native banks, sophisticated mercantile operations, and, in the cruelest blow, Chinese

[7] Fairbank, "The creation of the treaty system," pp. 249–61; Fairbank, *China*, pp. 183–86; Hao, *The comprador in nineteenth-century China*, pp. 15–88. Also see discussion in chapter 4.

merchants even took advantage of efficient foreign steamship service to undercut foreign merchants; the trade of Shanghai exemplified their dominance. In 1866, Chinese merchants controlled the trade in Chinese commodities produced for domestic consumption (sugar, vegetable oils, hemp, tobacco) and shipped on foreign vessels between Shanghai and Chinkiang, the Yangtze River port at its junction with the Grand Canal, and they accounted for the majority of the textiles and opium shipped on foreign vessels from Shanghai to Kiukiang, a Yangtze port upriver from Chinkiang, and for much of the tea shipped from Kiukiang to Shanghai. They also controlled the trade in domestic commodities (silk piece goods, sugar, rice, vegetable oil, paper) between Shanghai and Tientsin, the port that served Beijing and North China, and by 1867 they handled most imported goods (textiles, opium) sent from Shanghai to Tientsin in foreign vessels. Chinese firms operated thirty branch offices in Tientsin; their headquarters were in Canton and Hong Kong (20), Swatow (7), and Foochow (3). Those firms headquartered in Tientsin ordered most imported goods directly from Shanghai firms, rather than from local branches of foreign firms, and twelve of these Chinese firms had their own agents in Shanghai.[8] Foreign firms, therefore, faced formidable competition from multilocational, hierarchically structured Chinese firms with headquarters in leading metropolises, such as Hong Kong and Shanghai, and branches in lesser metropolises along the coast.

Chinese merchants competed fiercely because they had built well-oiled domestic wholesaling networks for the leading imports (opium, cotton textiles) and exports (tea, raw silk) well before foreign merchants arrived at the treaty ports. By the late 1860s, Chinese merchants controlled most opium wholesaling at the treaty ports, with the exception of Shanghai, the key import center. After some success buying tea upcountry through compradors during the 1860s and initial forays to buy tea at the Yangtze ports of Hankow and Kiukiang, foreign firms retreated to purchasing tea from Chinese brokers on the Foochow market during the 1870s. Although foreign merchants dominated raw silk exports into the 1920s, they never sustained a presence after 1860 outside Shanghai, the major export center. Chinese silk merchants in Shanghai, including compradors for foreign firms and independent silk brokers, occupied the terminus of an interior wholesaling network, and they sold to foreign firms. The changes in volume and value of the leading imports and exports help explain the failure of foreign firms to penetrate markets outside the principal ports of Shanghai, Foochow, and Canton (table 5.1). Small economic gains of Chinese peasants between 1860 and 1910 kept per capita consumption of cotton textiles stagnant and left little margin for foreign merchants to boost exports. Sales of British textile firms to China

[8] Le Fevour, *Western enterprise in late Ch'ing China*, pp. 48–52; Liu, *Anglo-American steamship rivalry in China, 1862–1874*, tables 12, 15, pp. 64, 67, and pp. 108–10; Rawski, "Chinese dominance of treaty port commerce and its implications, 1860–1875," pp. 461–62.

Table 5.1. *Foreign commodity trade of China, 1867–1934*

	1867	*1884*	*1894*	*1904*	*1914*	*1924*	*1934*
				Exports			
				£ (millions)			
	18	19	21	36	51	145	108
				Percentage			
Raw silk	30	27	26	26	17	15	5
Silk goods	4	7	7	5	5	3	4
Tea	64	42	24	12	10	3	7
Beans & bean products	0	0	2	3	18	24	1
Eggs & egg products	0	0	0	1	2	4	6
Vegetable oils	0	0	1	2	3	6	6
Raw cotton	1	1	6	10	3	5	3
Other	1	23	34	41	42	40	68
Total	100	100	100	100	100	100	100
				Imports			
				£ (millions)			
	22	21	26	51	80	190	207
				Percentage			
Opium	50	35	20	10	6	0	0
Cotton textiles	25	37	44	51	43	22	4
Rice	2	0	6	2	4	6	6
Sugar	1	1	6	5	5	7	3
Cigarettes & tobacco	0	0	0	1	3	5	3
Kerosene & gasoline	0	0	5	8	6	6	6
Other	22	27	19	23	33	54	78
Total	100	100	100	100	100	100	100

Source: Hsiao, *China's foreign trade statistics, 1864–1949*, tables 1–3, 9a, pp. 21–124, 190–92.

never escaped the range of £5–7 million from 1870 to 1904 (fig. 4.3). Price falls severely pressured the profit margins of trading firms: from 1867 to 1894, the annual value of imported opium fell 51 percent and the price per pound plunged 53 percent; the volume of tea exports surged 69 percent from 1867 to 1886, but total value declined 29 percent; and the volume of raw silk exports rose almost threefold from 1872 to 1912, whereas the price per pound dropped 57 percent. These price falls required that firms boost their efficiency, but foreign traders had high-cost structures of expensive Western staff and relied on compradors to trade in China.[9] To their dismay, additional treaty ports after 1860 failed to offer competitive advantages against Chinese merchants, who successfully blocked foreign merchants in the domestic components of the import and export trades before the Suez Canal opened in 1869 and the telegraph reached treaty ports in 1871. Chinese merchants then used these technological innovations to challenge foreign firms in the international components of the import and export trades. They moved quickly, according to a satirical article written for *Harper's Magazine* in 1878 by Thomas Knox, an American merchant. By that date, Chinese importers and exporters operated in treaty ports with agencies in London, Marseilles, New York, and San Francisco, and even if Chinese firms did not have agencies, they established credit relations with foreign manufacturers to purchase goods directly. A full panoply of specialized trade and financial firms and organizations owned and managed by Chinese, including banks, insurance companies, steamship companies, and boards of trade, supported this challenge to foreign firms on more specialized forms of trade that required greater capital.[10]

The greatest competitors to foreign firms, paradoxically, operated in their midst. They chose the richest, most sophisticated Chinese merchants as compradors because they had the capacity to enlarge the firms' trade. Compradors continued to accumulate capital after 1860 through merchant activities for foreign firms, and when foreign banks entered treaty ports, they also hired compradors. The great firms, including Jardine, Matheson & Company, Dent & Company, and Butterfield & Swire, hired the wealthiest compradors whose large capitals lubricated the credit channels of wholesaling networks and native banks and who supplied the most complex market information. Compradors often left foreign firms to continue merchant trading as independent firms. By the 1870s, some compradors, including Jardine's Tong Kingsing, invested extensively in large enterprises such as steamship companies.[11] These compradors and other treaty port merchants had gentry status, a prerequisite for merchants who increased their specialization and scale of business and engaged in greater interregional and international trade. Their literati membership aided negotiations with local, provincial, and central government officials, and participation in literati information networks kept them informed about government policies. Some gentry used wealth gained from

[9] Bernhardt, *Rents, taxes, and peasant resistance*; Gardella, "The boom years of the Fukien tea trade, 1842–1888," pp. 40–50; Hao, *The commercial revolution in nineteenth-century China*, pp. 193–94; Hsiao, *China's foreign trade statistics, 1864–1949*, tables 2–3, 9a, pp. 52–54, 117–21, 190–92, appendix C, p. 297; Huang, *The peasant family and rural development in the Yangzi Delta, 1350–1988*; Li, *China's silk trade*, pp. 91–93, 157–62; Ma, "The modern silk road"; Ma, "The modern silk road," (unpublished paper), appendices 2, 4; Owen, *British opium policy in China and India*, pp. 214–79, 329–54; Pomeranz, *The making of a hinterland*; Rawski, *Economic growth in prewar China*, table 2.11, p. 97; Sargent, *Anglo-Chinese commerce and diplomacy*, diagrams D, F, pp. 330–31; Spence, "Opium smoking in Ch'ing China," pp. 152–53.

[10] Eiichi, "'The traffic revolution'"; Knox, "John Comprador," p. 431.

[11] Hao, *The comprador in nineteenth-century China*, pp. 89–105.

government positions to move into merchant occupations, and merchants increasingly purchased degrees to acquire literati rank.[12] Gentry merchants, therefore, formulated a sophisticated strategy that incorporated wide information networks and communities of trust to lower their risks of operating. The combination of gentry status and merchant increased the number of bridges to other social networks of capital, and this complemented their tendency to organize firms along family and clan lines.

The banking challenge

Bankers also had an incentive to acquire gentry status because all three levels of the Chinese banking hierarchy – pawnshops, native banks, and Shansi banks – had close connections to government. Family members often owned chains of pawnshops that served primarily as retail lenders to the farm population; government deposits represented low-cost sources of funds, and bank taxes could reach high levels. Native banks owned by gentry merchants, including compradors in the treaty ports, occupied the next level. These wholesale banks made commercial and industrial loans, issued and discounted commercial paper, remitted funds, and accepted deposits, primarily from wealthy gentry and Chinese businesses. Native bank orders were issued to Chinese merchants to pay foreign merchants; the latter deposited them in foreign banks, who willingly accepted bank orders because their compradors guaranteed them. The circle closed when foreign banks requested payments from native banks, who collected from Chinese merchants after they sold goods. Native banks also accepted checks drawn on foreign banks and issued bills of exchange to Chinese merchants to purchase goods in the interior. Some families and clans controlled networks of native banks, allowing the banks to engage in intra- and interregional financial transactions. Shanghai native banks, which had the greatest capital, captured the largest bank territories after 1860 because the owners' family and clan structures bridged to networks of merchants and bankers across China; they offered effective, cheap means to transfer capital and provide investment funds. Shanghai banks' "chop loans," so called because they carried the "chop," or guarantee, of the comprador, were short-term loans from foreign banks and large merchant firms. Because these loans often were continued indefinitely, they became surrogate deposits that allowed native banks to expand lending. Paradoxically, this transfer of capital from Europe and North America to China augmented lending by native banks to Chinese merchants, thus helping them expand, specialize, and compete better against foreign merchants. Although native banks depended

[12] Chang, *The income of the Chinese gentry*, pp. 149–95. Small-scale wholesalers and retailers would not require gentry status, and they could not afford to purchase degrees in any case. Wealthy gentry also would not bother to enter the small-scale mercantile activities, because they offered limited profit potential.

on foreigners for capital, the power of these banks came from control of information and lines of capital to areas of demand and supply in China. Shansi banks specialized in interprovincial remittances of merchant and government funds and occupied the highest level of the bank hierarchy. A small number of wealthy gentry merchants with close ties to the Qing government in Beijing controlled these banks. These merchants required those contacts because they remained vulnerable to "squeeze" from officials who deposited funds in Shansi banks and maintained regulatory control. As of about 1900, just thirty banks operated a wide network of 470 offices, especially in commercial cities and treaty ports, but in the late nineteenth century the growing capital of native banks allowed them to encroach on the interprovincial finance of Shansi banks. Because they never needed to compete vigorously for business, Shansi banks acquired few skills and market ties to fend off native banks.[13]

From 1860 to 1910, nevertheless, China's poverty and stagnant economy hindered the Chinese banks' expansion and shift to higher capitalization and specialization; as late as 1899, the capital of the largest Shanghai native banks did not exceed £10,000. Capital levels increased after 1912, when the growth of the economy accelerated, and this coincided with the emergence of modern commercial banks organized with stockholders and limited liability on the model of European and American banks. Between 1912 and 1921, the number of these modern banks rose from 9 to 123, with 343 branches; they underwrote huge volumes of loans issued by the Chinese government. Native banks thrived into the 1930s as industrialists added capital to that of the merchants, and banks underwrote factories and infrastructure.[14] From the 1860s to the 1930s, Chinese banks successfully repelled domestic challenges from foreign banks; the latter faced a number of difficulties: a formidable language hurdle; low-cost Chinese banks; complex, multiple currencies without a central bank; a legal system that did not protect financial contracts; and reliance on personal ties to deal with local, provincial, and national bureaucracies. Following the example of trading firms, foreign banks hired compradors to access the China market, but this prevented them from directly competing with Chinese banks because compradors also owned banks.

Chinese merchants hesitate

The British government imposed open markets on China in the treaty settlements, presumably to benefit its merchants, yet Chinese merchants made no sustained efforts to "appeal to force" to have their government counteract

[13] Chang, *The income of the Chinese gentry*, pp. 151–80; Hao, *The commercial revolution in nineteenth-century China*, pp. 77–80; McElderry, *Shanghai old-style banks (ch'ien-chuang), 1800–1935*, pp. 33–53.

[14] McElderry, *Shanghai old-style banks (ch'ien-chuang), 1800–1935*, pp. 33, 142–47; Rawski, *Economic growth in prewar China*, pp. 135–68.

British measures. This failure of Chinese merchants to attempt to subvert the treaties did not arise from a commitment to free trade. The richest merchants benefited from the salt and grain-tribute monopolies that the Chinese government granted, and they curried favor with the government to acquire banking franchises. The quick success of Chinese merchants in competing with foreigners for domestic components of foreign trade at new and old treaty ports, instead, probably convinced them that an "appeal to force" offered few benefits. Their approach to major export and import trades, however, implies that they hesitated to push these achievements too far. Gentry merchants and high gentry officials with large land-holdings might have developed tea plantations, employed abundant low-wage labor, and invested in capital-intensive processing technology to compete with British plantations in India and Ceylon for European and American tea markets, but they failed to take those steps.[15] Similarly, Chinese silk merchants made little attempt to improve sericulture and cocoon production on peasant farms and kept most filature factories safely in treaty ports. When domestic production of opium became legal after 1860 and output swelled after 1870, Chinese merchants could have organized opium farmers into low-wage, efficient suppliers for overseas Chinese markets in Southeast Asia, instead of conceding those markets to Indian opium. Because Chinese merchants in Southeast Asia controlled much of the sale of opium, those in China had secure outlets through Nanyang networks. Finally, Chinese merchants made only feeble efforts to start factory production of cotton yarn and cloth before 1910 and instead left the market to foreign factories, yet Chinese merchants had ample capital, a vast supply of low-wage workers, and access to foreign textile technology.[16]

The reluctance of Chinese merchants to transform these trades into competitive enterprises before 1910 seems paradoxical, given their capital, mercantile skill, information networks, and market opportunities. The government did not directly support agricultural improvements for merchants, presumably because the Qing state had weak organizational and financial resources. Sophisticated, gentry merchants, nevertheless, could have devised strategies to compensate for that neglect. Most agricultural improvements in the United States during the nineteenth century, for example, commenced as private actions: farmers improved land, organized agricultural societies, and lobbied

[15] Chang, *The income of the Chinese gentry*, pp. 127–47. High officials in the provincial and central governments had large incomes to build up extensive land-holdings. Their positions allowed them to resist surtaxes imposed by local officials, yet they could extract full rent from their tenants. The evidence for tea markets is from: Gardella, "The boom years of the Fukien tea trade, 1842–1888," p. 42, table 13, p. 72, and pp. 70–75; Sargent, *Anglo-Chinese commerce and diplomacy*, diagram G, p. 332.

[16] Allen and Donnithorne, *Western enterprise in Far Eastern economic development*, pp. 60–68; Chao, *The development of cotton textile production in China*; Li, "The silk export trade and economic modernization in China and Japan"; Liu, *Anglo-American steamship rivalry in China, 1862–1874*; Rawski, *Economic growth in prewar China*, table 2.10, pp. 93–94.

for governmental agencies at the state and federal level to support agricultural programs. Failure to invest long-term capital in fixed enterprises was the common thread in this hesitation of Chinese merchants. In the absence of a legal system that protected property rights, investors remained vulnerable to official squeeze and to uncertainty about changes in government officials. Investments in tea plantations, opium farms, raw silk farms, and silk filature and cotton textile factories, inevitably brought enterprises into contact with all regulatory layers of the local, provincial, and national bureaucracy. In contrast to high returns on investments in trade and finance or in large land-holdings rented to tenants, sunk capital investments in production looked risky from a long-term perspective.[17]

Chinese merchants, however, were not averse to the principle of fixed-capital investments. They invested substantial capital in leading steamship companies that foreign firms founded, such as the Shanghai Steam Navigation Company of Russell & Company in 1862 and the China Coast Steam Navigation Company of Jardine, Matheson & Company in 1872; nevertheless, top officials in Jardine's recognized a pattern to fixed capital investments of Chinese merchants. Besides selected cooperative ventures with foreign firms, they mostly invested in government-supervised, merchant-managed enterprises such as China Merchants' Steam Navigation Company (1873) and Kaiping Mines (1878). This organizational form was part of the self-strengthening movement of political and intellectual leaders who turned towards modern enterprises after 1870 to build the wealth and power of China. Gentry merchants, including compradors, in treaty ports led the private contributions, but incongruities in this organizational form quickly surfaced. Official meddling, the vulnerability of enterprises to demands for funds from bureaucracies, and the temptation of officials and merchants to milk firms of funds undercut their capital structures. By the 1880s, gentry merchants began to reduce their role as the unprofitability of enterprises became apparent. Provincial officials took more control and invested greater amounts of government funds, but this did not improve performance. This solution to the industrial transformation of China failed; no great industrial corporations built on production efficiency and market dominance had emerged by the late nineteenth century.[18]

The focus on problems of large, fixed-capital investments, however, misses the more consequential dilemma China faced after 1860: limited investment in small-scale industry. The status of industrial firms in China in 1912 reveals

[17] Chan, *Merchants, mandarins, and modern enterprise in late Ch'ing China*; Chang, *The income of the Chinese gentry*; Danhof, *Change in agriculture*; Feuerwerker, *China's early industrialization*; Li, "The silk export trade and economic modernization in China and Japan," pp. 85–99; Mann, *Local merchants and the Chinese bureaucracy, 1750–1950*.

[18] Chan, *Merchants, mandarins, and modern enterprise in late Ch'ing China*, pp. 235–43; Chang, *The income of the Chinese gentry*, pp. 149–95; Fairbank, *China*, pp. 217–21; Feuerwerker, *China's early industrialization*, pp. 5–28; Le Fevour, *Western enterprise in late Ch'ing China*, pp. 57–59; Liu, *Anglo-American steamship rivalry in China, 1862–1874*, pp. 24–36, 138–41.

the tragic consequences of that failure. Among 20,749 so-called factories, only 363 employed mechanical power; thus, most "factories" were workshops. Mechanics, retailers, unspecialized wholesalers, and professionals with limited amounts of capital invested in small manufacturing firms, such as processing mills, wagon shops, and machine shops, but few of these potential investors held gentry status because they could afford neither the time to go through the schooling and examination system nor the money to purchase degrees. Small-factory owners, therefore, had little defense against official depredation and harassment from the gentry and no protection for property rights, the same problem that gentry-merchant enterprises faced. Small and large factories, consequently, were risky fixed-capital investments. Limited development of small factories that served intraregional markets prevented China from entering this crucial phase of early industrialization. Bustling factory villages would have offered job opportunities for surplus agricultural labor, markets for farmers, and incentives for farmers to raise productivity, and vibrant intraregional markets would have undergirded the transformation to large-scale manufacture for interregional markets. At least through 1910, China remained almost devoid of bustling industrial satellites of metropolises, factory villages, and mining and lumber towns, such as existed in the eastern United States. China started this transformation after 1895, but it did not gain momentum until after 1910 as the political economy opened more to risk-taking investments; even then, the industrial share of the labor force reached only 0.4 percent by the early 1930s.[19] Thus, Chinese intermediaries, as well as foreign merchants, could not develop high levels of specialization in commodity trading and sophisticated finance before World War II.

Nanyang Chinese intermediaries

Foreign merchants and financiers also faced formidable competition from Chinese intermediaries in the Nanyang, the core of which comprised Malaya, the Straits, and the Indies, while its full extent included bordering areas of the Philippines, Indochina, Siam, and Burma (map 3).[20] Emigrants from South China, especially Fukien and Kwangtung Provinces, targeted these areas for centuries before foreign merchants and financiers arrived. The "overseas Chinese" of the nineteenth century consisted predominantly of males, but more females began to emigrate during the latter part of the century. This emigration had several consequences: agglomeration of males in plantation and mining areas; remittance flows back to China; circular migration as people returned to China; and intermarriage, muting Chinese identity. The actual

[19] Feuerwerker, *China's early industrialization*, table 2, p. 5; Meyer, "Emergence of the American manufacturing belt"; Rawski, *Economic growth in prewar China*, labor force figure from table 2.2, pp. 72–73; Rothenberg, *From market-places to a market economy*.

[20] Fairbank, *Trade and diplomacy on the China Coast*, p. 50.

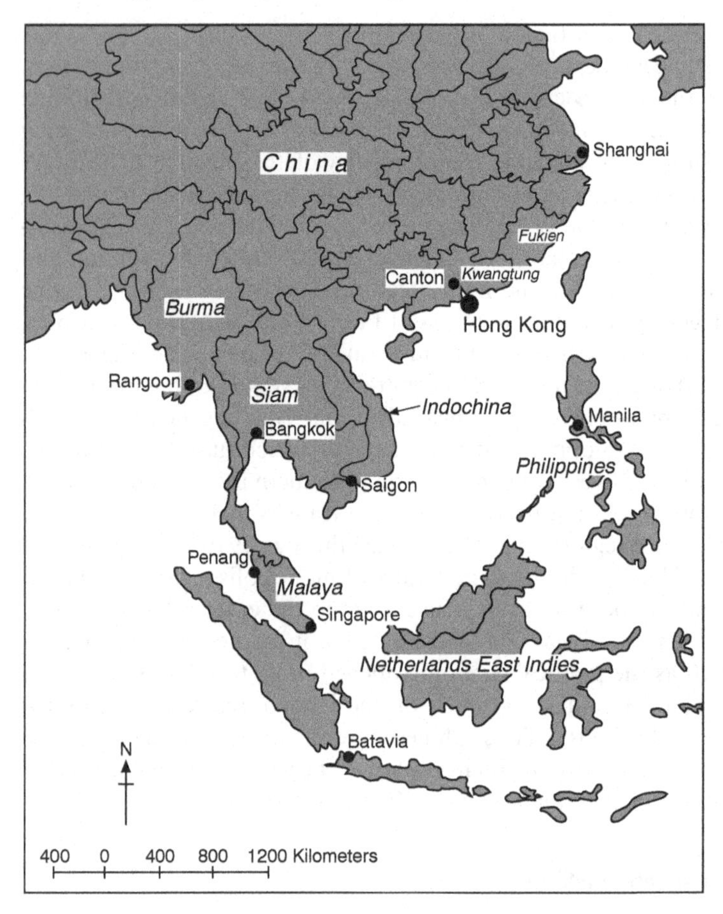

Map 3. Hong Kong and the Nanyang trade centers, *circa* 1870.
Source: author.

proportions of foreign-born Chinese in countries of the Nanyang rarely exceeded 1–2 percent, but the mix of residents with Chinese ancestry varied dramatically. The best data exist for Malaya; by the early twentieth century, up to one-third of its population had Chinese ancestry. Siam (Thailand) had the next largest proportion, possibly 10–20 percent at the turn of the century. Shares in other countries remained modest, but the importance of the Chinese did not rest on numbers. The Chinese diaspora to the Nanyang created labor supplies, markets, and flows of capital and commodities that were united through Chinese merchants and financiers who utilized family, clan, and ethnic ties to forge a formidable network of intermediaries.[21]

[21] Purcell, *The Chinese in Southeast Asia*.

The base of Chinese merchants

Ethnic groups such as the Chettiars from India or the Bugis from the Celebes Islands traded in portions of the Nanyang, but none approached the depth and breadth of Chinese trade networks. The family formed a redundant information network, but most merchants extended their activities to clans and broader Chinese ethnic networks. The multiple, non-redundant bridges across social networks of merchants maximized information flow, conveyed trustworthiness, and provided participants with powerful means to enforce sanctions. These networks reduced the risks of operating within each country where the Chinese were an alien minority, and their bonds aided merchants when they negotiated with national leaders, such as in Siam, or with colonial officials, such as in Burma, Malaya, the Netherlands East Indies, Indochina, and the Philippines. Their trustworthiness also lowered risks of trading among countries because they could not rely on colonial powers to defend their interests. Chinese merchants built their control over trade in the Nanyang on a base of retailers who served Chinese settlers, plantation workers, and miners. Unspecialized Chinese wholesalers supplied these Chinese retailers with consumption goods targeted towards immigrants, and in turn, the wholesalers captured the export trade of local economies. These unspecialized wholesalers maintained links to highly capitalized Chinese merchants in Saigon, Bangkok, Singapore, Penang, and Rangoon who traded with locally based foreign import and export firms or with large-scale Chinese firms engaged directly in international trade. Foreign firms could not compete effectively with Chinese firms in many of the trades; Siam and Malaya illustrate this structure of trade.

British and other foreign merchants entered Siam after the treaty of 1855 opened trade to them, anticipating that they could displace Chinese merchants whom the government had solicited to support royal trade. Foreign firms deployed large sailing vessels and steamships to drive much of the junk trade out of business and quickly gained control of shipping services, but they needed compradors to enter Siamese trade. As with the China trade, prosperous Chinese merchants in Siam, principally Bangkok, were logical candidates. High-cost foreign firms never effectively penetrated the domestic market because Chinese wholesalers headed well-oiled mercantile organizations with low labor costs and bridges across social networks from Bangkok through towns down to village retailers that gave them superior market information and efficient exchange of capital. Chinese firms in Bangkok augmented their dominance with the expansion of Chinese tin mining after 1860 and of Chinese pepper and sugar plantations up to the 1880s. Foreign firms remained active in some large-scale import and export trades, but they worked through Chinese wholesalers for the rest of their business. Compradors increased their control over trade and accumulated large amounts of capital in the service of foreign firms by the end of the nineteenth century. Chinese merchants in Malaya built their commercial prominence on continual infusions of coolie labor from China that numbered as many as 100,000 to 200,000 annually from 1881 to 1896; the operational headquarters of this trade stayed in Hong Kong, and Singapore and Penang served as redistribution points. These laborers often went to tin mines owned by Chinese firms and funded by Chinese merchants in Singapore, Malacca, and Penang. When rubber plantations expanded after 1900, coolies entered that sector, but Indians formed a somewhat larger share. From 1870 to 1900, Chinese merchants, led by large Singapore firms, operated an efficient credit machine that embraced advances of food and manufactured imports to mines and gambier and pepper plantations and payments with commodity outputs. They captured almost full control of Malaya's domestic trade and of some commodity exports, and they dominated imports of Chinese consumption goods. According to Thomas Knox, the American merchant who examined the China trade in 1878, this interpretation of Siam and Malaya applied to most Nanyang trade. Foreign merchants at each of the leading ports – Saigon, Bangkok, Rangoon, Singapore, Penang, Batavia, and Manila – dealt with Chinese wholesalers who purchased export commodities in the interior and controlled the distribution of imports. Hong Kong, in his view, was the pivot in the Chinese trade of the Nanyang with China.[22]

Chinese merchants cooperate with governments

The British and, to a lesser extent, French and Dutch promotion of a liberal international trade regime that maintained a somewhat *laissez-faire* approach created conditions for Chinese merchants to transform their advantages of emigrant markets and a powerful network of families, clans, and ethnicity into success.[23] These advantages clarify the Chinese merchants' capacity to defeat foreign firms in smaller-scale, less-specialized trades, but they do not explain their disproportionate dominance of domestic trade, even though the Chinese formed a minority of the population (except in Singapore where they were the majority). Colonial regimes, including the British in Malaya and Singapore, the Dutch in Indonesia, and the French in Indochina, followed strikingly similar policies between the 1860s and 1890s; they incorporated Chinese merchants into the revenue collection apparatus of the state. Siam, an independent monarchy, followed the same tack, implying that the motivation for such a policy did not rest solely on colonial exploitation. These states shared the dilemma of underdevelopment – extraction of revenue from a poor popula-

[22] Huff, *The economic growth of Singapore*, pp. 43–68; Knox, "John Comprador," pp. 433–34; Purcell, *The Chinese in Southeast Asia*, pp. 257–93; Skinner, *Chinese society in Thailand*, pp. 91–118; Wong, *The Malayan tin industry to 1914*, pp. 1–69.

[23] Elson, "International commerce, the state and society," pp. 142–51; Latham, *The international economy and the undeveloped world, 1865–1914*, pp. 65–94.

tion to run a government that had limited bureaucratic capacity; their solution was the "revenue farm," which rooted prior to 1860. The government awarded monopolies to firms or syndicates that gave them rights to sell goods, provide services (gambling), or collect taxes. They won monopolies by submitting the highest bids, and winners paid their bids to the government, guaranteeing it revenue. Opium farms, focused on sales to Chinese emigrants, accounted for the greatest revenue sources. In Siam and Singapore, whose populations had large proportions with Chinese ancestry, opium farms contributed 40–50 percent of government revenue.[24]

Chinese merchants and their syndicates tapped large amounts of capital from families and clans, and their international networks provided them with greater resources than merchants could acquire from domestic trade alone. This allowed them to outbid competitors to win revenue farms. Nevertheless, a paradox remains: colonial regimes and the Siamese monarchy permitted an alien minority to acquire extensive capital to compete with the native mercantile elite, and, in colonial regimes, this meant that the alien group also competed with firms from the home countries. These governments devised a brilliant tactical maneuver; as an alien minority, Chinese merchants could be expelled, thus they were unlikely to support subversion by Chinese laborers or to align themselves with native rebels. Because revenue farms represented thinly disguised taxes, Chinese merchants became identified as exploiters of peasants; therefore, rebel groups would not align with them. Colonial regimes also coopted the landed elite by granting them power to exploit peasants through manipulation of land ownership and direct coercion, and the Siamese monarchy also coopted the nobility by allowing them to enforce a client–patron relationship based on paternalism and interdependence. At the turn of the century, colonial regimes and the Siamese monarchy increasingly turned the government apparatus into modern, bureaucratic, centralized states whose power and efficiency to tax the population allowed them to dispense with revenue farming. This eliminated a profitable business of Chinese merchants, but leading ones had already accumulated substantial capital. These states continued to coopt Chinese merchants and financiers by granting them participation in funding government activities, and the landed elite and the nobility remained coopted through their incorporation into the highest levels of the government bureaucracy.[25] Within the embrace of the

[24] Butcher and Dick, *The rise and fall of revenue farming*; Ingram, *Economic change in Thailand, 1850–1970*, pp. 176–77; Murray, *The development of capitalism in colonial Indochina (1870–1940)*, pp. 75–77; Purcell, *The Chinese in Southeast Asia*, pp. 265–69, 430–31; Skinner, *Chinese society in Thailand*, pp. 118–25; Rush, *Opium to Java*; Trocki, *Opium and empire*, pp. 70–78.

[25] Adas, *The Burma Delta*; Brown, "Chinese business and banking in South-East Asia since 1870," pp. 187–88; Elson, "International commerce, the state and society: economic and social change," pp. 141–86; Murray, *The development of capitalism in colonial Indochina (1870–1940)*; Osborne, *The French presence in Cochinchina and Cambodia*; Rush, *Opium to Java*; Skinner, *Chinese society in Thailand*, pp. 91–125.

government, therefore, Chinese merchants and financiers acquired dominance of domestic trade within each state of the Nanyang, except Burma, even though they had the status of an alien minority, and they leveraged this with their international network to challenge foreign firms and defeat them in the greatest trade within Asia.

Dominance of the international rice trade

Large foreign firms, such as Jardine, Matheson & Company, had capital to fund the redistribution of rice among Asian countries and superior access to information about demand and supply through their headquarters in Hong Kong, branch offices, and correspondents at all major ports of Asia; but these advantages proved insufficient to control the international rice trade of Asia. With the exception of Burma, Chinese merchants captured most of this trade. The explanation of this paradox penetrates to the core of their competitive position. This rice trade required immense capital that commenced with advances to innumerable peasant farmers, and the circuit was completed when buyers in consumption countries paid for the delivered rice. First-stage intermediaries typically operated as brokers or independent merchants, and even if they were agents for rice mills or exporters, organization of this early stage had structural similarities. Intermediaries required a labor-intensive purchase and credit system, with a premium placed on low-cost operation, to deal with small-scale, fragmented production units. They granted credit to farmers indirectly through credit they furnished village retailers who received paddy in payment for goods, and rice buyers took paddy in payment for goods they supplied retailers. These transactions gave rice merchants entrée to the distribution of imported products such as food and manufactures. Rice had to be warehoused before aggregation into bulk quantities for shipment to rice mills, who purchased paddy from upcountry dealers and milled it. Rice mills exported rice or sold it to exporters, and the latter also might purchase paddy directly for export. With high fixed costs, foreign firms could not control paddy outside their headquarters or branch offices in ports, given this fragmented, intensely competitive production, credit, and supply system. The use of comprador offices for interior purchases, warehousing, and transportation of paddy introduced prohibitive costs, including comprador expenses and commissions and costly management of a fragmented supply system. Because the international rice market set prices, firms could not recoup these expenses through monopolies or oligopolies. Foreign firms could compete to purchase paddy at ports if they achieved economies of scale in milling, transportation, and bulk sale; they constructed this winning combination by buying in Burma and selling in Europe. Huge British investments in the Burma delta transformed it into a vast production machine; between 1865 and 1890, Europe took two-thirds of Burmese rice. British firms initially shipped sizable quantities of paddy to Liverpool and London rice mills, but soon they established large steam-powered mills in Burma. As late as 1898, European mills in Burma dominated production, whereas Chinese mills accounted for under 10 percent, reflecting the unimportance of Chinese merchants in earlier stages of the rice trade; this confirms the significance of that portion of trade. Chettiars from the Madras presidency of India were the dominant alien minority; they developed close ties to the Burma delta before Chinese merchants could establish themselves. Chettiar moneylenders provided most large loans for production and distribution, and the more numerous Burmese creditors focused on small loans.[26]

Prior to 1860, the rice trade in the Nanyang and East Asia comprised small shipments that balanced normal fluctuations in supply and demand and episodic jumps that coincided with demand surges during natural disasters and wars. After 1860, this trade commenced a sustained rise as mining and plantation areas in Malaya, the Netherlands East Indies, and the Philippines expanded and as demand for rice in China and Japan increased; Siam and the Mekong delta of Indochina constituted two huge supply sources. Chinese rice merchants captured control of the wholesaling of rice from the purchase of paddy from farmers to the sale to mills or export firms. Before 1880, Western steam rice mills in Bangkok and Saigon/Cholon attained modest success in buying paddy at the mills, but by the late 1880s, Chinese steam mills accounted for about two-thirds or more of the mills. Dominance of the supply of paddy to rice mills did not guarantee that Chinese merchants would control exports. Large Western firms might have employed branch offices at major ports that reported to their headquarters in Hong Kong, Singapore, or Shanghai to trade rice, but they failed. Chinese merchants copied that office structure and captured the entire trade from paddy farmer to consumer because the rice trade had a fragmented market structure of ports, interior towns, mines, plantations, and rural deficit areas throughout Asia. Western firms could not achieve sufficient economies of scale to compensate for their high-cost office structure. Singapore and Hong Kong were pivotal redistribution points in the Asian rice trade. Chinese rice firms headquartered in Bangkok used Singapore branch offices, and they joined with local merchants to redistribute rice to Malaya and the Netherlands East Indies. Most rice imported to Singapore came from Siam, whereas the volume of rice from Indochina remained small; supplies from Burma rose after 1880. Chinese merchants in Bangkok and Saigon/Cholon redistributed rice to China, Japan, and the Philippines through offices in Hong Kong or direct shipments. According to Thomas Knox, the American merchant, the dominance of Chinese merchants and this structure of the rice

[26] Adas, *The Burma Delta*, pp. 110–18; Cheng, *The rice industry of Burma, 1852–1940*, pp. 48–70, tables 4.5, 8.1, 8.2, pp. 85, 201, 203; Coclanis, "Distant thunder"; Ingram, *Economic change in Thailand, 1850–1970*, pp. 71–74; Latham and Neal, "The international market in rice and wheat, 1868–1914."

trade solidified by the late 1870s, if not earlier. By the first decade of the twentieth century, some Chinese rice merchants had accumulated sufficient capital that they started to specialize in the large-scale rice trade to Europe; in 1909, a Siamese firm began selling rice directly in European markets.[27]

Chinese merchant families and clans in the Nanyang who controlled the international rice trade had the most extensive bridges to social networks of capital in Asia, linking them to merchant firms in all major metropolises and to unspecialized trade of the Nanyang down to rural villages. During the late nineteenth and early twentieth centuries, they leveraged these advantages to build large merchant firms and to found banks. Between 1912 and 1940, Chinese families in Malaya, Singapore, and Siam, whose wealth came from international trade that included rice and rubber, shipping, and import and export businesses, shifted substantial capital into banking. Chinese banks numbered five in Malaya and the Straits Settlements, two in Siam, and two in the Netherlands East Indies. They established branches in Burma, the greatest source of surplus rice in Asia, and most branched to Hong Kong and China. Rather than compete directly with Western banks, they served their own trade and industrial operations and those of other Chinese firms. Chinese banks accumulated capital outside the reach of Western banks, who could not compete effectively for the business of Chinese firms.[28] Chinese merchants and bankers in China and the Nanyang, thus, built competitive enterprises based on low-cost management and organizational structures and extraordinary access to information through family and clan networks. Those networks also provided frameworks to gain trust, thus reducing risks of intermediary exchange both within and across uncertain political economies over which they had little control.

Foreign firms react to competition

Foreign firms faced a watershed in their reactions to competition during the 1860s. Chinese merchants reasserted control over the domestic commerce of China, even at the treaty ports, and they increasingly challenged foreign firms in import and export trades. Because the China trade did not expand much, foreign merchants could not readily react to the competitive challenge by deploying their greater capital into larger-scale, more-specialized commodity trades. Meanwhile, the gradual extension of the telegraph to Asia during the 1860s, with growing numbers of telegraphic short-cuts, reduced time-lags for

[27] Brown, "Chinese business and banking in South-East Asia since 1870," p. 174; Huff, *The economic growth of Singapore*, pp. 63–65, 102–05; Ingram, *Economic change in Thailand, 1850–1970*, pp. 70–71; Ingram, "Thailand's rice trade and the allocation of resources," table 2, p. 107; Knox, "John Comprador," pp. 433–34; Latham and Neal, "The international market in rice and wheat, 1868–1914"; Murray, *The development of capitalism in colonial Indochina (1870–1940)*, pp. 449–54; Skinner, *Chinese society in Thailand*, pp. 102–05.

[28] Brown, "Chinese business and banking in South-East Asia since 1870," pp. 174–83; Godley, *The mandarin-capitalists from Nanyang*, pp. 9–12.

information flows among Asia, Europe, and North America. As credit cycles shortened and price uncertainties lessened, firms required less capital to engage in import and export trades. The number of banks in China climbed from two in 1858 to twelve by 1865, and large London and New York insurance firms expanded their operations through agents in Asia, furnishing small firms greater access to low-cost credit and shipping insurance. This allowed unspecialized small trading firms in the treaty ports of China and ports of the Nanyang to shift to brokerage, and new, small firms entered business. Medium- and large-size trading firms, with high fixed costs in personnel and infrastructure, faced devastating competitive challenges.[29]

Failures

At the start of the 1860s, Rathbone Brothers & Company of Liverpool, England, looked confidently to the future. Their thriving commission business spanned Asia, the Americas, and Europe, and to manage this trade better, they shifted their headquarters to London at the end of the decade. Nevertheless, their fateful decision to close the Shanghai office in 1852 meant that they transacted commission business in Asia through correspondents in ports. This eliminated their direct participation in social networks of capital of sophisticated trading firms headquartered in Hong Kong and Shanghai and severed their bridges to other networks of capital in Asia. Their bridges from London to those metropolises reached correspondents who filtered information, making the Rathbones, as principals, dangerously dependent on other firms, as agents. They also followed a catastrophic business plan when they continued to operate as an unspecialized commission firm and increased participation in some trades to gain greater profits during the 1860s. The Rathbones sold diverse manufactures in China, traded in Indian cotton, exported tea and silk from China, dabbled in the Japan trade, and engaged in small trades in the Nanyang (rice, sugar, hemp). They even purchased vessels to offer shipping services, which became their second largest business between 1864 and 1874, but remained an adjunct to their commission trade. As banks undercut profitable trade in bills of exchange, they shifted surplus funds to short- and long-term securities but stayed out of the banking business. Their expansion of diverse trades involved greater capital investment at reduced profit margins while Chinese merchants and small-scale brokerage firms in Asia undercut them. To compete better, the Rathbones needed to reduce overhead costs and specialize, but they refused to do that. By the late 1870s, they were on a trajectory of declining profitability, and their commission business almost ceased during the early 1880s; by the end of the next decade the firm was moribund.[30]

In contrast, Augustine Heard & Company, the American commission

[29] See earlier discussion, and Lockwood, *Augustine Heard and Company, 1858–1862*, pp. 104–09.
[30] Marriner, *Rathbones of Liverpool, 1845–73*.

house, built a powerful organizational presence in the China trade during the 1850s. The firm moved its headquarters from Canton to Hong Kong in 1856 and established large branch offices in Shanghai and Foochow and agencies in Amoy and Ningpo during the 1850s. This gave the firm superb access to information about the China trade through participation in two great merchant agglomerations (Hong Kong, Shanghai) and through bridges to networks of capital in other major ports. Its "price circulars" provided Heard's with bridges to commercial houses in London and other major ports in Europe, the United States, and Asia. This approach to accessing information and managing the China business, nevertheless, did not guarantee success because Heard's accumulated high fixed costs with large Western staffs of well-paid employees and supporting staffs of Chinese compradors and assistants. During the late 1850s, this organizational machine effectively managed a diverse commission business: imports of cash, specie, bills of exchange, opium, and cotton goods; exports of silk and tea; shipping and ship chartering; and banking and acceptance business in the ports. At the start of the 1860s, however, Heard's participated heavily in businesses where competition from Chinese merchants, small Western brokers, and new banks would flare hotter. Heard's foolishly attempted to compete with low-cost Chinese firms in the rice trade and quickly accumulated a huge loss, and the firm dabbled in the steamship business on the Yangtze. It soon faced the reality of heavy competition and never committed to shipping beyond small-scale operations along major routes in East Asia. Heard's continued as a commission firm, but by the middle of the 1870s, it had declined to insignificance.[31] It failed to recognize that a large-scale firm with high fixed costs must specialize in intermediary activity that did not face fierce competition from low-cost, unspecialized, mostly Chinese, firms.

Trade services: path to success

The expansion of small brokers and unspecialized Chinese merchants who had limited capital created opportunities for well-capitalized firms to provide trade services, rather than compete in commodity trades. Low-cost Chinese junks dominated shipments of inexpensive, bulk commodities, but steamships offered competitive transport for high-value commodities, such as silk, tea, and manufactures. Firms needed substantial capital to own and run steamships, and regular service along the Yangtze and the China coast required at least four ships on each route, waterfront sites, wharves, and warehouses. During the early 1860s, about forty British and American firms in Shanghai operated steamships on the rivers and coast of China, but these offered irregular service. In 1862, Edward Cunningham, managing partner of Russell & Company in

[31] Lockwood, *Augustine Heard and Company, 1858–1862.*

Shanghai, organized the Shanghai Steam Navigation Company to provide regular service on the Yangtze and along the North China coast. The ownership and capital structure of about $1.4 million, following well-honed intermediary principles, comprised a cooperative venture of Russell's partners, small British merchant firms, and Chinese merchants, including three former compradors of Russell's. By 1867, Shanghai Steam Navigation had twelve ships and monopolized steamship service on the Yangtze, servicing Hong Kong, Ningpo, and Tientsin; five years later it had expanded service to other ports using a fleet of nineteen ships. That same year, well-heeled British firms commenced operations to garner a share of the large profits: Butterfield & Swire founded the China Navigation Company to service the Yangtze trade; and Jardine, Matheson & Company started China Coast Steam Navigation Company to service the North China coast trade. In 1873, the China Merchants' Steam Navigation Company, organized under the principle of Chinese government supervision and merchant management, began operations.[32]

Nevertheless, the meager growth of the Chinese economy from 1860 to 1910 did not offer endless expansion opportunities to steamship companies; steamship trade over a broader territory of Asia provided alternatives. Individual tramp steamers that moved among large and small ports offered stiff competition for small cargoes of high-value or bulky commodities. The smaller capital required for these vessels, compared to regular steamer service, was a wedge for Chinese merchants to form small steamship companies and compete for inter-port trade. During the 1880s, for example, steamers of Thio Thiau Siat started to dominate the trade of Malayan and Netherlands East Indies ports with close ties to Singapore. A steamship line that offered regular service among major ports of East and Southeast Asia, such as Shanghai, Hong Kong, Saigon, Bangkok, and Singapore, offered another alternative. However, a grander scheme combined that trade with the Europe–Asia trade: Alfred and Philip Holt founded the Ocean Steam Ship Company in 1865 to compete for the China trade, and by 1870, a year after the opening of the Suez Canal, the Holt firm had forged agreements with major merchant firms, including Butterfield & Swire, with offices in Shanghai and Hong Kong, and Walter Mansfield, with offices in Singapore, to serve as shipping agents to acquire cargo and passengers. This shrewd strategy gave the Holt firm entrée to the three leading centers of merchant firms in Asia, and their agents had an extensive array of bridges to other trading centers in the region.[33] The Ocean Steam Ship Company exemplified the shift to trade services based on large-scale capital investments, but this represented a narrow specialization in trade services, one always susceptible to competition from equally well-capitalized firms.

[32] Liu, *Anglo-American steamship rivalry in China, 1862–1874*.

[33] Godley, *The mandarin-capitalists from Nanyang*, pp. 11–12; Hyde, *Blue Funnel*.

Butterfield & Swire followed a different tack; it started in merchant trade and shifted to trade services. In the early 1860s, John Swire & Sons, a minor commission house in Liverpool, England, operated diversified trade with the United States, Europe, Australia, and Japan, but the contagious enthusiasm of British merchants to trade with China following the opening of new treaty ports inspired Swire's to commence textile shipments to China and Japan, and by the late 1860s it started to import tea and silk. In 1866, Swire's founded Butterfield & Swire to manage its Asian business and a year later opened a branch office in Shanghai. By 1870, the firm consolidated British operations at a London headquarters, gaining access to information and business opportunities in that global metropolis, and added branches in Hong Kong, Foochow, and Yokohama (Japan); later, it opened branches in Swatow, Amoy, Hankow, Tientsin, and Kobe (Japan). The branches in the pivotal nodes of Shanghai and Hong Kong, which they entered early, provided social network bridges to other ports in Asia and to their London headquarters. Swire's sweeping access to information set it apart from Rathbone Brothers & Company, which attempted to run its Asian business from London; nonetheless, Swire's confronted the conundrum of the China trade: limited growth opportunities in a stagnant economy and fierce competition from small brokers and Chinese merchants. Between the 1870s and the 1890s the firm shrewdly kept costs under control, specialized in a few trades, and exited from those that experienced declining profitability, such as textiles and tea, thus preventing erosion of its capital.[34] Butterfield & Swire diverged from Alexander Heard & Company, which never reduced costs, and from both Heard's and the Rathbones, who stayed in too many trades and failed to leave them before losing capital.

Swire's quickly recognized that trade services offered lucrative opportunities. Starting with its investment in the Holts' new line, the Ocean Steam Ship Company, in 1865, the firm developed a range of interests: it became shipping agent for the line through its branches in Hong Kong and Shanghai by 1870; it accumulated insurance agencies at its Asian branches after 1870; it founded China Navigation Company to compete for Yangtze trade by 1872 and merged that with Coast Boats Ownery, which engaged in China coast trade, by 1883; and it started Taikoo Sugar Refinery Company in Hong Kong to draw on raw sugar from the Philippines and Java in 1881. Swire's focused on trade services after 1880, including a shipping line, shipping agencies for other lines, insurance, sales agencies in China for Standard Oil's kerosene, and agencies for banks. This produced swelling profits from 1868 to 1900, in sharp contrast with the failures of the Rathbones and Heard's. Swire's shrewdly perceived that low-cost Chinese merchant firms, with superb domestic and international social networks of capital in Asia, competed formidably in

[34] Marriner and Hyde, *The senior John Samuel Swire, 1825–98*, pp. 1–57.

unspecialized, small-scale commodity trades, and the impoverishment of China and the Nanyang meant that few commodity trades would grow and offer opportunities for economies of scale from specialization. Chinese firms, however, had insufficient capital to specialize in trade services. For some services, such as sales and distribution agencies for foreign manufacturers and insurance companies, Swire's drew on its social networks that bridged from Asia to London, whereas Chinese firms initially had no access to those networks; that came later.[35] Jardine, Matheson & Company, in contrast to Swire's, committed enormous capital to commodity trades, placing itself in a treacherous position; failure to move expeditiously would dangerously erode that capital.

The trade giant transforms

Jardine Matheson, like other firms, dreamed of opportunities at the new treaty ports; it opened agencies or branches at Tientsin, Newchwang, and Swatow in 1861 and at Hankow in 1862. That organizational expansion seemed astute as Jardine's opium sales to China, the cornerstone of its profits since the Canton days, rose from 1860 to 1872; yet, Jardine's virtually ceased trading opium by 1873, an astounding reversal. The roots of this decision can be traced back to the mid 1850s when David Sassoon & Sons of Bombay and Hong Kong decided to increase their investment and specialization in the opium trade through backward vertical integration in India, boosting advances to Indian dealers and reducing costs of operation there, and bulk sales in China. Although Jardine's operated sophisticated, low-cost services in currency exchange, shipping, and insurance, they refused to react to competition by following Sassoon's strategy. It had an office in Hong Kong, but Bombay was Sassoon's chief operational site; then its senior officers shifted to London by the 1870s and viewed Asian trade from that end. This made the firm heavily reliant on agents in Asia who filtered information about the opium trade in China. When the price of opium in China started to fall after 1867, Sassoon's kept its capital tied up in the trade, but domestic opium producers were expanding and offering stiffer competition to foreign importers. In contrast, Jardine's prescient decision to leave the opium trade followed from its superb social networks of capital: its management structure placed the most sophisticated officers in the Hong Kong headquarters, and secondarily, in Shanghai; and Jardine's network bridges reached from the headquarters to the treaty ports of China where its top-flight compradors utilized their networks inside China to keep the firm appraised of the market context for the falling price of opium. To react to competition from Sassoon's and the Chinese domestic opium traders, Jardine's had to invest extensive fresh capital in the opium

[35] Marriner and Hyde, *The senior John Samuel Swire, 1825–98*.

trade, increasing its specialization in that business. Its decision to leave the trade swiftly, therefore, avoided the drain on capital and freed vast amounts for redeployment. That decision was repeated in other commodity trades; by the early 1880s, Jardine's seldom traded cotton goods on its own account and sold most textiles on consignment at auction, and it commenced a withdrawal from the silk and tea trades.[36]

Although Jardine, Matheson & Company probably debated the decision to shift from commodity trades to trade services by the late 1860s, it formulated an explicit policy around 1872. In that year Jardine's ended significant involvement in the opium trade, freeing huge amounts of capital, and simultaneously, Francis Johnson, Jardine's manager of the Shanghai branch, led efforts to form the China Coast Steam Navigation Company; Jardine's supplied two-thirds of the capital. The firm had operated steamships for over a decade in Chinese waters and between India and China, but steamship companies now became a major business. Excluding capital employed in commodity trades and loans, Jardine's committed the enormous sum of £605,507 to trade services by 1885, distributed among sugar refining (10 percent), wharves and warehouses (29 percent), steamship lines (37 percent), and banking and insurance (24 percent) (table 5.2). Its emphasis on financial services was based on earlier adjuncts to its commodity trades. The Canton Insurance Company dated from the Canton days, but starting in the early 1860s, Jardine's enlarged services as an agent for other insurance companies; by the early 1870s, it served at least eight firms. Banking services originated in Jardine's traditional provision of credit to smaller foreign and Chinese merchants, but it declined to participate in founding the Hongkong and Shanghai Bank in 1864, possibly because Dent & Company, Jardine's bitter rival, took an active role. At the time, Jardine's had sufficient capital to operate as its own bank, but by 1877, it reversed course and William Keswick, taipan of the firm in Hong Kong, joined the board of the Hongkong Bank when capital requirements for Jardine's activities exceeded its resources. The firm's sophisticated networks that bridged to leading Chinese merchants, who in turn bridged throughout China, including provincial and national bureaucracies, gave Jardine's insights that fixed capital investments were risky; therefore, following the lead of major Chinese merchants, it scrupulously avoided those investments outside the treaty ports. Jardine's, however, recognized opportunities in loans to the Chinese government, especially if customs revenue secured them. By 1875, it granted loans for administrative purposes, and by the mid 1880s, Jardine's often cooperated with the Hongkong Bank on loans to the government. The Sino-Japanese War (1894–95) revealed weaknesses of the self-strengthening movement and convinced Jardine's to cease acting as a financier on its own. After that date, the firm formalized cooperative relations with the Hongkong

[36] Jackson, *The Sassoons*, pp. 37–59; Le Fevour, *Western enterprise in late Ch'ing China*, pp. 25–59, 152–71; Roth, *The Sassoon dynasty*, pp. 50, 108–41.

Table 5.2. *Investments of Jardine, Matheson & Company, 1885*

Activity	£	*Percentage*
Processing		
China Sugar Refining	29,068	4.8
Luzon Sugar Refining	31,712	5.2
Wharves and warehouses		
Hunt's Wharf Property	56,584	9.3
Jardine Piers and Godowns	87,583	14.5
Hong Kong & Whampoa Dock Company	21,583	3.6
Shanghai & Hongkew Wharves	8,193	1.4
Steamship lines		
Hong Kong, Canton & Macao Steamship Company	10,497	1.7
Indo-China Steam Navigation Company	216,326	35.7
Banking and insurance		
Hongkong and Shanghai Bank	122,766	20.3
Canton Insurance Company	9,526	1.6
Hongkong Fire Insurance Company	11,669	1.9
Total	605,507	100.0

Source: Hsiao, *China's foreign trade statistics, 1864–1949*, table 9a, pp. 190–92; Le Fevour, *Western enterprise in late Ch'ing China*, p. 175.

Bank; Jardine's would leave finance to specialized institutions.[37] Jardine, Matheson & Company, therefore, commenced a transformation before its capital base was eroded and successfully implemented a massive reorganization over three decades following 1860, in contrast to the Rathbones and Heard's who lingered too long in declining trades. Rather than attempt to compete with new, specialized commodity brokers and swelling ranks of ever more specialized Chinese merchants, Jardine's deployed its large capital to specialize in trade services for them, thus maintaining its prominence in Asian trade and finance. Ultimately, Jardine's engaged in finance indirectly because that sector, especially banking, increasingly became a specialized intermediary activity in Asia after 1860.

Banks displace merchants in finance

The poverty of China and Southeast Asia before 1860 inhibited indigenous capital accumulation; therefore, interest rates stayed high in trading ports and

[37] Le Fevour, *Western enterprise in late Ch'ing China*, pp. 60–140; Liu, *Anglo-American steamship rivalry in China, 1862–1874*, pp. 135–50; King, *The Hongkong Bank in Late Imperial China, 1864–1902*, pp. 52, 335.

reached extortionate levels in rural areas. Because banks remained undercapitalized and merchants had limited capital to provide as credit or loans, foreign merchants lubricated the finance of exports and imports. Big Agency Houses, such as Jardine, Matheson & Company and Dent & Company, dominated the provision of credit, loans, currency exchange, and the acceptance business of bills of exchange for large-scale trades, and medium-size firms, such as Augustine Heard & Company and Russell & Company, filled the next tier of finance; even small trading firms participated somewhat in trade finance. Consequently, specialized financial institutions such as banks captured little business in that poor economic setting.

Eastern exchange banks

As the dominance of the East India Company in India faded, Eastern trade expanded, and trade rivalry intensified during the 1850s, opportunities opened for British banks to enter trade finance in competition with Agency Houses. London boards of Eastern exchange banks included leading British bankers and merchants with impeccable ties to the financial and mercantile communities, and they formulated strategies to optimize deployment of capital in London and among Asian countries. The London headquarters provided access to the cheapest capital in the world and the most efficient markets for investing and lending. A typical progression consisted of branch offices in Bombay and Calcutta, followed by offices in Hong Kong and Shanghai; by 1859, the Chartered Bank of India, Australia and China, the Commercial Bank of India, and the Oriental Bank had offices in Hong Kong and Shanghai. They followed that expansion mode because trade flows among Britain, India, and China generated extensive demands for currency exchange and bills of exchange to settle international accounts. The astute choice of offices in major mercantile centers embedded these banks in pivotal social networks of capital in Asia that provided access to sophisticated information about trade, finance, and politics. Information flowed between branches in Asia and the London headquarters, and after the telegraph reached most Asian ports by 1871, they used telegraphic fund transfers to settle accounts rather than relying on the cumbersome acceptance of bank drafts and bills of exchange. Branches had orders to fund local lending with local deposits, but capital and surplus funds were invested in safe British securities, providing a sound, growing stock of capital to underwrite expansion. Starting in the late 1850s, Eastern exchange banks in the treaty ports of China and elsewhere in Southeast Asia supplied credit to small, specialized brokers in the import and export trades who challenged the commodity trades of medium- and large-size firms, eroding their trade and financial business. These banks benefited from specializing in the complex transactions necessary to equilibrate international financial flows, but that did not constitute their chief advantage; Agency Houses also internalized that activity within their organization.

Banks gained a competitive edge by raising large amounts of capital on the London financial market to underwrite business and deploying all of it to support financial transactions. Only the largest Agency Houses, such as Jardine, Matheson & Company, could devote equivalent capital to finance because trading firms also needed capital for commodities, shipping, warehouses, and other components of long-distance trade. During the 1870s, even the greatest Agency Houses could not keep pace with the capacity of exchange banks to add capital for expanding financial services.[38] These Eastern exchange banks, mostly headquartered in London (though some started in India), reached to East and Southeast Asia to control financial transactions. Their branches had substantial decision-making authority, but the nerve center remained in London; in contrast, the Hongkong Bank housed its decision-making center within the Far East.

Hongkong and Shanghai Bank

The prospectus of the Hongkong and Shanghai Banking Company, dated 1864, stated that "The Scheme of a Local Bank for this Colony, with Branches at the most important places in China, has been in contemplation for a very long period." Leading trading firms in the Far East had experienced wrenching changes as resurgent Chinese merchants and small specialized brokers in the treaty ports of China eroded their trade and as Eastern exchange banks competed with them to finance trade. Some big houses presciently recognized that they must react to competitive threats by founding a new organizational entity that specialized in finance. The directors comprised trade rivals from British, German, American, and Parsee (India) firms, testimony to their capacity to cooperate when faced with severe competition in trade and financial business. The Hongkong Bank entered a fiercely competitive market: about ten British, Indian, and French banks operated branches in Hong Kong by 1864–65, and many had branches and agencies in other Far Eastern ports; five bank agencies operated in Hong Kong; and eleven banks, including six headquartered in India and five in London, planned to establish operations in the Far East. Confronted with those rivals, the founders of the Hongkong Bank leveraged their huge capital and created one of the world's largest banks capitalized at £1.1 million. They aimed to finance the most important trade of the Far East, trade with China and, secondarily, that with Japan. Consistent with that goal, they formulated a management structure within the first few years that comprised a Hong Kong headquarters, with responsibility for global exchange of funds and broad oversight, a Shanghai branch in charge of

[38] Baster, *The international banks*, pp. 160–63; Cain and Hopkins, *British imperialism*, pp. 107–40; Jones, *British multinational banking, 1830–1990*, pp. 13–62; King, *The Hongkong Bank in Late Imperial China, 1864–1902*, p. 335; Lockwood, *Augustine Heard and Company, 1858–1862*, pp. 106–09; Marriner, *Rathbones of Liverpool, 1845–73*, pp. 201–22; MacKenzie, *Realms of silver*, pp. 28–70.

finance for Central and North China and with oversight of the Japan branch, and a Yokohama branch directly in charge of finance in Japan. The directors astutely established a branch office in London, even though they were wary that it might try to dominate the affairs of the bank, and they arranged a credit line of £1.6 million with London and County Bank, one of Britain's largest. The bank quickly established agents, including offices of some directors, that operated in most of China's treaty ports and in financial-trading centers in Southeast Asia, India, the Americas, and Europe (table 5.3).[39]

The Hongkong and Shanghai Bank, therefore, vaulted into the ranks of the world's most sophisticated banks with unparalleled access to capital; a branch in London, pivot of global trade and finance; headquarters in Hong Kong, hub of the Far East; and far-flung merchant-agents that kept the bank informed of global trade and finance. Within the Far East, it operated through Chinese compradors, similarly to trading houses and other large banks, and its prestige and size allowed the bank to attract the leading Chinese merchants and financiers to its ranks. These compradors provided superb access to information about trade and finance and were conduits for granting loans to Chinese merchants and banks within China and throughout the Nanyang. No other Asian bank had such sweeping bridges to the social networks of capital within Asia and across the globe. As business expanded at metropolitan centers with agencies or merchant-agents, they were converted either to branches or agencies, and the dates of founding these offices from 1864 to 1918 convey the history of the transformation of Asian trade and finance (table 5.4). Besides global oversight, the Hong Kong headquarters supervised agencies in treaty ports from Foochow to Canton, whereas the Shanghai branch supervised those in Central and North China, and the creation of agencies in North China after 1880 underscored the growing importance of that trade. The headquarters closely monitored agencies in Saigon and Bangkok as those ports became centers of the rice trade, especially with China and Japan; the Saigon office also handled investments in sugar plantations in Vietnam. The importance of the India–China trade forced the Hongkong Bank to quickly upgrade the status of its offices at the pivotal metropolises of Calcutta and Bombay. The bank inserted itself into trade finance in the Nanyang as plantations and mines expanded in those economies after 1880, and branch officers increased commitments of the bank's capital to finance economic activity in their territories of operations, such as natural resource development, imports, exports, and infrastructure (ports, railroads), besides handling exchange operations for the bank. The Singapore branch, which served as the control center for trade finance in Malaya, led financing of the rubber industry and the Penang branch provided money for locally based trading and

[39] King, *The Hongkong Bank in Late Imperial China, 1864–1902*, pp. 52–102 (quote, p. 73), table 3.4, p. 89. The estimate of the capital of the bank in pounds sterling was converted from Hong Kong dollars using Tom, *The entrepot trade and the monetary standards of Hong Kong, 1842–1941*, appendix 15, pp. 151–52.

Table 5.3. *Earliest branches, agencies, and agents of Hongkong and Shanghai Bank, 1864–1866*

Branches and bank agencies	
Hong Kong	
Shanghai	
London	
Yokohama	
Agent offices	**Agents**
India	
Bombay	B. and A. Hormusjee
Calcutta	McKillop, Stewart and Co.
	Nanyang (Southeast Asia)
Singapore	Borneo Co., Ltd.*
Bangkok	Messrs. Pickenpack, Thies and Co.
Saigon	Messrs. Behre and Co.
Manila	Russell, Sturgis and Co.
China	
Foochow	Gilman and Co.*
Amoy	Tait and Co.
Ningpo	D. Sassoon, Sons and Co.*
	Davidson and Co.
Swatow	Bradley and Co.
Kiukiang	Augustine Heard and Co.*
Hankow	Gibb, Livingston and Co.
Japan	
Yokohama	Macpherson and Marshall
Australia	
Sydney/Melbourne	Union Bank of Australia
France	
Paris	La Société Général de Crédit Industrial et Commercial
United States	
San Francisco	Bank of California
Chile	
Valparaiso	Messrs. Th. Lachambre and Co.

Note:
* Member of Board of Directors
Source: King, *The Hongkong Bank in Late Imperial China, 1864–1902*, table 3.5, p. 95.

Table 5.4. *Branches and agencies of Hongkong and Shanghai Bank, 1918*

	Year opened		Year opened
London	1865	YOKOHAMA	1866
San Francisco	1875	Kobe/Hiogo	1869
New York	1880	Nagasaki	1891
Lyons	1881	Taipei	1909
Hamburg	1889		
		SINGAPORE	1877
HONG KONG	1865	Penang	1884
Foochow	1867	Malacca	1909
Saigon	1870	Kuala Lumpur	1910
Amoy	1873	Ipoh	1910
Bangkok	1888	Johore	1910
Canton	1909		
		CALCUTTA	1867
SHANGHAI	1865	Rangoon	1891
Hankow	1868		
Tienstsin	1881	BOMBAY	1869
Peking	1885		
Hongkew	1909	COLOMBO	1892
Tsingtau	1914		
Harbin	1915	MANILA	1875
Vladivostok	1918	Iloilo	1883
		BATAVIA	1884
		Sourabaya	1896

Note: Offices named in capitals supervised offices that appear inset below each of them.
Source: King, *The Hongkong Bank in the period of imperialism and war, 1895–1918*, table 2.1, p. 92.

import/export firms, but the bank expanded cautiously because the Chartered Bank of India, Australia and China had captured an early financial lead. By the 1880s, the Hongkong Bank strengthened its financial bridges to North America and Europe, testimony that Hong Kong was the hub of Asian trade and finance *vis-à-vis* the most developed nations of the world; those bonds would strengthen over the next century.[40]

The choice of Hong Kong as headquarters of the Hongkong and Shanghai Bank proved astute. Most great trading firms that founded it had either their

[40] King, *The Hongkong Bank in Late Imperial China, 1864–1902*, pp. 132–81, 347–50, 509–23; King, *The Hongkong Bank in the period of imperialism and war, 1895–1918*, pp. 90–146; Lim, Nooi, and Boh, "The history and development of the Hongkong and Shanghai Banking Corporation in peninsular Malaysia," pp. 356–58.

headquarters or one of their most important branches there. At the time of its founding in 1864, Hong Kong traders still focused mostly on the China trade, and foreign and Chinese merchants intermediated trade flows that bound China with India and the Nanyang. As trade of the Nanyang with China and Japan grew, the ties of Hong Kong traders to the Nanyang deepened. The bank's headquarters, therefore, had exceptional access to information to evaluate both the stagnation and looming problems of the leading China trades of tea, silk, and opium and the opportunities in the Nanyang as the rice trade, tin mining, and sugar plantations expanded from the 1860s to the 1880s. The Hong Kong directors raised the capital of the bank from £1.1 million to £1.4 million in 1883, but the London committee of the bank expressed serious reservations about the increase. They viewed the Britain–China trade problems from the London end and missed the opportunities in the Nanyang that the Hong Kong directors recognized. Concurrently, the bank raised its credit line with London and County Bank from £1.6 million to £2.5 million, vastly boosting its financing capacity. The Hong Kong directors also had their eyes on the growing appetites of governments, especially China's, for loans. They had long experience maneuvering among government trade regulations in Asia and in their home countries, and the headquarters and the London branch kept them attuned to British imperial policy in Asia. The bank started agencies in Tientsin in 1881 and Peking in 1885 to cultivate closer ties with the Chinese government and became banker to the Imperial Maritime Customs, giving it access to crucial information about the volume and availability of customs revenue as collateral for Chinese government loans. From 1895 to 1914, the bank leveraged these advantages to maintain leadership in consortia granting government loans and railway finance, and it advised the government on currency and banking policies. It perfected a globe-spanning organization that few other banks could duplicate in competing for this lending: the Hong Kong headquarters supervised negotiations in China over terms of loans, and the London branch operated as the merchant bank and negotiated loan syndications at that global financial center. The headquarters also gave the bank unparalleled capacity to operate successfully through a major financial crisis, the decline of 61 percent in the price of silver relative to gold from 1875 to 1903. The London branch handled gold, silver, and sterling trades at that global market, similarly to other Eastern exchange banks. Its nerve center in Hong Kong, however, gave the Hongkong Bank a competitive advantage through social networks of capital that bridged to its branches and other firms in the key business centers in Asia. The head office used its superb information to devise sophisticated tactics to acquire silver funds in the East, whereas many Eastern exchange banks failed because they tried to direct their branch strategies in the Far East from the London headquarters.[41]

[41] King, *The Hongkong Bank in Late Imperial China, 1864–1902*, pp. 261–302, 332–66, 451–99, 535–62, tables 8.3, 9.4, pp. 283, 309.

Multinational banks from continental Europe, including Germany, France, and Russia, expanded their presence in the Far East after 1890 and acquired a competitive edge in their colonial territories or in "spheres of influence" in China; nevertheless, the Hongkong Bank had an entrenched position throughout the Far East and the backing of widespread British colonial influence. By 1914, some indigenously owned modern banks emerged in the Far East, but their limited capital made them weak competitors in large-scale finance against the Hongkong Bank and other Eastern exchange banks. The years from 1914 to 1945 proved unsettling for the Hongkong Bank as wars, trade disruptions, depression, and turmoil and political change in China upset established financial relations. The bank, however, achieved dominance in the Far East and wove itself tightly into the regional economies, and its ties to Chinese merchants and financiers in China and the Nanyang would prove a critical resource in the era after World War II.[42]

Competition and cooperation

The treaty settlements of 1858 and 1860 opened China more to foreign traders and financiers, and European governments subsequently accelerated imperial expansion in the Nanyang to acquire raw materials for their factories and foodstuffs for their populations. The dreams of foreign traders and financiers that they would dominate intermediation within Asia, nevertheless, collapsed quickly. Asia's impoverishment meant that each peasant represented a minuscule amount of demand, and large volumes in trade and finance consisted of aggregations of these tiny demands. This placed a premium on low-cost intermediary business, but the expensive staffs and heavy fixed costs of foreign firms made them uncompetitive because they could not achieve economies of scale. Chinese merchants and financiers had low-cost operations, and they leveraged their families and clans into strong local networks and bridges to other social networks that conveyed rich information about demand and supply, built trust, and provided mechanisms to enforce sanctions for malfeasance within and across the countries of the Nanyang and East Asia. Decisions of leading foreign firms to react to this competitive challenge and specialize in trade services, especially for Chinese firms, imply that Chinese intermediaries had triumphed by 1870. Few commodity trades offered opportunities to specialize because Chinese firms controlled access to domestic production and distribution, and the huge commodity trades of plantations and mines in the Nanyang either became the province of a few big traders or were internalized within large plantation or mining corporations. Foreign merchants and financiers cooperated with Chinese firms: compradors allowed foreign firms entrée to the Chinese social networks of capital, and commodity and finance capital

[42] Jones, *British multinational banking, 1830–1990*, pp. 96–98; King, *The Hongkong Bank between the wars and the bank interned, 1919–1945*.

moved between less-specialized, less-capitalized Chinese merchants and financiers and more-specialized, more-capitalized foreign firms. As Chinese firms accumulated capital and specialized, they challenged foreign firms to react to that competition by boosting their capitalization and specialization. This process dominated from 1860 to 1940, but the impoverished state of Asia always retarded capital accumulation. Chinese and foreign firms, thus, competed *and* cooperated, and Hong Kong was the great meeting-place of these two massive social networks of capital.

6

Trade and finance center for Asia

> In Hong-Kong the Chinese houses are increasing annually, while the English and other foreign ones are decreasing . . . Nineteen-twentieths of the population of Hong-Kong are Chinese, and there is not a branch of business into which they have not entered. They have shipping and commercial houses, steam-ship, banking, and insurance companies.[1]

Hong Kong pivot

Commodity trade

By the 1860s, visitors to Hong Kong knew they had arrived at a Chinese city, and that character continued as the Chinese proportion of the population stayed around 95 percent through the 1920s; their in-migration and natural increase powered its population growth (fig. 6.1). Because Hong Kong operated as a trade and financial center, businesses in those sectors signified its status as a metropolis. The number of non-Chinese provide a crude indicator of those workers for the foreign group. Between 1861 and 1901, that number rose almost sevenfold from about 3,000 to 20,000, and then stayed at that level until 1931 (fig. 6.1). These counts, however, overstate the actual number affiliated with trade and finance because they include dependants, government officials, and military personnel. Fewer than a third of the non-Chinese worked in foreign firms, implying that about 6,000 were employed in trade and finance by the start of the twentieth century. Nevertheless, this group mostly comprised upper-level managers and office workers because the Chinese supplied the laborers. The number of Chinese traders and financiers ranged from about 500 to 1,000 in 1871 to over 3,000 by 1881 and 1891.[2] Hong Kong, thus, had a sizable group of pivotal intermediaries of capital that may have reached 10,000 or more by the late nineteenth century.

[1] Knox, "John Comprador," p. 432.

[2] Kwan, *The making of Hong Kong society*, table 3.5, p. 67; Tom, *The entrepot trade and the monetary standards of Hong Kong, 1842–1941*, appendix 2, p. 105.

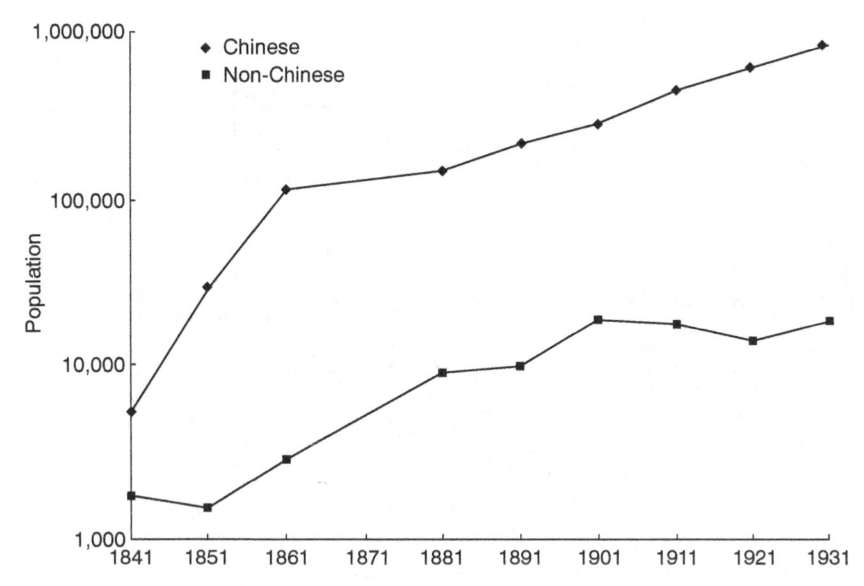

Fig. 6.1. Population of Hong Kong, 1841–1931.
Source: Tom, *The entrepot trade and the monetary standards of Hong Kong, 1842–1941*, appendix 2, p. 105.

Hong Kong played a growing role in the trade of the Far East as the number of ships and aggregate tonnage entering and clearing rose substantially from the 1860s to the 1930s; most of the cargoes were carried by clippers and steamships, not Chinese junks (fig. 6.2). It was not a transshipment point for plantations and mines in the Nanyang because their large, bulk cargoes exited directly to world markets from ports near those resource sites or they were aggregated first at Singapore. Hong Kong trading firms retained close ties to London, but direct trade between them expanded little from the 1870s to the 1930s; it loomed much larger as redistributor of British exports to Asia than as exporter to Britain (fig. 6.3). Hong Kong's entrepot role as exporter to China far surpassed its imports from that country (fig. 6.4). Between the early 1870s and late 1920s, exports to China rose about tenfold and imports climbed about fifteenfold, a notable expansion of Hong Kong's trading firms as intermediaries between China and the rest of the world.[3]

The firms of Hong Kong were linchpins in the coolie trade, a traffic in human labor that resembled a commodity business. It garnered substantial

[3] These measures are in Hong Kong dollars, which remained stable relative to the Shanghai tael; this implies a substantial real increase in the position of Hong Kong as a trade center for China. The value of trade measured in pounds sterling expanded about half as much because the Hong Kong dollar fell that amount compared to the British currency during the period. Tom, *The entrepot trade and the monetary standards of Hong Kong, 1842–1941*, appendix 15, pp. 151–52.

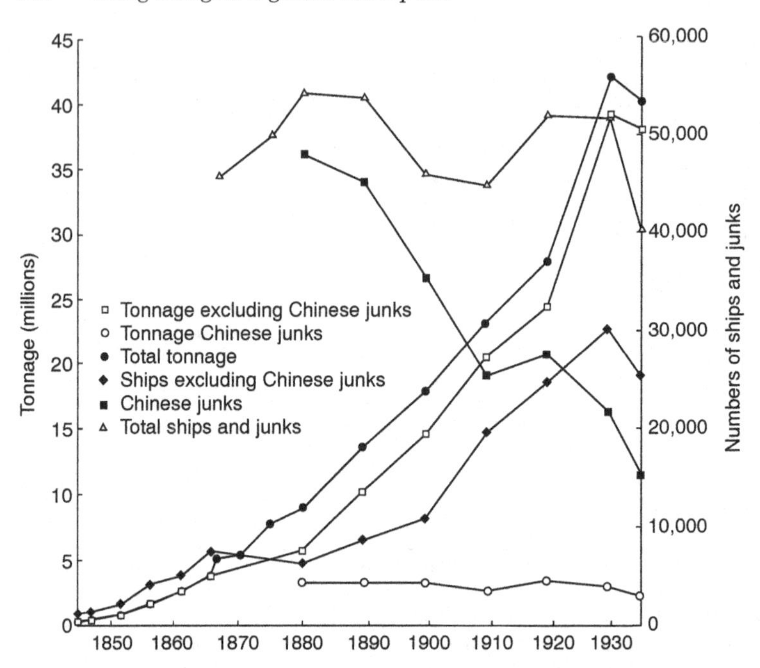

Fig. 6.2. Tonnage and number of ships and junks entering and leaving Hong Kong, 1844–1936.
Source: Tom, *The entrepot trade and the monetary standards of Hong Kong, 1842–1941*, appendix 4, p. 107.

profits for shipping firms, brokers, and labor recruiters, contributing to capital accumulation in Hong Kong. Most coolies came from the poor provinces of Kwangtung and Fukien, home of many Chinese merchants in Hong Kong. During the 1860s, about 100,000 coolies embarked from Hong Kong, and in the following decade the total more than tripled. Between the 1880s and 1900–09, the total embarking each decade stayed around 700,000, and from 1910 to 1919 almost one million left. During the 1870s and 1880s, the number returning approximated the number leaving, but after that, returners exceeded those embarking. The United States was one of the largest single destinations of coolies, and the Nanyang also took vast numbers, including to tin mines and rubber plantations in Malaya, tobacco and rubber plantations in Sumatra, and spice and sugar plantations in Java. The social networks that Chinese firms built or reinforced across the Nanyang and with North America through the coolie trade strengthened their hub position within the information network of Overseas Chinese; Hong Kong firms were the dominant supplier of Chinese goods to those emigrants and their offspring.[4]

[4] Tsai, *Hong Kong in Chinese history*, pp. 23–35.

Fig. 6.3. Hong Kong trade with Britain, 1843–1938.
Source: Tom, *The entrepot trade and the monetary standards of Hong Kong, 1842–1941*, appendix 18, pp. 157–58.

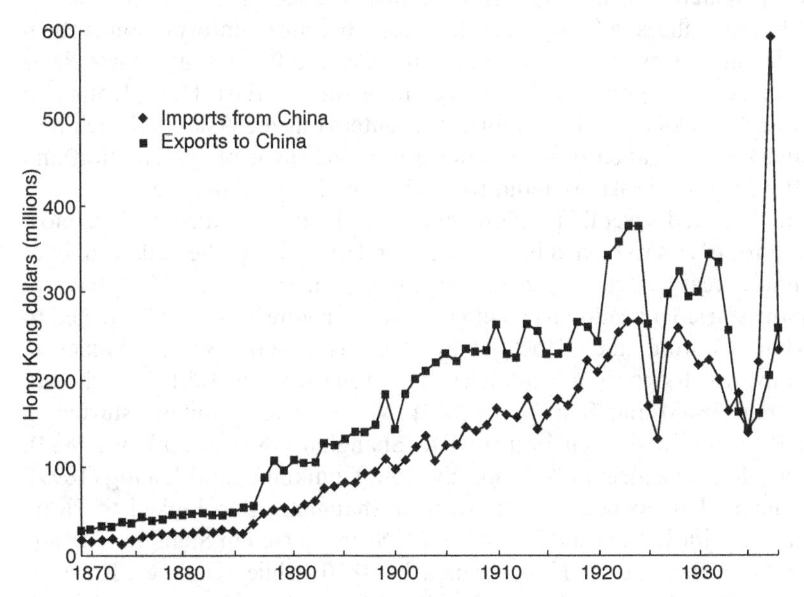

Fig. 6.4. Hong Kong trade with China, 1869–1938.
Source: Tom, *The entrepot trade and the monetary standards of Hong Kong, 1842–1941*, appendix 19, pp. 159–60.

Controller of exchange of capital

Collection, warehouse storage, and commodity redistribution made Hong Kong a pivot of trade, but this exchange was one of physical movement, not necessarily of decisions about the allocation of commodity capital. Once the telegraph reached Asia in the early 1870s, firms could gather information and execute decisions about capital exchange separately from the physical movement of commodities and passengers, but this bond between information flow and physical movement did not completely break. Passenger travel for face-to-face exchange of complex information always remains important, and this overlapped with commodity shipments during the nineteenth and early twentieth centuries. Although specialized passenger ships existed, business travelers took fast steamships that carried high-value commodities and mail, such as those of the Blue Funnel Line (Alfred Holt and Company) and Peninsular and Oriental Steam Navigation Company.[5] Hong Kong's prominence as pivot of Asian trade and finance, nevertheless, did not rest on commodity shipments; Shanghai and Singapore were also great commodity ports. Instead, Hong Kong housed the headquarters or leading branches of the largest, most heavily capitalized trade and financial firms that directed the exchange of commodity and financial capital. They came to dominance before 1860 on the basis of their control over the greatest trade of Asia, the China trade. Firms that participated in it had powerful incentives to establish headquarters or major branch offices in Hong Kong to access specialized information embedded in leading firms and to cooperate on trade and finance, and these firms had branches throughout Asia. As the hub of the Far East, Hong Kong had the strongest links to trade and financial centers outside Asia, and each new headquarters or branch office enhanced it as the pivot of information and expertise to make decisions about the exchange of capital in Asia.

Firms followed several standard processes of office expansion. The most frequent route was to open a headquarters in Hong Kong, then add branches in China's treaty ports, especially Shanghai. Firms such as Olyphant & Company started a headquarters in Hong Kong before 1860 and branched to Shanghai, Canton, and Foochow by the early 1860s. Smith, Archer & Company, which opened a headquarters in Hong Kong in 1861, branched to Yokohama (1865) and Shanghai (1872). Melchers & Company started in Hong Kong in 1866, then branched to Shanghai (1877), Hankow (1884), Canton (1891), Tientsin (1896), and Swatow, Chinkiang, and Ichang (1899). Major firms that opened headquarters in Shanghai often looked to Hong Kong as a major branch site. Butterfield & Swire started in Shanghai in 1867 and founded their Hong Kong branch by 1870, while Kirchner, Boger & Company started in Shanghai in the late 1860s and its Hong Kong branch

[5] Farnie, *East and west of Suez*; Hyde, *Blue Funnel*, pp. 179–80.

opened in 1867. Numerous leading Parsee firms that controlled trade between India and China from the India end relocated their headquarters to Hong Kong. Other firms entered the Nanyang trade: Landstein & Company opened a headquarters in Hong Kong in 1865 and developed business with Indochina, including the transport of rice and cotton from Saigon. These firms from Europe, the Americas, and India operated in the import and export business, but many shifted to trade services (insurance and ship brokerage) in the same way as the great trading houses, such as Jardine, Matheson & Company. By about 1900, Shewan, Tomes & Company operated a huge import and export trade, shipping and insurance services, and shipping lines from their headquarters in Hong Kong and branch offices in Kobe, Shanghai, Tientsin, Canton, London, and New York, and it had agencies throughout the other treaty ports of China, in Manila, and in the Straits Settlements.[6] The decisions of the trade giants to choose Hong Kong as their Asian headquarters during the 1840s and 1850s, therefore, locked in the pattern of forming a headquarters or establishing the leading Asian branch in Hong Kong after 1860. Firms had to participate in the social networks of capital within Hong Kong to compete effectively in Asian trade and finance, and the branch structure of these firms to metropolises and smaller ports across Asia made Hong Kong the hub of bridges to other networks of capital. These foreign firms, nevertheless, did not stand alone; Hong Kong's position as the principal meeting-place of the foreign and Chinese social networks of capital that formed in the 1840s continued to strengthen after 1860.

Chinese and foreign networks meet

Chinese merchants did not choose Hong Kong as their trade and financial center for Asia simply because it was a British colony; Singapore, too, held that status. They also operated from the treaty ports of China and functioned relatively freely as merchants in Bangkok, Saigon, and other ports of the region. Instead, Chinese merchants agglomerated in Hong Kong for the same reasons as foreign merchants; both groups needed access to each other's social networks of capital. Chinese merchants filled the ranks of unspecialized exporters and importers in Hong Kong, and at least sixty-five major firms operated in Hong Kong by 1859. The largest-capitalized, most-specialized firms engaged in the pivotal trade termed "Nam Pak Hong," the south–north trade between the Nanyang and China, and they boosted their capital and specialization after 1860 and dominated the entrepot trade of Hong Kong by the mid 1870s. Their numbers swelled between 1876 and 1881: the count of Nam

[6] Bard, *Traders of Hong Kong*, pp. 51–108; Feldwick, *Present day impressions of the Far East and prominent & progressive Chinese at home and abroad*, pp. 550–72; Jones, *Two centuries of overseas trading*, pp. 17–19, 159–87; Wright, *Twentieth-century impressions of Hongkong, Shanghai, and other treaty ports of China*, pp. 210–28, 618–20.

Pak Hong firms rose from 215 to 395, while the number in the general category of Chinese traders jumped from 287 to 2,377; Chinese bullion dealers, which did not operate as specialized firms in 1876, numbered 34 by 1881. At that date, Chinese constituted seventeen of the twenty largest ratepayers in Hong Kong, indicating their prominence in the economy. By the last several decades of the century, some firms shifted into trade services, including banking and insurance. As of 1886, Hong Kong housed twenty small Chinese banks that operated in other business centers through agencies, and they accessed Shansi banks in China as conduits for long-distance transfers of capital.[7] Chinese merchants, as collectors of commodities from the Nanyang and China and redistributors of foreign imports and other commodities produced in Asia, had access to the lowest wholesale prices in Hong Kong; leading traders, therefore, needed either a headquarters or a major branch office there.

The Chinese mercantile community in Hong Kong occupied the nexus of social networks of capital based on ethnic, clan, and family bonds that bridged to Chinese merchant communities across Asia; this network had two vital parts based on Cantonese and Hokkien-Teochiu dialects respectively. Numerous Cantonese merchants moved to Hong Kong during the 1850s, when turmoil wracked Kwangtung Province, but they maintained access to markets in South China through other Canton merchants. The migration of many Cantonese merchants to Shanghai during the 1850s gave Hong Kong Cantonese entrée to the emporium for Central and North China. Because some Cantonese in Shanghai were compradors in foreign firms, those in Hong Kong had an inside track to deal with foreign firms in Shanghai. Hokkien (Amoy) and Teochiu (Swatow) merchants, as well as Cantonese, had long-standing mercantile bridges through their dialect/clan networks to compatriots in Nanyang ports, such as Bangkok, Batavia, and Singapore. Because merchants from these dialect/clan groups were compradors to foreign firms in Hong Kong, they were well positioned to participate with foreign firms in the Nanyang and China trades.[8]

From 1869 through the end of the century, the Tung Wah Hospital was the hub of the social networks of capital of Chinese firms in Hong Kong. Leading merchants sat on the board of twelve, and every major guild was represented. Between 1869 and 1896, the board consistently included three compradors, one California trader, two Nam Pak Hong (Nanyang–China) merchants, one rice merchant, one opium trader, one mixed goods merchant, one yarn dealer, one pawnbroker, and one miscellaneous member. A founder and first chairman of the board, Liang An, was comprador for Gibb, Livingston &

[7] King, *The Hongkong Bank in Late Imperial China, 1864–1902*, p. 504; Kwan, *The making of Hong Kong society*, table 3.5, p. 67, and p. 107; Lee, "Chinese merchants in the Hong Kong colonial context, 1840–1910", Sinn, *Power and charity*, pp. 1–89.

[8] Tsai, *Hong Kong in Chinese history*, pp. 22–35.

Company, a prominent foreign firm, this appointment signifying the bonds that linked Chinese and foreign networks in Hong Kong. Guild members came from different regions in China, making them powerful hubs who forged cooperative relations within the guilds and non-redundant bridges to other networks within China and across the Nanyang. The hospital remained an open organization rather than an exclusive club, providing widespread communication among Chinese merchants. Because the hospital became heavily involved in issues of Chinese emigration, it was a clearing house of information about the far-flung Overseas Chinese in the Nanyang, Australia, and North and South America. Chinese merchants in Hong Kong represented diverse ethnic, clan, language, and regional groupings in China, but they did not fragment into opposing assemblies. Instead, merchants maintained multiple bridges within the Hong Kong mercantile community, and their bridges reached throughout Asia and to the Americas; this made Hong Kong the pivot of the Chinese social networks of capital. Chinese merchants in Hong Kong followed trajectories similar to foreign firms. Ko Man Wah founded the firm of Yuen Fat Hong in 1853 as a specialist in the import and export of rice and other Chinese food products. It branched to the major ports of East Asia and the Nanyang, including Singapore, and acquired ownership of five rice mills in Bangkok. Beginning in the 1880s, the firm expanded into trade services, including ship brokerage and insurance. Ng Li Hing started the firm of Goh Guan Hin around 1878; it operated as commission agent, rice and sugar importer, and exporter of marine edibles. Its export trade covered the Straits Settlements, Java, the Philippines, and South China. Ng Li Hing also chaired the Hongkong and Manila Yuen Shing Exchange and Trading Company, which focused on finance, and built branches in Shanghai, Amoy, Manila, Singapore, and Penang. Chinese merchants in Nanyang ports, as well as those in the treaty ports of China, especially Shanghai, branched to Hong Kong. Knox, the American merchant, had already noted, in 1878, the status of Hong Kong as the pivot of Chinese trade between the Nanyang and China.[9]

Because this trade remained fragmented and unspecialized through the 1860s, Chinese firms dominated on the basis of their low-cost operations and wide social networks of capital. Foreign (Western and Parsee) firms focused on the China trade, either between India and China (opium) or between Europe and China. The actions of Butterfield & Swire, the trade firm with headquarters in London, suggest that foreign firms may have reevaluated business opportunities in the Nanyang during the 1870s, a period when the output of plantations and mines first reached significant levels. Following the opening

[9] Feldwick, *Present day impressions of the Far East and prominent & progressive Chinese at home and abroad*, pp. 572–601; Knox, "John Comprador," pp. 433–34; Sinn, *Power and charity*, pp. 1–120, appendix 3, p. 273; Sinn, "A history of regional associations in pre-war Hong Kong"; Tsai, *Hong Kong in Chinese history*, pp. 22–35; Wright, *Twentieth-century impressions of Hongkong, Shanghai, and other treaty ports of China*, p. 229.

of Swire's chief Asian branch in Shanghai in 1866, they added an office in Hong Kong in 1870. The Shanghai branch contributed the greatest share of the profits of Eastern branches between 1875 and 1900, demonstrating their focus on the China trade and trade services, but the Hong Kong branch regularly accounted for 25 percent or more of the profits by the late 1880s.[10] Transformation of the Hongkong and Shanghai Bank provides the clearest indication of the spreading tentacles of Hong Kong firms into the Nanyang and throughout Asia. The quick establishment of its agencies in Asia between 1864 and 1866 demonstrates that leading firms viewed Hong Kong as the pivot for Asian trade and finance (table 5.3), and the opening of bank branches and upgrading of merchant agencies measures the timing of the deepened reach of Hong Kong firms into Asia (table 5.4). The bank almost immediately moved into Japan with offices at Yokohama (1866) and Kobe (1869) and commenced its financial reach into the Nanyang in 1870, when the Saigon office opened. As plantations and mines expanded, the pace accelerated; the Manila office opened in 1875, followed by Singapore (1877), and the 1880s witnessed offices at Iloilo (1883) in the Philippines, Penang (1884) in Malaya, Batavia (1884) in the Netherlands East Indies, and Bangkok (1888) in Siam. The distribution of net profits of the bank in 1887 shows that this expansion achieved notable results (table 6.1). The Hong Kong office, excluding those under its immediate supervision, dominated the bank's business, followed respectively by the Shanghai and Yokohama offices. Although the bank had recently upgraded offices in the Nanyang, profits from Singapore, Penang, and the Philippines suggest a lucrative business, but Saigon had a loss that year. A snapshot of trading areas of Chinese merchants in 1915 (table 6.2) that comports with the organizational structure of branches of the Hongkong and Shanghai Bank as of 1918 (table 5.4) reiterates the dominance that firms in Hong Kong achieved over Asian trade and finance. Numerous Chinese firms (84) operated throughout East and Southeast Asia, but most firms specialized in regions of the Far East. Hong Kong firms bound metropolises of the Nanyang (Southeast Asia; 186 firms), including Manila, Saigon, Bangkok, Singapore, and Penang, with China and Japan; the leading firms operated in the rice trade.[11] The trade areas of Hong Kong firms penetrated all regions of China through the treaty ports, and their biggest trading partners were the merchant agglomerations in Shanghai and Singapore, the greatest metropolises of Asia outside Hong Kong.

The agents and correspondents of Chinese native banks headquartered in Hong Kong, which served Chinese merchants, spanned Asia and reached to the United States. Starting in 1912, modern Chinese banks opened in the district termed Central, the financial heart for the foreign banks in Hong Kong.

[10] Knox, "John Comprador," pp. 433–34; Marriner and Hyde, *The senior John Samuel Swire, 1825–98*, table 2, pp. 194–95.

[11] Faure, "The rice trade in Hong Kong before the Second World War."

Table 6.1. *Net profits of Hongkong and Shanghai Banking Corporation, July–December, 1887*

Branches and agencies	Net profits ($ '000s)	
Hong Kong (head office)		+334.0
New York	+1.7	
San Francisco	+58.0	
Lyons	−10.9	
Bombay	+33.2	
Calcutta	−36.4	
Saigon	−3.5	
Amoy	+13.4	
Foochow	−19.0	
Hong Kong office*	+297.5	
Shanghai branch		+200.0
Peking	−3.5	
Tientsin	+18.8	
Hankow	−10.6	
Shanghai office*	+195.3	
Yokohama branch		+100.0
Hiogo	−6.7	
Yokohama office*	+106.7	
Singapore branch		+78.0
Penang	+4.2	
Singapore office	+73.8	
Manila branch		+12.2
Iloilo	+9.1	
Manila office	+3.1	
Batavia branch		+16.7
London branch		nil
Total		+741.0
Published profit (net of contingencies)		+598.7

Note: *Residual: branch returns less net profits attributed to agencies listed
Source: King, *The Hongkong Bank in Late Imperial China, 1864–1902*, table 9.9, p. 328.

Table 6.2. *Trading areas of selected Chinese exporters and importers in Hong Kong, 1915*

	Number of firms	Total number of firms
Australia and United States		
Melbourne, Sydney, San Francisco, and Honolulu		239
Other America		9
Havana	3	
Panama	2	
Peru	4	
India (Calcutta)		1
South Africa		1
Japan		30
East and Southeast Asia		84
Southeast Asia		186
Manila and Spain	31	
Haiphong	2	
Annam (mostly Saigon)	36	
Cambodia	1	
Siam (mostly Bangkok)	19	
Sandakan	4	
Java	5	
Singapore	73	
Penang	15	
China		152
Haikow (Hainan)	6	
Yunnan Province	24	
Peihai (Kwangtung Province)	1	
Swatow	12	
Amoy	22	
Foochow	11	
Shantung Province (Tsingtao, Chefoo)	3	
Tientsin	7	
Hankow	8	
Shanghai	58	
Canton (numerous, but not detailed)	?	

Source: Tsai, *Hong Kong in Chinese history*, pp. 31–32.

A Chinese returner from San Francisco founded the first modern bank, the Bank of Canton; it started branches in Canton and Shanghai and focused on foreign exchange and remittances between the United States and China. The Lau family, leading rice dealers with operations in French Indochina, formed the Chinese Merchants Bank in 1918 and opened branches in Saigon, Canton, Shanghai, and New York by 1921. The Bank of China housed a branch in Hong Kong that supervised a sub-branch in Canton and managed the businesses of the bank in South China by 1919. That same year, the Bank of East Asia also started; its founders headed prominent Chinese firms in Hong Kong that controlled commerce between the Nanyang and China and did business with Japan and the United States.[12]

Chinese and foreign intermediaries in Hong Kong maintained their integral position in Asian trade and finance during the decades prior to 1940. Shares of Hong Kong's imports and exports by country from 1918 to 1938 continued trends in place by the 1860–80 period. China typically ranked first as a trading partner, Japan held a high position, and the Nanyang countries of the Straits Settlements, French Indochina, the Netherlands East Indies, and Siam were significant, accounting collectively for 20–30 percent of trade. Hong Kong firms solidified their grip on the Nanyang trade within Asia, such as between the Nanyang and China/Japan, and between the Nanyang and the United States, but Singapore firms dominated the direct trade of the Nanyang with Europe. Old trades of Hong Kong with Britain and India stayed modest, and far below their importance during the middle decades of the nineteenth century, whereas trade with the United States occupied a prominent place, foreshadowing the years after 1945. About one-third of Hong Kong's trade was with countries outside Asia, indicating that its traders and financiers had a global reach, but most of their business focused on Asia. Hong Kong consistently ranked between third and eighth in the world as a financial center between 1900 and 1935; it outranked all Asian financial centers before 1925, and only between that date and 1935 did it slip slightly below Shanghai or Yokohama.[13] The growing disruption of commerce in Asia, especially with China and Japan, during the 1930s probably caused that slippage, and the Hongkong and Shanghai Bank also experienced problems. The challenge for Hong Kong intermediaries would be to resume their prominence following World War II.

Shanghai: metropolis of Central and North China

The treaty settlements of 1858 and 1860 allowed new ports on the Yangtze River and along the North China coast, and Japan opened to foreign trade in

[12] Sinn, *Growing with Hong Kong*, pp. 4–15.

[13] Reed, *The preeminence of international financial centers*, table A.11, pp. 131–38; Tom, *The entrepot trade and the monetary standards of Hong Kong, 1842–1941*, appendix 8, pp. 111–12.

1858 through commercial treaties with Britain and the United States. From the perspective of the great China Houses headquartered in Hong Kong, these events accelerated the transformation of Shanghai from a branch-office regional metropolis to the multiregional headquarters for Central and North China, Korea, and Japan. They created a hierarchical control structure that proceeded from their Hong Kong headquarters through the Shanghai branch to lower-level branches in other treaty ports. To Shanghai Chinese merchants, however, the organizational structure of foreign firms ratified existing trade networks. As the most highly capitalized, specialized firms, China Houses in Hong Kong led the expansion to Japan. During the treaty negotiations in 1858, the Shanghai branches of Jardine, Matheson & Company and Dent & Company assisted the emissaries and negotiators, and Jardine's, consistently with its powerful position in Asian trade, established a branch office in Yokohama by 1859.[14]

Comprador management

Compradors of the Shanghai branches of the largest British and American firms served as "heads" of the firms' other compradors in branch offices at ports on the Yangtze and along the North China coast. The activities of Jardine's Tong King-sing demonstrate that the Shanghai comprador operated at the broad policy and strategic level, though always under supervision of the headquarters in Hong Kong. Tong traveled frequently to other treaty ports to develop new business opportunities, gather sophisticated market intelligence, settle disputes with Chinese merchants, and, as a supplement to written correspondence, meet with compradors he supervised. He also supervised the firm's banking, insurance, and shipping operations. Although Tong handled some purchases and sales of commodities, staff of the branches at Shanghai and other treaty ports or the headquarters in Hong Kong handled most routine buying and selling. The Foochow branch reported directly to Hong Kong because the large-scale purchase of tea had become a routine, specialized business. The headquarters continued to finance and supervise the shipment of opium from India to Shanghai during the 1860s, whereas the Shanghai branch redistributed opium to ports on the Yangtze and along the North China coast.[15] The comprador administration and opium business structure of the great trading houses demonstrate that Hong Kong served as the high-level decision-making center for Asian trade, and Shanghai operated as the multiregional office center for Central and North China and Northeast Asia. This hierarchical management structure reveals an almost seamless bonding of Chinese and foreign social networks of capital. Yet, foreign firms

[14] Barr, "Jardines in Japan," pp. 154–58; Lubbock, *The opium clippers*, pp. 351–52.

[15] Hao, *The comprador in nineteenth-century China*, pp. 83–88, appendix A, p. 227; Hao, *The commercial revolution in nineteenth-century China*, pp. 133–35.

Table 6.3. *Foreign population of treaty ports, 1870*

Treaty port	Foreign residents
Original five ports	
Shanghai	2,074
Foochow	196
Canton	181
Amoy	115
Ningpo	92
New coastal ports	
North	
Chefoo	165
Tientsin	104
Newchwang	65
South	
Swatow	69
Yangtze ports	
Hankow	88
Kiukiang	48
Chinkiang	37
Total	3,234

Source: Rawski, "Chinese dominance of treaty port commerce and its implications, 1860–1875," table 1, p. 453.

continually faced the principal–agent dilemma over the authority of Shanghai compradors; the latter acquired extraordinary power as filters of specialized business information gathered from their own firms and from supervision of compradors in other ports.

Shanghai dominated as China's trade, financial, and shipping center under the overall hegemony of Hong Kong, headquarters for Asian commerce. The number of foreign residents in treaty ports in 1870 provides a surrogate for the managers and staff of foreign firms (table 6.3). Shanghai stood ten times larger than the second largest port, Foochow, the specialized center of tea exports. Canton, a virtual satellite of Hong Kong, was ranked third under Foochow, and Ningpo's small foreign population coincided with its status as a satellite of Shanghai. The large size of Chefoo and Tientsin implies that trade with North China ports flourished. The Yangtze ports of Hankow, Kiukiang, and Chinkiang also had modest foreign enclaves, but Hankow, the commercial metropolis of Central China, housed a larger group of foreign firms.

Nevertheless, the extreme concentration of foreigners in Shanghai, compared to the river and coastal ports, symbolized the foreign firms' failure to wrest control of domestic distribution of imports and collection of commodities for export from Chinese merchants.[16]

Commodity port

During the 1860s and 1870s, foreign firms developed shipping from a pattern of loosely organized sailing times into formal shipping lines with regular schedules that offered freight and passenger service on the Yangtze River, along the China coast, and with Hong Kong and Japan. The infrastructure of shipping, warehouses, wharves, and ship-repair services made Shanghai an efficient, low-cost center to physically exchange vast quantities of commodities with the markets in Central and North China, but this expansion of shipping services did not mean that Shanghai grew significantly as a commodity port after 1860. From the 1860s to the 1890s, the real value of imports and exports of China expanded little (table 5.1), and poverty in China limited the growth of Shanghai to servicing the slow, aggregate expansion of the domestic economy as the population increased. The shift of shipping from junks, which went unrecorded, to foreign and Chinese steamships, which customs statistics recorded, contributes an upward bias to trade figures.[17] Estimates of the total population of Shanghai and of foreign males, a surrogate for the size of the foreign commercial sector, suggest that the port activity of Shanghai did not grow extensively until after 1890 (fig. 6.5). The modern institutionalization of Chinese commodity shipping that had relied on small-scale junks, warehouses, and wharves powered much of the expansion of foreign trade services at Shanghai.

It stood unsurpassed as China's commodity port and typically served more as a redistributor of foreign imports than as an exporter; about 50 percent of China's imports and 40 percent of its exports passed through Shanghai between 1870 and 1930 (table 6.4).[18] Its share of exports fell as goods increasingly left North China through Tientsin, Tsingtao, Antung, Darien, and Harbin. The decline of Canton *vis-à-vis* Shanghai, which had commenced in the 1850s, continued as this satellite of Hong Kong witnessed its shares of imports and exports collapse by 1930.[19] The commodity trade of Shanghai swelled after 1890 as its share of trade was maintained with the rise in the exports and imports of China (tables 5.1 and 6.4). Shanghai's share of the tra-

[16] MacPherson and Yearley, "The 2½% margin"; Rowe, *Hankow*, pp. 17–157.

[17] Murphey, *The outsiders*, pp. 197–220.

[18] These figures represent the formal measurement of trade through treaty ports, thus biasing the data upwards for Shanghai in 1870. The addition of treaty ports after that date raised the total measured trade and reduced values for Shanghai closer to its "true" share.

[19] Hsiao, *China's foreign trade statistics, 1864–1949*, table 7a, pp. 168–79; Inspectorate General of Customs, *China*, p. 549.

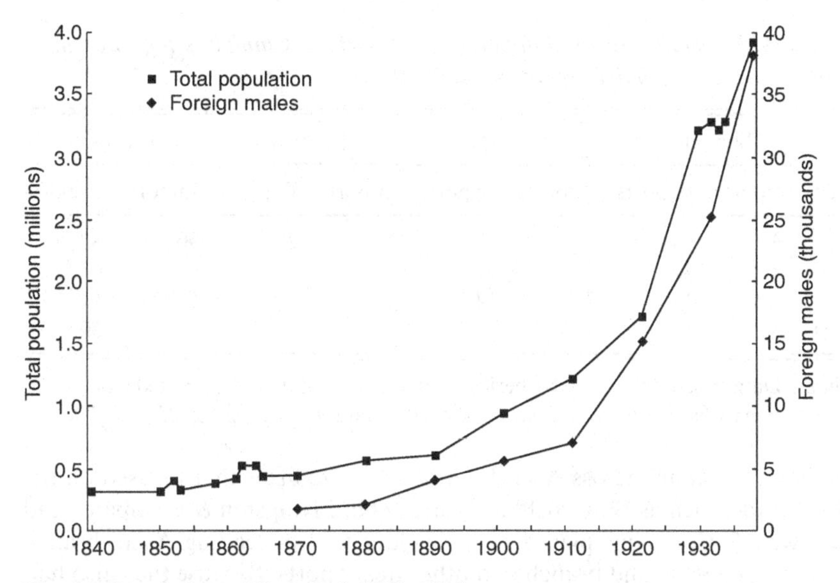

Fig. 6.5. Population of Shanghai, 1840–1936.
Source: Murphey, *Shanghai*, graphs 1–2, pp. 22–23.

ditional exports of raw silk and tea and imports of opium and cotton textiles fell, whereas it increasingly exported raw commodities from the Yangtze valley, including vegetable oils, egg products, hides, and, after 1910, textiles and miscellaneous manufactures. Imports shifted towards industrial raw materials and intermediate goods such as machinery. These industrial imports and exports mostly related to factories in Shanghai as it captured about 50 percent of the meager manufacturing expansion that commenced in China after 1890, but the transformation of Shanghai to industrial metropolis rested on shaky grounds. In 1933, textile factories accounted for 56 percent of the 214,736 manufacturing employees, a share similar to a specialized textile city, such as Manchester, England, or Lowell, Massachusetts, rather than to an industrial metropolis such as London or New York.[20]

Foreign firms complement Chinese firms

For foreign firms that specialized in trade and trade services, China's impoverishment provided few incentives to establish sophisticated offices in Shanghai even after 1890. The greatest trade-services firm in Asia, Jardine, Matheson & Company, maintained a major branch in Shanghai but kept its headquarters in Hong Kong to access the wider array of bridges to social

[20] Lieu, *The growth and industrialization of Shanghai*, pp. 13–14, table C-V, p. 348; Murphey, *Shanghai*, tables 1–2, pp. 119–20.

Table 6.4. *Percentage of total imports and exports of China passing through Shanghai, Canton, and Tientsin, 1870–1930*

	Shanghai		Canton		Tientsin		Total	
Year	Imports	Exports	Imports	Exports	Imports	Exports	Imports	Exports
1870	79	60	9	20	2	2	90	82
1890	54	40	9	17	1	5	64	62
1910	44	45	7	14	7	2	58	61
1930	53	34	4	7	8	9	65	50

Source: Data based on three-year period centered at each date. Computed from Hsiao, *China's foreign trade statistics, 1864–1949*, tables 1, 7a, pp. 22–25, 168–79.

networks of capital across Asia. Intermediate-size firms that focused on the China trade, such as Butterfield & Swire, Gibb, Livingston & Company, and Carlowitz & Company, placed their headquarters or leading Asian branch offices in Shanghai and branches in other treaty ports. Because they also had moderate amounts of business outside China, they located prominent branches in Hong Kong and sometimes a branch in Japan. Smaller firms headquartered in Shanghai, such as A. R. Burkill & Sons or Telge & Schroeter, focused on the China trade, but they could not afford extensive branch networks. Foreign firms that traded with both the Nanyang and China or provided trade services to that broader region chose Hong Kong, rather than Shanghai, as their headquarters. By the early twentieth century, most offices of trading and trade-services firms in Shanghai operated either as branches of Hong Kong firms or of firms headquartered in Europe, North America, and Japan. These branches engaged in the import and export business, but they chiefly served as agents of insurance and manufacturing firms based in their home nations.[21] They drew on their national social networks of capital to dominate the agency business, whereas Chinese firms could not as effectively access those networks.

Shanghai merchant networks based on family, clan, and dialect groups remained rooted in Kwangtung and Chekiang (near Shanghai) provinces, but the relative importance of the Kwangtung base, which gave Shanghai merchants contacts with South China and were reinforced through offices in Hong Kong, declined after 1880 as tea exports, the Cantonese specialty, fell. The significance of the Chekiang base, which provided merchants contacts with Central and North China, rose; by 1920, most leading Shanghai compradors came from the surrounding region. Chinese merchants maintained local ties

[21] Feldwick, *Present day impressions of the Far East and prominent & progressive Chinese at home and abroad*, pp. 387–456; Wright, *Twentieth-century impressions of Hongkong, Shanghai, and other treaty ports of China*, pp. 602–64.

through memberships in organizations such as the Shanghai Chinese Chamber of Commerce and service on boards of directors of firms such as Wah Shing Insurance Company.[22] Shanghai compradors transmitted insights about the regional economies of China to foreign firms, and compradors shared this information among themselves. Successive service of one comprador in different foreign firms and participation of family members in various firms reinforced their network ties. Rather than seeing this as detrimental, foreign firms willingly hired compradors with such ties, implying that those firms recognized the benefits from access to information and trust embedded in Chinese merchant networks, and that trustworthiness reduced the costs and risks of intermediary activity in an uncertain Chinese political economy. This market information, the potential for cooperative ventures, and market demand and supply of China articulated in the Shanghai wholesale mart required that foreign firms engaged in imports and exports and acting as agents for manufacturers, insurance, and other firms agglomerate with Chinese merchants in Shanghai. That explains why most foreign firms that focused on the China trade had a headquarters or Asian branch office in Shanghai.

American multinationals that arrived in growing numbers after 1880 to sell to the Chinese market established marketing and sales distribution systems that mirrored the wholesale network of Chinese merchants in Shanghai. The New York or London office supervised the chief branch office at Shanghai. Some firms, such as the British-American Tobacco Company, headquartered in New York, built divisional and territorial office structures with their own sales representatives housed at strategically located warehouses. This firm sold cigarettes through Chinese compradors and regional agents and managed its distribution network from the Shanghai branch. Oil companies diverged from this organization, but they also copied the Chinese mercantile structure. In 1885, Standard Oil Company commenced a major expansion in China when it opened an office in Shanghai to market in Central and North China; it located an office in Hong Kong to sell to South China in 1894 and branched to other treaty ports by 1908. Standard Oil initially sold kerosene through compradors and leading Chinese wholesalers, but changed strategy and established a distribution network directly under its control by 1914. The firm rationalized an organizational structure for China with a "head" office in Shanghai, district offices that covered each region, including one in Hong Kong for South China, and branch offices in leading treaty ports. This distribution network internalized the structure of the Chinese wholesale network within Standard Oil, and it complemented this with external wholesalers for

[22] Bergere, *The golden age of the Chinese bourgeoisie, 1911–1937*, pp. 144–46; Feldwick, *Present day impressions of the Far East and prominent & progressive Chinese at home and abroad*, pp. 459–62; Wright, *Twentieth-century impressions of Hongkong, Shanghai, and other treaty ports of China*, pp. 525–72.

smaller markets; other oil companies replicated its approach.[23] The correspondence between the marketing and sales distribution networks of American multinationals, whether internal or external to the firm, and the Chinese mercantile networks testifies to the depth and resilience of the social networks of capital within China that radiated from Shanghai.

Because Shanghai served as the dominant wholesale mart of China, its firms generated prodigious demands for financial capital to underwrite trade. Those opportunities attracted the early branches of foreign exchange banks, including the Oriental Banking Corporation (1848), the Mercantile Bank of India, London and China (1854), the Chartered Bank of India, Australia and China (1857), and the Hongkong and Shanghai Bank (1865). China's impoverishment and slow economic growth, however, retarded expansion of foreign banks in Shanghai until the economic upturn around 1890. Within just over a decade, the Deutsche-Asiatische Bank (1889), the Yokohama Specie Bank (1892), the Banque de l'Indo Chine (1899), and the predecessor of National City Bank of New York (1902) opened offices; and the pace of expansion accelerated so that Shanghai housed twenty-seven foreign banks by 1937. They financed foreign trade, trade-services firms, and infrastructure in the treaty ports, such as commercial buildings, wharves, warehouses, and factories. Shanghai native banks owned by leading merchants, many of them members of the Ningpo *pang*, attracted foreign banks who used native banks as conduits to lend capital throughout China. Numerous compradors of foreign banks came from the ranks of the Ningpo *pang* bankers, cementing bonds between foreign and native banks; compradors sometimes served several banks, and family members occasionally served different banks.[24] This cross-service of comprador family members, coupled with ownership of native banks, created a seamless web for capital to move among foreign and native banks in Shanghai.

Between 1900 and 1930, Chinese financiers in Shanghai strengthened their dominance of banking in China. In addition to funding the growth of Shanghai, these bankers created a powerful, unified financial community that gave them leverage to provide capital to the government and private sectors in China. The Ningpo *pang* shrewdly maintained their practice of redundant internal ties and building bridges to other social networks of capital in China; this expanded network, with Shanghai banks at the hub, dominated banking in China. The Ningpo *pang* gradually incorporated native bankers from nearby Shaohsing to form a larger Chekiang *pang*, and this group enlarged

[23] Cheng, "The United States petroleum trade with China, 1876–1949"; Cochran, "Commercial penetration and economic imperialism in China"; Wilkins, "The impacts of American multinational enterprise on American–Chinese economic relations, 1786–1949."

[24] Bergere, *The golden age of the Chinese bourgeoisie, 1911–1937*, pp. 142–43; Jones, *Shanghai and Tientsin*, p. 75; McElderry, *Shanghai old-style banks (ch'ien-chuang), 1800–1935*, pp. 39–51; Wright, *Twentieth-century impressions of Hongkong, Shanghai, and other treaty ports of China*, p. 540.

itself through the formation of the Shanghai Bankers Association (modern banks) in 1915 and Shanghai Native Bankers Association in 1917; these associations also incorporated other regional cliques such as those from Kwangtung. Chekiang-group banks forged numerous interlocking directorates; in 1931, for example, six Shanghai bankers served on the boards of five or more leading local banks and fifteen served on the boards of three or more banks. The Chekiang *pang* encompassed a network of personal and economic relations that bridged business lines and joined native and modern bankers with industrialists, merchants, and shippers. These financiers and their powerful business network, nevertheless, reached its apogee during the late 1920s. After that, Chiang Kai-shek's Kuomintang government increasingly engaged in expropriation and extortion of the Shanghai business elite to retain power as they battled the Communists under Mao Tse-tung and coped with the disorder sweeping China under the Japanese threat.[25] To regain power, the Shanghai business elite would reorganize in Hong Kong after 1945.

Singapore: metropolis of Southeast Asia

Meeting-place for Nanyang merchants

Singapore merchants, located about 2,500 kilometers southwest of Hong Kong, never seriously challenged merchants in Canton (before 1840) or Hong Kong (after 1840) for control of the China trade. Traders in Hong Kong, Canton, Shanghai, and other China ports served as hubs through which Singapore merchants accessed information about trade and participated in business networks. Merchant firms in Singapore were rooted in, or tied to, the same Agency Houses of London, Bombay, and Calcutta as Canton, and later, Hong Kong firms. Until the demise of the East India Company's monopoly over the China trade in 1833, British firms in London, India, and Canton used Singapore as a transshipment point to circumvent prohibition on direct trade between Britain and China. Singapore was one component of their global mercantile networks, not a competitor of Canton or Hong Kong, where these firms had other offices. Between 1819, when Raffles founded Singapore as a free port, and 1824, when the Dutch acknowledged it as a permanent British possession, it became the metropolis of Southeast Asia, the territory that included Burma, Siam (Thailand), Malaya, Sumatra, and the Indonesian archipelago. Restrictive Dutch policies throughout the Indonesian archipelago raised mercantile costs; to reduce them, merchants flocked to Singapore to establish offices and others came to trade. The agglomeration lowered exchange costs as access to information about Southeast Asian markets improved, and merchants became the repository of information about

[25] Coble, *The Shanghai capitalists and the Nationalist government, 1927–1937*, pp. 13–27, 261–69; McElderry, *Shanghai old-style banks (ch'ien-chuang), 1800–1935*, pp. 51–53.

markets in China, India, and Europe through their network ties. Singapore's quick rise represented mostly a reorganization of highly developed, fragmented trade in Southeast Asia, rather than new economic growth or the subterfuge of transshipment trade between Britain and China.[26]

British merchants dominated the foreign group at Singapore, but the arrival of Nanyang merchants solidified its pivotal status. Bugis from the Celebes, who controlled small-scale, unspecialized trade of the archipelago east of the Malay peninsula, rushed to Singapore to escape onerous trade duties in Dutch ports. They collected goods, such as sarongs, gold, beeswax, and antimony from the islands and exchanged them at Singapore for opium, salt, tobacco, and manufactures. Merchants in small ports in eastern Sumatra forwarded coffee, rice, and gambier to Singapore in exchange for Siamese salt, British cottons, Indian textiles, opium, and raw silk, and Indian merchants at Penang shifted operations to Singapore. Chinese merchants in Malaya, the Straits, and the East Indies represented the key addition to Singapore trade; some settled there while most reoriented their trade from other ports to Singapore. This placed Singapore merchants at the pivot of the network of Chinese traders in Malacca, Penang, Bangkok, Manila, and Batavia and other Javanese ports, who exported the production of those areas and imported Chinese consumption goods to Chinese settlements, and they forged bridges to Chinese merchant networks at Canton (before 1840) and Hong Kong (after 1840). The arrival of numerous unspecialized traders at Singapore allowed highly capitalized European firms to achieve economies of scale in importing opium and textiles from India and textiles from Europe and exporting products from Southeast Asia. By 1824, twelve European firms operated in Singapore, most as agents for London and Calcutta Agency Houses. The total trade of Singapore languished, implying that the transshipment trade between Britain and China and trade within Southeast Asia grew little; termination of the East India Company's monopoly over the China trade in 1834 ended the subterfuge of transshipment. Continued migration of Chinese from Kwangtung and Fukien provinces to Singapore from the mid 1820s to the mid 1840s boosted the Chinese share of the population to about 60 percent. Merchants among these immigrants enhanced bridges to merchants at Canton (Kwangtung Province) and Amoy and Foochow (Fukien Province) and to merchants from the same provinces who moved to Siam and the Malay peninsula. Singapore became a mercantile meeting-place; highly capitalized British commission houses supplied credit to unspecialized Chinese merchants, who controlled trade within Southeast Asia and served as intermediaries to other traders, such as the Bugis and Malays.[27]

[26] Greenberg, *British trade and the opening of China, 1800–1842*, pp. 87–88, 96–99; Turnbull, *The Straits settlements, 1826–67*, pp. 233–55; Turnbull, *A history of Singapore, 1819–1988*, pp. 1–32.

[27] SarDesai, *British trade and expansion in Southeast Asia, 1830–1914*, table 5, pp. 52–53; Trocki, *Opium and empire*, p. 57; Turnbull, *The Straits settlements, 1826–67*, pp. 163–84; Turnbull, *A history of Singapore, 1819–1988*, pp. 13–40; Viraphol, *Tribute and profit*, pp. 210–22.

The 72 percent surge in Singapore's trade from 1825–36 to 1840–51 had several causes. During the 1840s, tin and gold mining expanded in Malaya, and the influx of Chinese miners increased demand for opium and Chinese consumption goods. Chinese merchants who controlled this trade dealt through Singapore and achieved economies of scale handling the larger trade. Increasingly, Penang and Malacca trade passed through Singapore; Penang merchant offices often became branches of Singapore firms, and by the early 1860s half of its largest trading firms had Penang branches. As friction with the Dutch decreased after 1840 and they opened more free ports, Singapore traders penetrated deeper into the Indonesian archipelago, but Dutch efforts to achieve gains from free trade came too late. The Singapore merchant agglomeration offered too many advantages: access to information through the British and Chinese networks of capital, highly capitalized merchant firms, and low-cost trade. The growth of China trade controlled by Hong Kong firms during the 1850s indirectly underwrote the 75 percent growth in Singapore's trade from the 1840s to the late 1850s as profits from China supported greater purchases of goods produced in Southeast Asia. New tin mines near Penang during the 1850s and the opening of trade by British firms with Burma after the Anglo-Burmese War (1852–54) and with Siam after the Bowring Treaty (1855) directly boosted Singapore's trade. The Siam opening aided Singapore because trade shifted from Siamese-Chinese merchants in Bangkok and other ports, who dominated because the government excluded foreign merchants, to British and Chinese firms headquartered in Singapore who operated with branches in Bangkok. By 1860, British and other European firms controlled large-scale trade with India and Europe, and they acquired a foothold in the growing specialized trades of Southeast Asia, including tin and rice. These European Singapore merchants totaled about 400, similar to the foreign group in Shanghai, but Hong Kong had over three times as many, indicative of its greater importance in Asian trade (table 4.1). Singapore Chinese merchants dominated small-scale trade throughout Southeast Asia and monopolized the trade in opium and Chinese consumption goods with dispersed Chinese communities.[28]

Merchants and financiers in Singapore intensified control over commodity and financial exchange within Southeast Asia between 1860 and 1940, but their territorial reach changed little even with the expansion of trade and the development of plantations and mines. These resilient bonds imply that structural relations among economic actors in Southeast Asia solidified by the 1860s. The fleeting foothold that British merchants in Singapore gained in the intraregional trade in tin and rice during the late 1850s collapsed under the onslaught of low-cost, Chinese merchant firms who employed ethnic, clan, and dialect bonds to forge network bridges that reached from Singapore to

[28] SarDesai, *British trade and expansion in Southeast Asia, 1830–1914*, table 5, pp. 52–53; Turnbull, *The Straits settlements, 1826–67*, pp. 22, 160–77.

major and minor ports for wholesale exchange and from these ports to wholesaler-retailers and on to village storekeepers. This substructure of intraregional trade within Southeast Asia built around the mercantile hub of Singapore had three related components. Chinese merchants in Singapore: controlled redistribution of foodstuffs among countries, especially rice; collected export commodities from the region and brought them to Singapore for sale to foreign firms who exported to Europe; and purchased manufactures from foreign importers in Singapore and redistributed them throughout the region. These small-scale trades in this poor region rested on credit that Singapore's Chinese merchants provided as advances on goods to wholesalers, retailers, and small commodity firms who paid with commodities. Chinese merchants had the most sophisticated information about local economies and about the financial viability and trustworthiness of business people to handle these high-risk transactions. Thus, the social networks of capital and low-cost operations of Chinese firms gave them competitive advantages over high-cost foreign firms with no access to those networks.[29]

Commodity trade

The value of the total import and export trade of Singapore rose about 3.3 percent annually in real terms from 1870 to 1937, testimony to the expanding influence of its trading firms in Southeast Asia.[30] That steady increase implies that the rise of bureaucratic states, the deeper impress of imperialism, and the arrival of large-scale, capital-intensive multinationals did not materially alter the growth path; perturbations in the pace of growth of Singapore's leading exports did not come until after 1915 (fig. 6.6). Merchandise exports, which reflect redistribution within the Nanyang of manufactured goods from Europe and North America, accounted for the largest share of exports. Rubber exports, which eventually formed the greatest share of commodities sent to industrial economies, did not jump above a low base until after 1915, and petroleum exports first spurted in the early 1920s. Tin and a set of sixteen tropical commodity exports rose at a steady pace but amounted to a tiny share of exports. Rice stayed mostly within Asia, and its share remained modest. This view from Singapore, pivot of Nanyang trade, suggests that the region finally achieved a burst of economic growth after 1915, but that this collapsed under the depression of the 1930s. The delayed surge of raw material commodity exports to industrial economies cautions against overstressing the impact of plantations and mines on the regional economy; they were islands in a sea of impoverished peasants.

Shifting trade flows between Singapore and other countries from 1871 to

[29] See discussion in chapter 5 and Huff, *The economic growth of Singapore*, pp. 49–68.
[30] Huff, *The economic growth of Singapore*, fig. 1.3, p. 12.

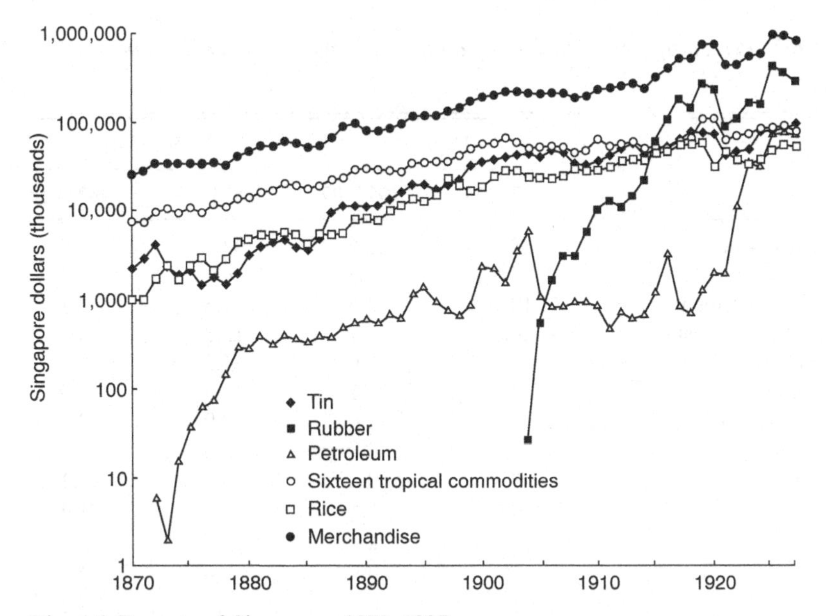

Fig. 6.6. Exports of Singapore, 1870–1927.
Source: Huff, *The economic growth of Singapore*, table A.1, pp. 372–77.

1927 offer another perspective on the position of its merchants within the regional and global economies (table 6.5), but a caveat is necessary; these flows measure physical movement, not control of exchange. Exports outside Asia as a share of total exports rose from 37 percent to 55 percent, and the United States, as the dominant market for tin (one-half to two-thirds) and rubber (two-thirds to four-fifths), accounted for all of that shift.[31] Imports from outside Asia as a share of total imports fell from 31 percent to 17 percent. The combined import and export trade of Singapore with countries outside Asia never amounted to more than about one-third of total trade. Singapore traders, similarly to those in Hong Kong, had a global reach, but both groups focused most of their business on Asia. Ignoring distortions caused by exports of rubber to the United States, the import and export trade of Singapore with Southeast Asia was in the range 45–55 percent of its total trade between 1871 and 1927. The Malay peninsula, the Straits Settlements, and the Netherlands East Indies, the heart of Singapore's trade hinterland, accounted for 60–75 percent of that trade. Trade with Siam, Indochina, and Burma continued strongly on the import side because they supplied rice for plantation and mining areas in the Nanyang. Imbalances in trade with Siam and Indochina emerged after 1900 as the value of exports fell far below imports, probably a

[31] Huff, *The economic growth of Singapore*, p. 80.

Table 6.5. *Percentage distributions of annual average exports and imports of Singapore, 1871–1927*

	Exports		Imports	
	1871–73	1925–27	1871–73	1925–27
Southeast Asia	48.1	38.6	43.4	71.0
Malay peninsula	4.5	12.5	6.1	18.3
Straits Settlements	10.4	4.6	11.3	2.1
Netherlands East Indies	20.6	11.7	15.3	34.6
Siam	6.4	2.6	4.5	7.3
Indochina	2.7	1.1	1.4	2.0
British Borneo	1.3	1.6	0.9	4.0
Burma	1.3	0.7	2.6	1.2
Philippines	0.8	0.2	0.9	0.1
Australia	0.0	3.5	0.5	1.4
East Asia	12.6	4.8	12.0	8.6
Hong Kong	9.6	0.8	9.9	2.7
China	2.9	0.9	2.1	3.2
Japan	0.0	3.1	0.0	2.6
South Asia	2.1	2.1	13.6	3.2
India	2.1	1.7	13.6	3.1
Ceylon	0.0	0.5	0.0	0.1
Europe and United States	31.0	50.8	27.5	15.7
United Kingdom	20.6	5.7	23.7	9.0
Europe	1.6	8.7	3.8	3.6
United States	8.8	36.5	0.0	3.1
Others	6.2	3.7	3.5	1.5
Total	100.0	100.0	100.0	100.0
Total $ (millions)	37.4	868.3	42.6	996.8

Source: Adapted from Huff, *The economic growth of Singapore*, tables 2.2, 2.3, 3.4, 3.5, pp. 50, 52, 81–82.

result of greater competition from Hong Kong traders for the sale of food (excluding rice) and manufactured goods to those countries. The Philippines was a minor partner of Singapore because Hong Kong merchants controlled that trade. Most of Singapore's trade with China and Japan physically passed through Hong Kong prior to 1900, and then commodities increasingly bypassed it; however, Hong Kong traders maintained their control over decision-making about this exchange. Singapore firms established branch offices in Hong Kong to handle their China trade because that city's foreign and Chinese firms made it the optimal place to access information about demand

and supply in China; and if firms focused on Central and North China, they placed a branch in Shanghai.[32] The decline in the share of Singapore's trade with East Asia between 1871 and 1927 mostly resulted from distortions of the rubber trade; the absolute value of East Asian trade rose. The historic link between Singapore and South Asia, principally India, weakened dramatically, whereas Singapore traders strengthened their bonds with industrial countries, especially the United States.

Transformation in control of capital

By the 1890s, growing intraregional trade in the Nanyang supported more unspecialized Chinese merchants, and leading Chinese firms in Singapore reacted to this competition by specializing in fewer lines of produce, such as rice or pepper and gambier. Shortly after 1900, they institutionalized this change as the Chinese Produce Exchange, and they formed the Singapore Chinese Chamber of Commerce in 1906 in recognition of their business prominence. Rising imports of manufactures after 1900 (table 6.5) allowed Chinese merchants, who dominated redistribution of manufactures in the Nanyang, to leverage their larger capital and encroach on the European firms' import trade by the 1920s. European Agency Houses reacted by acquiring agencies of the growing number of manufacturers in industrial countries that sought international distributors. These houses used their offices and contacts in industrial countries to acquire agencies, whereas Chinese firms did not have such bridges to European and North American networks of capital. Foreign firms competed fiercely with Chinese merchants for dominance of trade related to industrial commodities (tin, rubber, petroleum) produced for developed nations. As capital-intensive European mining firms gained greater control over tin mining in Malaya, the labor force dropped from 225,000 in 1913 to 88,000 in 1937. The consequent reduced demand by Chinese miners for specialty consumption goods caused Chinese merchants in Singapore to suffer losses on imports and on wholesale trade to Malaya. Agency Houses created management services, and European equipment and engineering firms supplied producer goods for mines, but Chinese merchants did not have the capital and contacts with European suppliers to compete for that trade. European merchants hoped to capture greater marketing of the swelling tin production; however, large mining firms internalized initial stages of marketing and looked directly to London and New York commodity markets for sales. Agency Houses even faced competition from firms that specialized in management services for mines.[33]

[32] See earlier discussion in this chapter and chapter 5. For evidence on foreign and Chinese firms during the early twentieth century, see Feldwick, *Present day impressions of the Far East and prominent & progressive Chinese at home and abroad*, pp. 819–41.

[33] Huff, *The economic growth of Singapore*, pp. 44–45, 75–90, 250–70.

European merchant firms in Singapore introduced the managing Agency House system when they floated rubber companies on the London stock exchange starting around 1900. These companies invested in rubber estates in Malaya, and the Agency Houses managed them. Yet, the production side remained on a smaller scale and more fragmented than in tin mining and smelting; as of 1913, only 83 British companies mined tin in Malaya, whereas 646 European rubber estates operated. Chinese merchants in Singapore also financed rubber estates in Malaya, but Agency Houses in Singapore extended their control over estate management to control over rubber trading because large production allowed firms to leverage their extensive capital to achieve economies of scale. Agency Houses devised a grading system for rubber, similar to London's, to trade it as an abstract commodity, not just in physical form. During the rubber boom following World War I, Agency Houses started rubber purchases from those estates they did not manage and from smallholder farms; they became rubber traders and took orders from international buyers in industrial countries. These Singapore firms offered some competition to the larger commodity exchange in London, and a few Chinese firms joined this trading. Europeans, however, could not achieve economies of scale in the Netherlands East Indies where smallholders produced rubber. Chinese merchants in Singapore leveraged their dominance of the East Indies trade into control of the lucrative import trade in rubber. Profits in Malayan rubber estates, control of the import trade in rubber from the East Indies, and milling rubber contributed to substantial capital accumulation among Chinese merchants in Singapore; they transferred it to manufacturing and banking investments. By the mid 1920s, several Chinese firms reached such a large scale that they ceased selling rubber to local European export houses and forged direct ties to London and New York rubber markets.[34]

Firms that provided management and agency services for large commodity producers faced the risk that greater scales of production units would encourage firms to internalize intermediary activities. That happened quickly with oil production in the Netherlands East Indies when large international oil companies, including Asiatic Petroleum Company, established by Royal Dutch Shell, and the Standard Vacuum Oil Company, associated with Standard Oil of New Jersey and Standard Oil of New York, internalized intermediary transactions for the simple, bulk commodity. Neither European Agency Houses nor Chinese firms were competitive against the low internal transaction costs. Singapore was ideal for storage facilities, transshipment, and marketing and management offices to handle sales to Asia, but capital linkages of the oil companies reached to their headquarters in industrial countries and to leading financial centers such as London and New York. Similarly, multi-

[34] Huff, *The economic growth of Singapore*, table 6.1, p. 182 and pp. 180–235; Wong, *The Malayan tin industry to 1914*, table 35, p. 214.

nationals such as Kodak, Singer, and Ford Motor Company that expanded to Southeast Asia placed sales branches in Singapore. Other firms with unbranded products or small sales awarded agencies to Agency Houses and other merchant firms.[35] Singapore, therefore, became an office management, marketing, and sales distribution center for multinationals to reach wholesalers, retailers, and consumers in Southeast Asia.

Singapore's large agglomeration of merchant firms that needed trade financing attracted branches of international banks as early as 1840, when the Union Bank of Calcutta opened a branch, but the arrival of branches of large British exchange banks signaled notable change. Branch expansion followed consistent sequences: after opening a branch in Singapore, banks moved to Penang because its merchants controlled Malayan trade under the supervision of Singapore firms; and other branches followed mostly after 1890 when plantation and mining production picked up. Leading exchange banks illustrate this: Chartered Mercantile Bank of India, London and China started a branch in Singapore (1856), followed by Penang (1860) and Kuala Lumpur (1909); and Hongkong and Shanghai Bank started its branch in Singapore in 1877, followed by offices in Penang (1884), Malacca (1909), and Kuala Lumpur, Ipoh, and Johore, all in 1910 (table 5.4). Starting in 1888, Dutch exchange banks opened branches in Singapore to service the trade of the Netherlands East Indies with Singapore and Malaya, where some branches were established. In 1902, First National City Bank of New York opened its first branch in Singapore, followed by Banque de l'Indo Chine in 1905. The choice of Singapore as the lead office for banking and subsequent branch expansion, thus, correlated highly with the social networks of capital of foreign and Chinese merchants in the Nanyang. Chinese banks lagged because the trade of Singapore's Chinese merchants mostly relied on credit to exchange commodities. Foreign exchange banks supplied that credit, and the small indigenous Chinese population in Malaya and the Netherlands East Indies hindered the formation of "native" banks. During the first decade after 1900, a few Chinese banks opened in Singapore to serve local people; leading Hokkien merchants founded the most important ones, the Chinese Commercial Bank (1912), Ho Hong Bank (1917), and Overseas-Chinese Bank (1919). Ho Hong Bank expanded throughout Malaya and to Batavia and Hong Kong. Singapore was not among the top ten world financial centers before 1940, unlike Hong Kong and Shanghai, but numerous Western bank branches and locally headquartered Chinese banks made Singapore a regional banking center for Southeast Asia. Its firms, nevertheless, did not have the reach of Hong Kong firms, which controlled the exchange of capital (commodities, money) over the Nanyang and China. The relative importance of the two metropolises is indicated by the foreign populations: in 1931, Europeans

[35] Huff, *The economic growth of Singapore*, pp. 237–70.

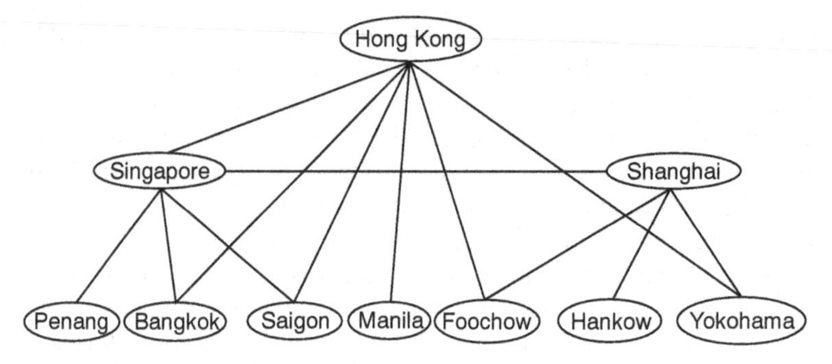

Fig. 6.7. Selective view of the system of metropolises in Asia, *circa* 1880–1920.
Source: author.

numbered 6,686 in Singapore, whereas Hong Kong's non-Chinese population was 19,369, almost three times larger.[36]

Metropolises of Asia

Merchants and financiers in Hong Kong, Shanghai, and Singapore dominated decision-making about the exchange of commodity and money capital in East and Southeast Asia by the 1860s, and they tightened their grip in subsequent decades. These agglomerations were not competitors because leading firms in each metropolis branched to one or more of the other cities. Hong Kong firms provided the greatest mercantile and financial links in Asia; therefore, it occupied the pinnacle of the hierarchy in Asia (fig. 6.7). These intermediary agglomerations always maintained strong ties with the leading global metropolises of London and New York, but merchants and financiers in Asian metropolises conducted the majority of their business as exchange within Asia, not between Asia and the industrial countries. Chinese merchants and, to a lesser extent, the financiers in all the metropolises undergirded this intra-Asian trade and financial network; they would operate similarly after 1945. The extension of economic development to the peasants of East and Southeast Asia would be the monumental transformation during the second half of the twentieth century. That change provided enormous opportunities for merchants and financiers from industrial countries, but Chinese intermediaries controlled the exchange from the bottom-up through their social networks of capital.

[36] Lee, *The monetary and banking development of Malaysia and Singapore*, pp. 66–78; Reed, *The preeminence of international financial centers*, table A.11, pp. 131–38. Singapore's European population is computed from Huff, *The economic growth of Singapore*, table 5.5, p. 158. Hong Kong's population comes from Tom, *The entrepot trade and the monetary standards of Hong Kong, 1842–1941*, appendix 2, p. 105.

7

Industrial metropolis

Hong Kong has become a regional manufacturing control centre and will continue to assume such a role in the future.[1]

Asian political and economic transformation

Hong Kong traders and financiers always had conducted business within complex environments, and since 1860 the imperial powers increasingly shaped the framework of business. Assertive economic and political moves of Japan that threatened Western dominance in Asia and the growing vigor of nationalist movements after 1900 hinted that traders and financiers would confront a transformed environment, although few observers anticipated the swift changes between the late 1930s and the 1950s. When Japan extended its invasion of China in 1937, all pretenses to continued normal business in Asia collapsed. The movement of Japanese military units to Guangdong (Kwangtung) Province and the fall of Guangzhou (Canton) in 1938 unleashed a flood of refugees to Hong Kong that totaled as many as 750,000 between 1937 and 1939, boosting its population to about 1.8 million. Under the harsh occupation of Japanese forces, the number fell to 0.5 million by the end of World War II, but it recovered to 1.0 million in 1946. Battles between the Communist Party under Mao Zedong (Mao Tse-tung) and the Nationalist Party under Jiang Jieshi (Chiang Kai-shek) for control of China set off another surge of refugees that totaled about 345,000 from 1949 to 1951. By the latter date, Hong Kong's population had returned to 1.8 million, setting the floor for subsequent growth. Shortly after the triumph of the Communist Party in 1949, traders and financiers in Hong Kong recognized that their influence over the China trade would not recover quickly.[2]

[1] Industry Department, *1996: Hong Kong's manufacturing industries*, p. 20.

[2] Endacott, *A history of Hong Kong*, pp. 288–307; Fairbank, *China*; Szczepanik, *The economic growth of Hong Kong*, pp. 25–27, tables 6–8, pp. 152–54; Tarling, *The Cambridge history of Southeast Asia*.

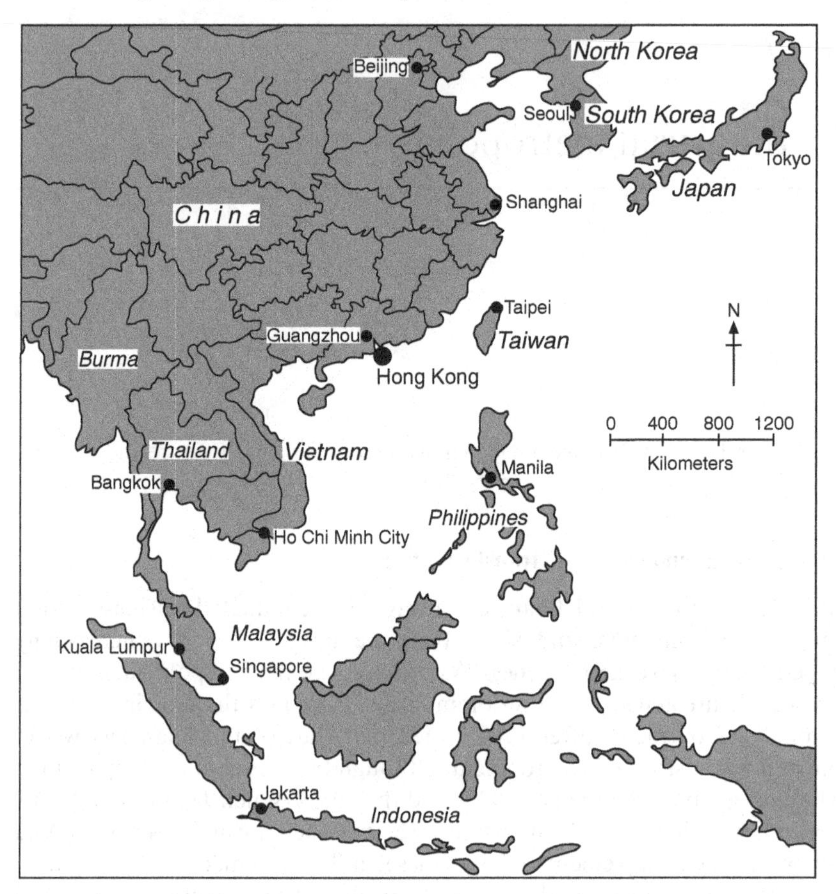

Map 4. Asian political units and metropolitan centers, late twentieth century. *Source:* author.

Following the defeat of Japan in 1945, reconstruction dominated politics in Northeast Asia for at least a decade. Freed from Japanese colonial control, Korea quickly became embroiled in Cold War politics and bitter war. After 1953, North Korea was isolated from Asian trade, except with China and the Soviet Union, whereas South Korea commenced recovery during the late 1950s; however, it remained economically weak as of 1960. Japan embarked on rapid economic growth by the early 1950s, and its export manufactures swelled, but demands for traditional raw commodities of Southeast Asia constituted its significant regional impact. The end of colonialism provided the unifying political theme in Southeast Asia between 1945 and 1960; yet, political units diverged (Map 4). Extraordinary paradoxes emerged, and the paths taken had significant consequences for economic growth and for traders and financiers in Asia.[3]

The Philippines and Taiwan followed diametrically opposite development trajectories. The Philippines quickly received independence from the United States in 1946 and embarked on rapid economic growth supported by American foreign aid, but the government did not institute serious land reform. The oligarchy controlled the land and processing industries, and import-substitution industries of the domestic elite and multinationals stayed inefficient behind protective tariff walls. By 1960, these deficiencies retarded growth, and the Philippines commenced episodic political and economic crises, with little change in the power of the oligarchy. Taiwan started under the brutal, corrupt occupation of the Nationalist Chinese government, but fortunes reversed, paradoxically, when Jiang Jieshi fled to Taiwan along with over 1 million mainlanders in 1949 and led an honest, competent government with strict military control. Backed by the United States, Taiwan instituted rural land reforms that gave peasants land ownership, and the government invested heavily in rural infrastructure. Agricultural productivity rose sharply, and this raised farm sector demands for small-scale, intermediate manufactures and low-cost consumer goods, setting off rural industrialization. Rural employment shifted to non-agricultural occupations, and by 1960 Taiwan housed a thriving small industrial sector; the government then switched to encouraging export industries.[4]

Indonesia and Malaysia shared a heritage of multinational control of large-scale plantations and mines under colonial protection, but their development paths diverged after 1945. Indonesia achieved independence in 1949, yet Dutch and other foreign firms continued to own 70 percent of the plantations and Chinese firms controlled 19 percent. The government suppressed dissent and invested little capital in rural infrastructure, leaving most peasants impoverished. By the late 1950s, widespread dissatisfaction and nationalist assertions led to expropriation of Dutch property and suppression of Chinese firms. The government started heavy industrialization aimed at self-sufficiency, but, without domestic savings, this generated inflation and economic collapse. In the late 1960s, a new regime commenced rural development that boosted farm productivity somewhat. Malaysia achieved independence almost a decade later than Indonesia. During the 1950s, agricultural investment went

[3] Fairbank, Reischauer, and Craig, *East Asia*, pp. 817–26, 882–86; Stockwell, "Southeast Asia in war and peace."

[4] Barrett and Whyte, "Dependency theory and Taiwan"; Dixon, *South East Asia in the world-economy*, pp. 200–206; Fairbank, *China*, pp. 339–41; Ho, "Decentralized industrialization and rural development"; Oshima, "The transition from an agricultural to an industrial economy in East Asia"; Oshima, *Economic growth in Monsoon Asia*, pp. 206–16; Owen, "Economic and social change," pp. 488–89; Stockwell, "Southeast Asia in war and peace," pp. 349–51.

into long-term projects to replant rubber trees and plant oil palm, but this did not pay off until the 1970s. Rural infrastructure investments that benefited peasant rice farmers lagged until the late 1960s, keeping the commercial economy bound to tin and plantation crops. The failure to promote peasant farming earlier suppressed incomes and rural industrialization; therefore, import-substitution manufactures soon saturated markets. The government promoted export processing zones funded by foreign capital in the 1970s, but peasant farmers still lagged.[5]

Thailand, in contrast, did not experience colonialism. Its farmers owned land and freely sold crops in open markets, but these advantages left peasants impoverished because the monarchy did not encourage infrastructure development and discouraged Chinese business investment. Major rural infrastructure investments commenced in the 1960s, yet the government continued to tax rice exports, depressing prices rice farmers received and subsidizing urban consumers; therefore, farmers had few incentives to raise productivity. Rural industrialization made little progress, and import-substitution manufactures, which faced meager demand, failed to spark development during the 1960s. Thai peasants did not starve, but those in Burma and Vietnam suffered grievously from internal chaos and war. Even as the British left Burma in 1948, it sank into recurrent violence and government coups that have kept peasants abysmally poor to the present. Except for brief periods, Vietnam experienced thirty years of war that left a devastated infrastructure and agricultural poverty. The Communist government tentatively opened the economy to market forces in 1979, but it did not seriously begin to free market prices and permit private firms until 1986. The long failure to free farmers to produce left the population impoverished.[6]

During the first quarter-century of Communist Party rule, Chinese peasants fared little better economically than other peasants in Asia, excluding those in Japan and Taiwan. In 1949, the Party embarked on rural land reform, but benefits dissipated when it commenced coercive measures that locked farmers, first into collective farms, and then into communes under the Great Leap Forward. Farmers lost all private property rights in land, animals, and equipment. In 1962, agriculture began a slow recovery from the disastrous economic policies of the Great Leap Forward and settled into collective agriculture that persisted for two decades. Agricultural investments raised production, but the government viewed agriculture as a source of capital to subsidize urban residents and support industrialization. Although the government

[5] Dixon, *South East Asia in the world-economy*, pp. 181–93; Oshima, *Economic growth in Monsoon Asia*, pp. 235–55; Owen, "Economic and social change," pp. 487–90; Stockwell, "Southeast Asia in war and peace," pp. 361–65, 374–80.

[6] Fforde, "The political economy of 'reform' in Vietnam – some reflections"; Ingram, *Economic change in Thailand, 1850–1970*, pp. 220–79; Owen, "Economic and social change," pp. 490–92; Stockwell, "Southeast Asia in war and peace," pp. 351–74.

promoted rural industrialization, this amounted to state funding of inefficient enterprises, rather than indigenous investment in businesses that met demands from prosperous agricultural and urban sectors. The Cultural Revolution created chaos from 1966 to 1976 and retarded development; at the time of Mao Zedong's death in 1976, peasants did not starve, but they stayed impoverished.[7]

Hong Kong traders and financiers faced a new political environment during the two decades after 1945, but the Asian economy retained trappings of the previous century. Japan was the only large country to grow rapidly, and the economic upturn in Taiwan generated tiny demands for trade and financial services. Raw material commodities dominated exports as they had since 1860, and most peasants remained impoverished; consequently, they demanded few goods and services. The potential of the vast China market stayed unrealized because the government failed to institute measures that set off vigorous agricultural and industrial transformation; thus, closure of the China market held little practical significance for Hong Kong traders and financiers.

Trade and finance languish

By the mid 1950s, the Inchcape family, a great global trading group, sold Gibb, Livingston & Company because it contributed few profits. The sale of this venerable trading company, that moved its headquarters to Hong Kong in the 1840s and frequently had members on the board of directors of the Hongkong and Shanghai Bank, epitomized limited trade opportunities in Asia during the 1950s and early 1960s. Hong Kong's exports to China surged in 1950 and 1951 in anticipation of the United Nations trade embargo during the Korean War, but then domestic exports (goods produced locally) and re-exports stagnated between $2.5 and $3.0 billion (Hong Kong dollars) for the rest of the decade. Re-exports to China stayed depressed after the Korean War because it looked to Soviet bloc countries for capital goods, and its leaders stressed self-sufficiency. Domestic exports commenced a sustained rise in the late 1950s, climbing by two-thirds between 1959 and 1963, but re-exports from Hong Kong stagnated and did not jump until after 1963. Continued impoverishment in Asia offered few opportunities for trading firms to control commodity exchange, and large multinationals that dominated raw material production, such as oil companies, often internalized initial intermediary sale on world markets. Imports to Hong Kong contributed the most to trade growth; they rose about 80 percent from the early 1950s to the early 1960s. South China, especially the Pearl delta, supplied 20–25 percent of imports, mostly food and construction materials to supply the burgeoning population

[7] Fairbank, *China*, pp. 343–405; Putterman, *Continuity and change in China's rural development*, pp. 3–31.

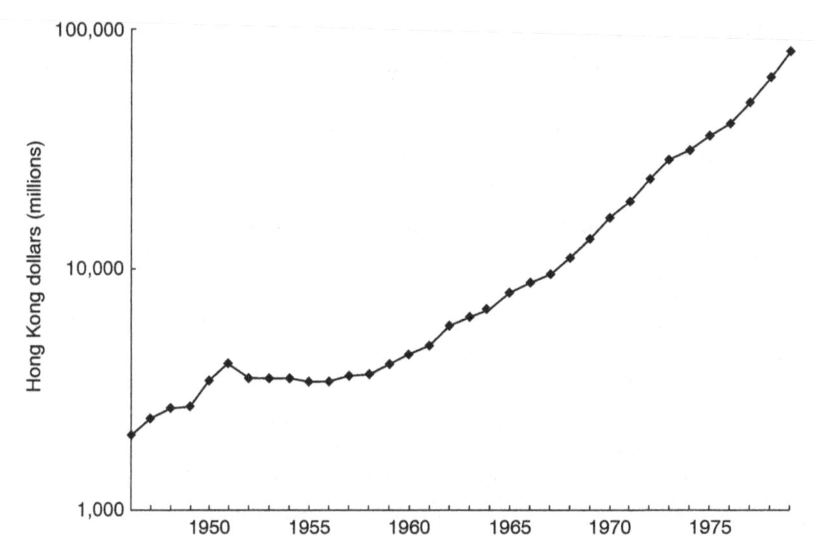

Fig. 7.1. Assets of the Hongkong and Shanghai Bank, 1946–1979.
Source: King, *The Hongkong Bank in the period of development and nationalism, 1941–1984*, tables 5.2, 15.11, pp. 197, 736.

of Hong Kong whose number soared from 1.7 million in 1950 to 3.1 million in 1961.[8]

Because much of Hong Kong's finance supported trade, absence of robust trade growth hindered financial expansion. Intermediation gains by banks barely compensated for reduced quasi "central bank functions," such as issuing currencies, as those governmental institutions formed in newly independent countries. Hong Kong, nevertheless, remained among the top ten international banking centers from 1955 to 1965, testimony to its financiers' prominence in Asia. Singapore did not make that group, and Tokyo was the only other Asian center in it; Hong Kong ranked just above Tokyo in 1955 and just below it in 1960 and 1965.[9] As the leading bank in Asia, expansion of the Hongkong and Shanghai Bank indicates the opportunities financiers faced. After an initial surge in assets associated with post-war recovery, they peaked in 1951; that remained a lid for the rest of the 1950s because most Asian countries failed to transform their economies (fig. 7.1). Assets turned markedly

[8] Census and Statistics Department, *Hong Kong annual digest of statistics, 1989*, table 2.1, p. 11; Census and Statistics Department, *Hong Kong review of overseas trade in 1978*; Fairbank, *China*, pp. 406–25; Jao, "Hong Kong's role in financing China's modernization," table 1.1, p. 15; Jones, *Two centuries of overseas trading*, p. 249; Szczepanik, *The economic growth of Hong Kong*, pp. 51–55, tables 7, 14, pp. 153, 158.

[9] Jao, *Banking and currency in Hong Kong*; Reed, *The preeminence of international financial centers*, table A.12, pp. 139–41.

upwards during the early 1960s and accelerated from the late 1960s to the 1970s as agricultural reforms took hold in some Asian countries and industrialization commenced. Japanese banks hid behind protectionist walls of government regulation and thrived in a large, rapidly growing economy, but great Asian exchange banks such as the Hongkong Bank had no safe harbor. Large American banks, including Bank of America and Chase, drew on the profitability of a robust post-war economy at home and expanded financing in Asia to serve American multinationals in manufacturing and raw material extraction and to look for new opportunities. Exchange banks had to react to this competition or face loss of market share, but they faced a dilemma; American banks gained competitive advantages from long-standing ties to their domestic corporations. Exchange banks had little hope of capturing that business; therefore, this pitted them against American banks for market share in an Asian economy with mostly impoverished peasants who witnessed slow income gains.

Reactions of the Hongkong and Shanghai Bank to this competitive threat from American banks foreshadowed Hong Kong's transformation as a financial center during the last half of the twentieth century. To protect itself within Asia, Hongkong Bank purchased Mercantile Bank of India in 1959, but the social networks of capital of that small, efficient bank, with its superb contacts and experience in Asia, replicated those of the Hongkong Bank. This purchase, however, removed the Mercantile from the clutches of American banks and forced them to find less experienced partners and to forge their own capital networks. Purchase of the British Bank of the Middle East, an experienced, profitable bank, in 1960 offered Hongkong Bank access to markets where it had little representation (although both banks had worked together since 1889), and added social networks, including bridges to other networks, that short-circuited the process of entering new markets. Robust expansion of the California economy during the 1950s offered a tempting counterpoint to slow Asian growth. Nevertheless, the Hongkong Bank of California, incorporated in 1955, stayed unprofitable because officials had little access to capital networks in the competitive, complex California market; they had to build those network bonds, a slow process. These changes in Hongkong Bank forced it to design an alternative organizational structure. By 1961, the bank inexorably created a structure that included the corporate head office in Hong Kong, with policy and supervisory power over global operations, and subsidiary upper-management offices in London to cover the Middle East and in San Francisco to monitor California, as well as a unit in Hong Kong that managed Asian banking and supervised branches in the region.[10]

Desultory business opportunities in Asia during the 1950s and early 1960s

[10] King, *The Hongkong Bank in Late Imperial China, 1864–1902*, tables 3.5, 5.5, pp. 95, 151; King, *The Hongkong Bank in the period of development and nationalism, 1941–1984*, pp. 482–564.

hurt the Hongkong and Shanghai Bank, but fortuitous local circumstances offered business opportunities that would temporarily move Hong Kong away from the century-long focus on Asian trade. The arrival of industrialists, manufacturing-led economic growth, and a large population increase dramatically boosted demand for loans to finance manufacturing and to purchase consumer durables. Hongkong Bank gained expertise in industrial lending and, by 1960, created a specialized unit, Wayfoong Finance, to service consumer loans. Branches that provided industrial and consumer lending generated 50–70 percent of the bank's total profits between 1951 and 1961.[11] An astounding paradox had transpired that set Hong Kong on a new course. Its foundation as the trade center for Asia rested on the forcible opening of China by the British after the Opium War of 1840–42, whereas after the closure of China in 1949, Hong Kong acquired firmer trappings of the great industrial metropolises of the nineteenth century, such as New York and London, that united industrial entrepreneurs, financiers, wholesalers, and labor.[12] As residents of a city-state, however, Hong Kong's manufacturers faced a dilemma; regional and national markets that undergirded the industrialization of American and European metropolises did not exist. They had to quickly enter international markets; fortunately, expert local exporters, trade financiers, and bankers with long experience in Asian and global markets aided them.

Industrial refugees

Hong Kong always housed typical metropolitan manufactures that comprised both exchange-related ones, including resource processing (sugar and chemical refining) and those that support intermediary activities of firms (shipbuilding and printing), and local consumer manufactures (apparel and food).[13] Firms that produced exchange-related goods faced familiar, secure markets linked to import and export trades, and trading firms kept them appraised of raw material supplies and demand for products; they had little incentive to branch into unfamiliar manufactures. Producers of local consumer goods likewise experienced no incentive to enter new industries, and expansion of their market areas outside Hong Kong required dramatic shifts in marketing, including using international traders to find consumer markets. Factories could not extend market areas to prosperous rural consumers because Hong Kong's population remained too small to have much impact beyond adjacent rural areas, and exploitation by elites and lack of public investment in rural

[11] King, *The Hongkong Bank in the period of development and nationalism, 1941–1984*, pp. 202–203, table 8.1, p. 327.

[12] For the nineteenth-century industrial metropolis, see Meyer, "The rise of the industrial metropolis."

[13] Meyer, "The rise of the industrial metropolis," pp. 736–37; Pred, *The spatial dynamics of U.S. urban-industrial growth, 1800–1914*, pp. 167–77.

infrastructure kept Asian peasants impoverished. The industrial transformation of Hong Kong, therefore, did not follow an inevitable course; it required a catalyst, the growing turmoil after 1920 that culminated in the triumph of China's Communist Party in 1949. Industrialists who moved to Hong Kong came from nearby Guangdong Province and Shanghai.

Starting in the mid 1920s, textile and clothing firms from Guangzhou and Foshan in Guangdong Province either opened branches in, or relocated to, Hong Kong, and following the Japanese attack in 1937 some Shanghai industrialists arrived. Although industry atrophied during the Japanese occupation of Hong Kong, firms from Guangdong and Shanghai formed an industrial base by 1947. At that time, a survey of 179 factories revealed that 87 percent started between 1926 and 1947; former residents of Guangdong headed most factories. After World War II, intensified battles between Communists and Nationalists sent hordes of immigrants to Hong Kong from groups with the most to lose under the Communists, urban classes such as the army, police, professionals, intellectuals, clerks, and commercial people. Because they brought little manufacturing experience, they could not instigate industrial development, although they could serve as workers. Entrepreneurs from nearby Guangdong and from Shanghai, the leading manufacturing center of China, acted as catalysts to accelerate the industrialization of Hong Kong; most had roots in textiles and, to a lesser extent, clothing manufacturing. These entrepreneurs came from the private sector rather than from firms with close ties to the government of Jiang Jieshi. Since the late 1920s, Shanghai industrialists who owned large firms had suffered under the Nationalists' statist policies, including expropriation and extortion; a move to Taiwan with the Nationalists had little appeal for these private-sector entrepreneurs.[14] Nevertheless, the Shanghai and Guangdong industrialists' choice of Hong Kong remains perplexing.

Hong Kong was open to Chinese refugees whereas other Asian countries shut their doors, but individuals with capital and strong business networks could circumvent restrictions. Other reasons for choosing it probably loomed larger. By 1949, most leading metropolises had limited appeal. Tokyo and other Japanese cities contained large, competitive textile industries, and their social, political, and economic environments were hostile to foreigners. Chinese firms in other metropolises, including Saigon, Manila, Jakarta, and Bangkok, coped with either revolutionary activity or heavy government regulation. As a British colony, Singapore held greater attraction than those cities, but Shanghai industrialists had few contacts with Singapore's traders and financiers because its firms chiefly controlled commercial exchange with Indonesia, Malaya, Burma, and Thailand. In contrast, Hong Kong's traders

[14] Leeming, "The earlier industrialization of Hong Kong"; Wong, *Emigrant entrepreneurs*, pp. 16–27.

and financiers, often through branch offices in Shanghai, had a long history of interaction with its industrialists. As many as 200 import and export firms based in Shanghai transferred their headquarters to Hong Kong by 1946; this move kept Shanghai industrialists within familiar social networks of capital. Decisions of Guangdong industrialists to move to Hong Kong had greater transparency because they had intimate contact with its importers and exporters. Guangdong and Shanghai industrialists risked founding factories within easy reach of the Chinese army because Hong Kong traders and financiers had deep knowledge of China through their business networks. The industrialists must have gained strong intimations about the future policy of China's Communist Party towards Hong Kong. So long as the British did not support local subversives and recognized the legitimacy of the new government, which Britain was one of the first to recognize, China would leave Hong Kong alone and use that city as a conduit to the "outside" world.[15]

Textile and clothing industrialists from Guangdong Province typically headed small firms, but cotton textile manufacturers from Shanghai who had substantial technical expertise and brought skilled workers along owned large, heavily capitalized firms. Most owners spoke English and sent their children to schools in the United States or United Kingdom; they drew on alumni networks to sell textiles outside Asia as those markets became important. Although these Shanghai firms could not bring equipment, their participation in social networks in Shanghai that linked foreign banks, native banks, trading firms (Chinese and foreign), and industrialists provided quick access to capital in Hong Kong because financial firms in the two metropolises had deep ties. Many Shanghai industrialists also had family members involved in finance; thus, they tapped large inflows of Chinese capital from China and elsewhere in Asia between 1947 and 1955, an amount that reached almost 40 percent of total income in Hong Kong over that period. Hongkong and Shanghai Bank, which had long-term ties to Shanghai industrialists through its local branch before World War II and close bonds with industrialists from Guangdong, probably contributed more than half of the bank capital for industrial finance during the 1950s, but that comprised a small share of total capital invested in manufacturing because it excluded most Chinese capital that flowed to Hong Kong and credit that export and import firms provided factories.[16]

Manufacturing surged as recent immigrant industrialists built factories and acquired machinery; between 1950 and 1955, the number of firms jumped from 1,478 to 2,437 and employment climbed from 81,718 to 110,574. The decline in average firm size from 55 workers to 45 over that period foreshadowed the next four decades as ever smaller firms typified the industrial struc-

[15] Vogel, *One step ahead in China*, pp. 44–47; Wong, *Emigrant entrepreneurs*, pp. 36–37, 75–76.

[16] Jao, "Financing Hong Kong's early postwar industrialization"; Wong, *Emigrant entrepreneurs*, pp. 26–126. The estimate of capital inflow is computed from Szczepanik, *The economic growth of Hong Kong*, table 46, p. 183.

ture. Between 1955 and 1960, the numbers of firms and employees doubled to 5,346 and 218,405, respectively, solidifying Hong Kong's status as an industrial metropolis. Guangdong and Shanghai industrialists built a vigorous textile industry even before the end of the Chinese civil war in 1949. As of 1950, textile firms employed 24,975 workers, amounting to 31 percent of the industrial labor force, and their number more than doubled to 54,759 by 1960, even as their share dropped to 24 percent. Within that decade, Hong Kong rapidly shifted towards clothing manufacturing; the meager total of 1,944 employees in 1950 soared to 51,918 a decade later, representing 24 percent of the industrial workers, and a few years after 1960 the clothing industry achieved a dominance that it did not relinquish over the next three decades.[17]

The local market could not absorb the swelling torrent of goods from factories in Hong Kong. Exporters first looked to Asia, which took 60 percent of total exports as late as 1957, whereas the United States and Western Europe purchased 7 percent and 16 percent, respectively. Export markets turned swiftly as Asia's share plummeted to 24 percent, and shares of the United States and Western Europe soared to 26 percent and 28 percent, respectively, by 1960. Disintegration of the China market could not have caused that switch because exports to China collapsed during the Korean War and the United Nations embargo on trade that lasted through 1953. This was at least four years prior to the switch away from Asia that commenced after 1957; Hong Kong's exports to China stayed low until 1979. As of the mid 1950s, Hong Kong's manufacturers successfully penetrated markets in the traditional trading region beyond China, including Indonesia, Malaya, Thailand, and Indochina.[18] However, those markets failed to grow and in most cases declined as turmoil and economic chaos gripped several countries, including Indochina and Indonesia, and import-substitution manufactures came into vogue across the region.

Hong Kong manufacturers faced the same barrier to Asian trade as European and American industrialists had done in the nineteenth century: Asia's impoverished peasants had limited capacity to purchase imported goods, and import-substitution manufactures also faced that obstacle. The large industrializing economy of Japan remained shut to most exporters, and tiny, growing Taiwan likewise remained closed to textiles and clothing because its government protected import-substitution manufactures. The capacity of Hong Kong's industrialists to switch to export markets outside Asia could not have taken place without the powerful global networks of its trading firms; and their ability to scan information about, and to penetrate, world markets would undergird the continued adaptability of industrialists. These traders

[17] Industry Department, *1995: Hong Kong's manufacturing industries*, tables 2.4, 2.12, pp. 24, 31.

[18] Jao, "Hong Kong's role in financing China's modernization," table 1.1, p. 15; Riedel, *The industrialization of Hong Kong*, tables 7, 8, p. 36; Szczepanik, *The economic growth of Hong Kong*, table 45, p. 180.

and industrialists formed a juggernaut in export markets outside Asia after 1960, but they could not directly alter the political economies of Asian countries. Opening those markets awaited transformations within each country that would allow peasants to rise out of poverty.

Asian industrialization

Industrial latecomers can draw on a reservoir of world experience with factory organization, production technologies, labor management, capital utilization, marketing, and distribution, but this knowledge does not automatically translate into a successful transition from impoverished agricultural peasantry to industrial economy. Specialists in development studies documented the failure of that transition in much of Africa, halting movements in Latin America, and limited progress in South Asia. These contrast with the "Asian miracle" economies, encompassing those from South Korea and Japan in the north to Indonesia and Malaysia in the south and west. Japan and the four "little dragons" (Hong Kong, Singapore, South Korea, and Taiwan) led this industrialization, and each started rapid industrial growth and entered export markets by the 1960s. Malaysia, Thailand, and Indonesia commenced significant industrial growth and export orientation after 1980, whereas the Philippines lagged, and China did not start major industrial change until after 1985. According to the World Bank, successful economies achieved superior accumulation of physical and human capital through sound development policies that included excellent macroeconomic management, financial reforms, educational investment, and agricultural policies to raise productivity. In most economies the government intervened systematically and through multiple channels; yet, they implemented diverse interventions, making it difficult to isolate positive or negative impacts of specific interventions on industrial growth.[19]

Since the industrialization of Japan and the four little dragons, however, much of this manufacturing change took the following paths: multinationals entered export processing zones to exploit low-wage labor, and governments subsidized these firms through tax benefits and free infrastructure; governments subsidized domestic industries to exploit this same labor for either import-substitution manufactures or exports; and governments invested in capital-intensive resource processing or intermediate goods industries. The large subsidies for these forms of industrial growth amounted to a redistribution of wealth from other sectors to manufacturing, and they offered rural peasants few benefits. Only laborers in export zones made higher wages, and

[19] Gereffi and Wyman, *Manufacturing miracles*; Vogel, *The four little dragons*, pp. 83–112; World Bank, *World tables 1993*; World Bank, *The East Asian miracle*; Yang, *Manufactured exports of East Asian industrializing economies*.

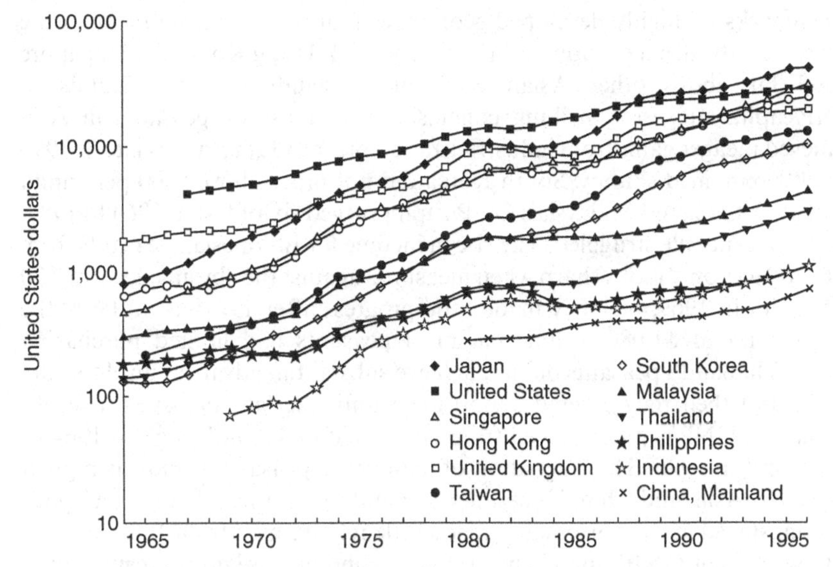

Fig. 7.2. Gross national product (GNP) per capita by political unit, 1964–1996. *Source:* World Bank, *World development indicators.*

multinationals demanded few inputs from host countries. Those that subsidized their export industries stimulated only minimal spread effects within the country, and the concentration on low-wage industries offered workers limited benefits. Import-substitution industrial policies typically forced rural peasants to pay higher prices for manufactures and receive lower prices for agricultural products in order to subsidize urban-industrial workers. Capital-intensive manufactures often were inefficient because domestic demand could not support them. These industrial approaches imposed a superstructure on the economy and society that left rural peasants, the overwhelming majority of the population, impoverished; domestic markets for manufactures stayed moribund.[20]

Timing the rise of the mass of rural peasants out of poverty offers clues to the economic transformation of Asia and its relation to industrialization. The level and change in gross national product per capita (GNP/capita) in current United States dollars provides comparable data, albeit crude, across economies from 1964 to 1996 (fig. 7.2). It is shown in logarithmic form; thus, parallel lines indicate equal percentage changes, and bends mark changes in growth rates. The United States, the United Kingdom, and Japan serve as

[20] Gereffi and Wyman, *Manufacturing miracles*; Reynolds, "Agriculture in development theory," pp. 20–21; World Bank, *The East Asian miracle*; Yang, *Manufactured exports of East Asian industrializing economies*.

benchmarks of highly developed economies that industrialized prior to the newly industrialized economies in Asia. By 1964, Hong Kong and Singapore stood far above other Asian economies, excluding Japan. Trends in GNP/capita suggest a striking conclusion: the mass of peasants in Asia achieved meager gains from industrialization prior to 1980, and as late as 1996 benefits remained skimpy. South Korea did not move above $500 per capita until after 1973; and Malaysia's GNP/capita surged sixfold from 1964 to 1980, but then its people struggled. The rise in income for Indonesian peasants from 1969 to 1981 only looks sharp when measured against the abysmal level of $70 per capita in 1969, but they made little progress after 1981; as of 1996, the GNP/capita of $1,080 meant that most peasants had limited purchasing power. Thailand's peasants did not achieve substantial advances until the late 1970s, and then they entered a slowdown until the late 1980s; as a result, Thailand's GNP/capita plunged from 20 percent of Hong Kong's in 1964 to 12 percent in 1996. The stagnation of Philippine peasants stands as a great tragedy in Asia, and China's peasants, accounting for the bulk of Asia's population, stayed at the bottom and made little relative gain from 1980 to 1996.

Low levels of GNP/capita in most Asian economies as late as the mid 1980s and limited subsequent gains suggest that industrialization is a superstructure on a mass of impoverished peasants. Economic progress in Malaysia, Thailand, and Indonesia only looks large when compared to failures in the rest of the less-developed world; export industrialization has provided few benefits. These economies first started rural development in the late 1960s and early 1970s, and those lags contributed to their small rise in GNP/capita from 1964 to 1996. The GNP/capita of Taiwan continues to exceed most Asian economies by a large margin (fig. 7.2), a product of its early, successful development policies to benefit rural peasants. South Korea made the next biggest investment in rural development, but a major push did not commence until the 1970s; and its inadequate investments in the rural economy retarded development even though exports surged. The failure of other Asian economies to invest significantly in rural development has impeded the rise of peasants out of poverty. When farmers boost incomes through increased productivity, they generate substantial multiplier effects in rural areas and stimulate local, small-scale manufactures that process farm products and provide intermediate producer goods for agriculture. Farm families' growing demands for consumer goods support local and regional industrialization, and the rise in farm productivity permits excess laborers to enter manufacturing. These gains to farm families only come if they receive market prices for products and can buy farm inputs and consumer goods at market prices. Rural development provides powerful growth impulses because most of the population benefits rather than only a small subset of urban-industrial workers. Development policies that extract capital from agriculture to subsidize urban-industrial workers short-circuit this growth process.[21] Meager rural development across much of Asia, therefore, meant that Hong Kong industrialists faced the same impoverished peasant markets that had always thwarted its traders and financiers. Lags in the climb out of poverty would keep manufacturers focused on markets outside Asia for decades; their success provides a window on the impact of government policy on export industrialization.

Industrial transformation of Hong Kong

Laissez-faire *or state power*

The debate over the reason for Hong Kong's rapid industrial expansion has raged between advocates of *laissez-faire* capitalism, on the one hand, and of state power, on the other hand. According to some critics, the state exerted power to favor business because that elite and the government form a seamless web of shared interests, but the political economy of Hong Kong has greater complexity than these simplifications would suggest. When the industrial surge gathered momentum during the 1950s, the business elite with the closest ties to the government, principally large British trading houses and banks, did not dominate that expansion. Chinese industrialists and trading firms and other foreign trading firms played significant roles, but they did not operate at the center of government power. Recently, the power of great British firms (hongs) has waned whereas small and large Chinese firms remain ambivalent about direct involvement with government, and Beijing leaders discourage Chinese government-owned firms in Hong Kong from intervening in its politics.[22] Government's role in Hong Kong's industrialization becomes clearer through comparison with Singapore.

Singapore's senior leaders had roots in the anti-colonial struggle within Britain that was tied to socialism and the drive to nationalize key industries after World War II. After the People's Action Party of Lee Kuan-yew gained power in 1959, Singapore embarked on policies to entice foreign investment but stayed away from the sweeping social welfare policies which were in vogue in Europe. During the two decades after separation from Malaysia in 1965, they enlarged the scope of industrial policies based on government direction and focused on attracting multinationals. The government introduced financial incentives such as tax subsidies, developed industrial parks with concessions for preferred tenants, and directly funded research institutes; and it implemented traditional supports, including funding councils to promote

[21] Nicholls, "The place of agriculture in economic development"; Oshima, "The transition from an agricultural to an industrial economy in East Asia."

[22] Cuthbert, "A fistful of dollars"; Schiffer, "State policy and economic growth"; Wong, "Business and politics in Hong Kong during the transition."

industry, the creation of housing and industrial training programs, and the funding of infrastructure investments (telecommunications and the airport). Multinational promotion strategies bore fruit; by the late 1970s, foreign firms accounted for just over half of employment and about two-thirds of value-added and capital investment in manufacturing. At that time, the government also identified strategic industries, including oil and petrochemicals, information technology, and biotechnology, and by the early 1990s, it directly acted as an entrepreneur in over 500 government-linked companies.[23]

The industrial policies of the Hong Kong government followed the traditional supports that Singapore provided, but Hong Kong took a different tack on direction of industry; for the most part, it refrained from tax incentives, subsidies, and industrial targeting. It offered land at concessionary prices, but this was available to all firms, with one exception: high-technology industries acquired special access to single-story buildings. The government dispensed research grants but generally refrained from full funding of research institutes. It offered institutional support and export promotion across a wide range of industries, rather than targeting a select group. Provision of public rental housing at subsidized rates remained the greatest investment of government in industry. It contributed substantial benefits to low-wage, labor-intensive industries, yet also benefited other sectors that employed low-wage workers (retail, personal services). Numerous entries and exits from industries and the rise and decline of entire sectors portrayed an unorganized and chaotic environment.[24] The absence of external guidance, and seeming chaos, nevertheless, concealed a production organization based on social networks among manufacturers and between them and traders and financiers. Most of these social networks in Hong Kong existed within the Chinese business sector, and it had global bridges to other social networks through Chinese and foreign firms. These networks provided information and assistance that supported the rapid adjustment of firms to global market conditions, and the absence of government subsidies and other guided interventions encouraged them to move quickly.

Growth and decline

Contemporary observers of Hong Kong in the 1960s proclaimed that its economy had transformed to manufacturing and that sector would dominate in the future. Its share of the labor force jumped from 30 percent in 1954 to 40

[23] Chiu, Ho, and Lui, "A tale of two cities rekindled," table 2, p. 109; Krause, "Hong Kong and Singapore"; Vogel, *The four little dragons*, pp. 74–77; Wong, "A comparative study of the industrial policy of Hong Kong and Singapore in the 1980s"; Yuan, "Hong Kong and Singapore."

[24] Chiu and Lui, "Hong Kong"; Krause, "Hong Kong and Singapore"; Schiffer, "State policy and economic growth"; Wong, "A comparative study of the industrial policy of Hong Kong and Singapore in the 1980s"; Yuan, "Hong Kong and Singapore."

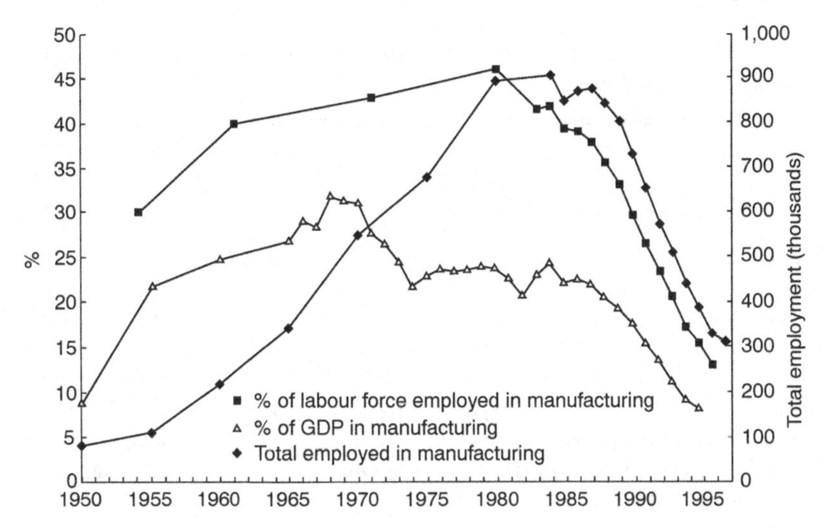

Fig. 7.3. Structure of manufacturing in Hong Kong, 1950–1997.
Source: Industry Department, *Hong Kong's manufacturing industries*, various years.

percent by 1961 and peaked near 46 percent about 1980; simultaneously, total factory employment rocketed upwards from 81,718 in 1950 to about 900,000 by the early 1980s (fig. 7.3). The designation of "newly industrializing country" that analysts awarded to Hong Kong in the early 1980s, therefore, seemed justified.[25] Employment, nevertheless, overstated the relative significance of local industrial production to the economy. Since 1950, manufacturing employment as a share of the total labor force remained substantially above the share of manufacturing in gross domestic product (GDP), and when manufacturing reached its peak share of total employment of about 46 percent around 1980, that sector accounted for only about 24 percent of GDP (fig. 7.3). Industry stagnated between 1980 and 1987 and then commenced a plunge that continued to the late 1990s: its share of the labor force fell to about 13 percent, its share of GDP collapsed to about 8 percent, and the number of factory workers dropped by two-thirds to about 310,000.

While the relative and absolute size of the industrial sector rose and fell, the average factory size in Hong Kong dropped relentlessly. The greatest shrinkage occurred between 1950 and 1975 as factory size fell from fifty-five workers to twenty-two, but the decline continued to a size of twelve employees, the level of a workshop, by the late 1990s.[26] The distribution of factories by size confirms this shift to small units. Those that employed fewer than ten workers

[25] Balassa, *The newly industrializing countries in the world economy*; Hopkins, *Hong Kong*; Riedel, *The industrialization of Hong Kong*; Szczepanik, *The economic growth of Hong Kong*.

[26] Data cited in this paragraph come from Industry Department, *1995: Hong Kong's manufacturing industries*, and unpublished data.

Table 7.1. *Percentage distributions of manufacturing employment by industry in Hong Kong, 1950–1997*

	1950	1960	1970	1980	1990	1997
Textiles	30.6	24.4	14.0	10.0	9.4	7.3
Clothing	2.4	23.8	28.8	30.9	34.5	24.8
Electronics	0.0	0.1	7.0	10.4	11.7	10.9
Plastics	0.3	8.3	12.9	9.7	7.3	4.0
Watches & clocks	0.0	1.1	1.8	5.5	3.7	2.6
Toys	0.0	3.3	7.2	6.2	3.4	1.1
Other	66.7	39.0	28.3	27.3	30.0	49.3
Total	100.0	100.0	100.0	100.0	100.0	100.0

Source: Industry Department, *1995: Hong Kong's manufacturing industries*, table 2.12, p. 31, and unpublished data.

constituted about two-thirds of all factories from 1975 to 1985, and then their share rose to 78 percent in 1996. Even these tiny workshops, nevertheless, could not escape the pervasive decline of manufacturing in Hong Kong; after 1990, the number of factories in every size group fell. Extremely large factories, those employing 1,000 or more workers, face extinction; in the late 1970s, about forty of these factories operated, but only twelve remained in 1996. Although tiny workshops (under ten employees) always constituted the overwhelming majority of factories, they accounted for only 12 percent of total employment in 1975, and this had risen to just 21 percent in 1996. At that date, most workers were in small-size (10 to 49 people) factories, which had 30 percent of total employees, and medium-size (50–199) factories, which had 25 percent of total employees. Extremely large (1,000 or more employees) factories accounted for only 6 percent of the workers in 1996, the same share they held in the early 1980s. By the standards of developed countries, therefore, Hong Kong houses mostly small factories, and they employ the majority of its manufacturing workers.

The industrial metropolis grows

The leading export industries of clothing, textiles, electronics, plastics, watches and clocks, and toys traditionally defined the industrial character of Hong Kong, apart from those industries that served its trade sector and local consumer demands. Collectively, they accounted for the majority of manufacturing employment (table 7.1) and an overwhelming share of exports (table 7.2) for most of the 1950–97 period. The numbers of workers in each industry moved synchronously with total manufacturing employment; they peaked

Table 7.2. *Percentage distributions of value of domestic exports by industry in Hong Kong, 1960–1997*

	1960	1970	1980	1990	1997
Textiles	19.3	10.3	6.7	7.5	6.0
Clothing	35.2	35.1	34.1	31.9	34.2
Electronics	0.0	8.7	19.7	25.9	26.7
Plastics	9.1	12.3	9.0	3.6	1.6
Watches & clocks	0.6	1.7	9.6	8.5	5.1
Toys	4.0	8.5	6.9	2.2	0.6
Other	31.8	23.4	14.0	20.4	25.8
Total	100.0	100.0	100.0	100.0	100.0

Source: Industry Department, *1995: Hong Kong's manufacturing industries*, table 2.13, p. 32, and unpublished data.

during the early 1980s, flattened from about 1980 to 1987, and plunged after that date (fig. 7.4). Changes in these industries' shares of total manufacturing employment, however, point to a new industrial structure. The textile share of employment declined inexorably from about one-third in 1950 to just 7 percent in 1997, and plastics, watches and clocks, and toys declined from peak shares during the 1970s to trivial levels in 1997; only electronics has maintained its share of total employment over the past two decades. The swift drop in clothing's share of employment from 35 percent in 1990, a level it had held since 1985, to 25 percent in 1997, combined with the sharp decline in the number of clothing workers, points to its impending demise as the signature industry of Hong Kong. The leading export industries reached their peak share of the value of domestic exports around 1980, when all other industries accounted for only 14 percent of total exports, but by 1997, these other industries had risen to 26 percent of total exports, a level they first reached during the 1960s, before declining (table 7.2). Industrial exports, however, did not become increasingly diversified within a declining export sector. Textiles, clothing, and electronics raised their share of domestic exports from 61 percent in 1980 to 67 percent in 1997; instead, the collapse of plastics, watches and clocks, and toys accounts for the falling share of the leading export sectors. The dramatic jump in the share of employment in industries outside these export sectors from 30 percent in 1990 to 49 percent in 1997 (table 7.1) suggests that local manufacturing is rapidly shifting away from exports towards those manufactures typical of an industrial metropolis, including commerce-serving, processing, and local-market products (table 7.3). They contribute little to total exports, and except for chemicals and packaging products, their exports as shares of gross output remain modest. Printing, which

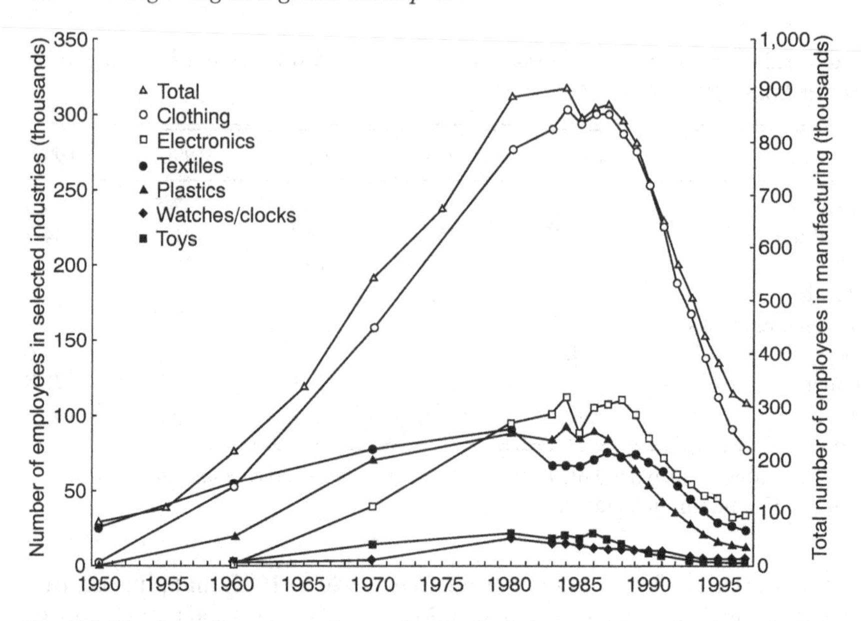

Fig. 7.4. Number of employees by total and selected type of manufacturing in Hong Kong, 1950–1997.
Source: Industry Department, *Hong Kong's manufacturing industries*, various years.

serves the huge trade, financial, and corporate sectors, as well as local demands for newspapers, books, and other printed matter, has become the second largest employer. Although its contribution to total exports remains small, printing increasingly serves the growing publishing (books, magazines, journals) sector of Hong Kong; it is now one of the four largest printing and publishing centers in the world and serves the Asia-Pacific market. This heightened importance of traditional metropolitan manufactures within overall industrial decline, coupled with the pervasive erosion of export industries from the mid 1980s to the late 1990s, implies that a common set of factors influences the industrial structure of Hong Kong.

The rise and decline of manufacturing in Hong Kong repeats the process that other industrial metropolises experienced. From the nineteenth to the early twentieth century, manufacturing expanded in the core of London, New York, and other large industrial metropolises, and as growth continued there, industrial satellites formed and manufacturing suburbanized because entrepreneurs sought lower-cost land and labor. When land, labor, and congestion costs in the core reached excessive levels, manufacturing shrank.[27] Industrial entrepreneurs in Hong Kong were not immune to these processes.

[27] Gras, *An introduction to economic history*; Haig, *Regional survey of New York and its environs*; Taylor, *Satellite cities*.

Table 7.3. *Hong Kong manufactures typical of industrial metropolises, 1985–1997*

	Industrial employment				Percentage of total value of domestic exports		Value of exports as percentage of gross output	
	Number ('000s)		Percentage of total					
	1985	1997	1985	1997	1985	1997	1985	1995
Commerce-serving								
Printing	30	46	3.5	14.8	1.3	2.3	25.9	19.0
Packaging products	11	5	1.3	1.8	0.7	0.7	54.6	45.3
Processing (entrepot)								
Chemicals	9	6	1.1	2.0	1.0	3.8	45.5	100.0
Local market								
Food and beverages	21	21	2.5	6.8	0.8	1.7	16.9	16.3
Metal products	48	17	5.6	5.5	2.3	1.6	35.7	29.1
Group totals/percentages	119	95	14.0	30.9	6.1	10.1	30.7	32.4

Source: Industry Department, *1995: Hong Kong's manufacturing industries*, tables 2.4, 2.10, 2.12–2.14, pp. 24, 29, 31–33, and unpublished data.

Table 7.4. *Growth rate of real payroll per person by economic group in Hong Kong, 1981–1996*

	Compound annual percentage growth in real payroll per person		
	1981–1988	1990–1994	1995–1996
Manufacturing	5.0	4.1	3.3
Wholesale, import/export, retail, restaurants, and hotels	3.8	2.8	0.1
Transport, storage, and communication	6.6	5.0	2.6
Finance, insurance, real estate, and business services	4.0	0.3	4.1
Community, social, and personal services	3.6	5.9	3.5

Source: Census and Statistics Department, *Hong Kong annual digest of statistics, 1989*, table 3.17, p. 45; Census and Statistics Department, *Hong Kong annual digest of statistics, 1995*, table 2.11, p. 23; Census and Statistics Department, *Hong Kong annual digest of statistics, 1997*, table 2.11, p. 23.

It achieved a dramatic rise in prosperity that raised GNP/capita relative to that of the United States from 19 percent in 1967 to 41 percent in 1984 to 87 percent in 1996 (fig. 7.2). Between 1981 and 1996, when manufacturing peaked and started to decline, most economic groups won sharp annual rises in real pay (table 7.4). Because factory owners competed for labor with other groups, manufacturing workers stood among the top gainers. This growing prosperity generated competition for space that affected workers and factory owners, causing rental prices for private housing units and for factory space in multistory buildings to climb rapidly, but these price increases must be standardized for rising incomes to measure the cost pressure on households and factories. Thus, the compound annual growth rate of real rental private housing prices was divided by the compound annual growth rate in real gross domestic product per capita (GDP/capita); this is termed the "housing cost pressure ratio". Housing cost pressure on factory workers in private housing did not rise during the peak years of manufacturing employment from 1980 to 1987 (housing cost pressure ratio=0.71), but when employment plunged from 1987 to 1996 (housing cost pressure ratio=2.82) housing cost pressures intensified as workers in economic sectors that expanded while manufacturing declined competed with those remaining in factory work. The government continues its policy of subsidizing housing to compensate for high land values in Hong Kong. The share of the population living in rental public housing has remained stable in the range 37–41 percent from 1979 to 1997. To measure the cost pressure on factory owners, the compound annual growth rate of real rental prices for space in multistory buildings, the quintessential factory space in Hong Kong, was divided by the growth rate of real GDP/capita; this is termed the "factory cost pressure ratio". Rental rates for factory space rose slower than GDP/capita from 1980 to 1987 (factory cost pressure ratio=0.88), the peak years of manufacturing employment, while they rose far faster from 1987 to 1996 (factory cost pressure ratio=2.34), when it commenced a plunge.[28] The enlarged supply of multistory factory space thrown on the market as manufacturing declined probably contributed to the lower cost pressure for factories than for apartment renters from 1987 to 1996.

Local industrial entrepreneurs remain wedded to their metropolis, the pivot of Asian trade, finance, and corporate management. These intermediary sectors helped industrialists export to distant markets, and this raised income levels. Growing demands for labor and land by intermediary sectors and rising household incomes, nevertheless, pitted them against factories; industrialists could not win that battle. This leaves a paradox. Industrial entrepreneurs specialized in labor-intensive manufactures, the type most susceptible to competition from countries with large supplies of low-wage labor. If they had invested in capital-intensive manufactures, they might have absorbed rising land costs because they account for small shares of total costs.

Labor-intensive choice

Hong Kong industrialists concentrated on labor-intensive manufacturing for export markets in the 1930s and reinforced that approach from the late 1940s through the 1950s. Firms used low-wage labor to penetrate consumer markets in developed countries that opened following import-tariff reductions after World War II, and they kept that approach through the 1970s. By then, rising wage levels in Hong Kong had forced manufacturers to boost capital investment per worker to stay competitive, but firms continued to invest much less capital per worker than the other "little dragons" (Singapore, Taiwan, and South Korea) into the 1990s.[29] Hong Kong's abundant supply of low-wage workers did not preordain this labor-intensive strategy. Because major trading companies had ample capital, they could have invested in machinery and employed low-wage, unskilled workers to tend automated equipment, but no evidence exists that they seriously tried that strategy.[30] Land costs may have prevented them, but other calculations probably loomed larger. These trading

[28] Computed from Census and Statistics Department, *Hong Kong annual digest of statistics, 1989*, tables 2.10, 7.2, 13.18, 14.12, pp. 17, 111, 184, 193; Census and Statistics Department, *Hong Kong annual digest of statistics, 1997*, tables 1.2, 6.17–6.18, 10.4, 17.2, pp. 4, 122–23, 188, 307.

[29] Chen and Li, "Manufactured export expansion in Hong Kong and Asian-Pacific regional cooperation"; Chow and Kellman, *Trade – the engine of growth in East Asia*, pp. 13–23, 63–83; Ho, *Trade, industrial restructuring and development in Hong Kong*, pp. 68–84; Nyaw, "The experiences of industrial growth in Hong Kong and Singapore."

[30] The strategy of trading companies moving into manufacturing and deploying large amounts of capital in machinery tended by low-wage workers has a long history. Boston merchants who invested in huge New England textile mills were prototypical examples; see Dalzell, *Enterprising elite*.

companies did not control distribution of products to final consumers; they sold goods to other factories, to wholesalers in consuming countries, or to large retail chains. Trading companies probably calculated that large capital investments in factories employing low-wage, unskilled workers entailed high risks because firms in other less-developed countries could duplicate that strategy, especially with government subsidies; Hong Kong's government refused to provide subsidies, eliminating that possibility. Although trading companies had experience with capital-intensive processing industries, those factories employed small numbers of skilled laborers. Trading firms did not have experience managing numerous low-wage, unskilled workers who utilized simple technology; this required sophisticated labor relations to maximize productivity. They decided that greater returns on capital came from exploiting their expertise in trading and letting local entrepreneurs manufacture.

Only textile industrialists from Shanghai had extensive capital, and they initially stayed in the same business; that left under-capitalized entrepreneurs from Guangdong Province to produce other export goods.[31] Although factories had little capital to invest in technology, the networks of capital of Hong Kong's global trading firms gave small factories extraordinary access to market information and orders. This allowed them to develop sophisticated strategies to respond rapidly to market changes, and their small size permitted them to manage low-wage workers effectively. They exited or refused to enter business lines that highly capitalized foreign competitors entered. A survey of small manufacturing firms (10–49 employees) in 1978 confirms this interpretation.[32] Most owners (88 percent) had roots in Guangdong Province, whereas Shanghainese accounted for only 7 percent of the firms. They started with little capital and over two-thirds (69 percent) of the owners took their first job in manufacturing, but their fathers had often worked as merchants (41 percent), which explains the firms' sensitivity to market-driven production; a significant minority (20 percent) of the owners' fathers had manufacturing-related backgrounds. Firms remained embedded in global networks of Hong Kong's trading firms: almost half (45 percent) of the factories received orders solely from those trading firms, and traders participated in another 12 percent of the orders. Trading firms served as contractors and sometimes offered technical advice and design and marketing suggestions. Direct subcontract relations outside the global networks of trading firms accounted for few orders; only 11 percent of the firms received orders directly from overseas. Local factories were highly interlinked; peers accounted for orders to 24 percent of the firms, especially in clothing and electronics. This subcontracting relation,

[31] Brown, "The Hong Kong economy," pp. 13–14; Leeming, "The earlier industrialization of Hong Kong."

[32] Sit, Wong, and Kiang, *Small scale industry in a laissez-faire economy*. The rest of the material in the paragraph is based on the data and discussion from this source; see especially tables 10.21, 11.11, 12.9, 14.5–14.6, pp. 260, 276, 294, 338, 340.

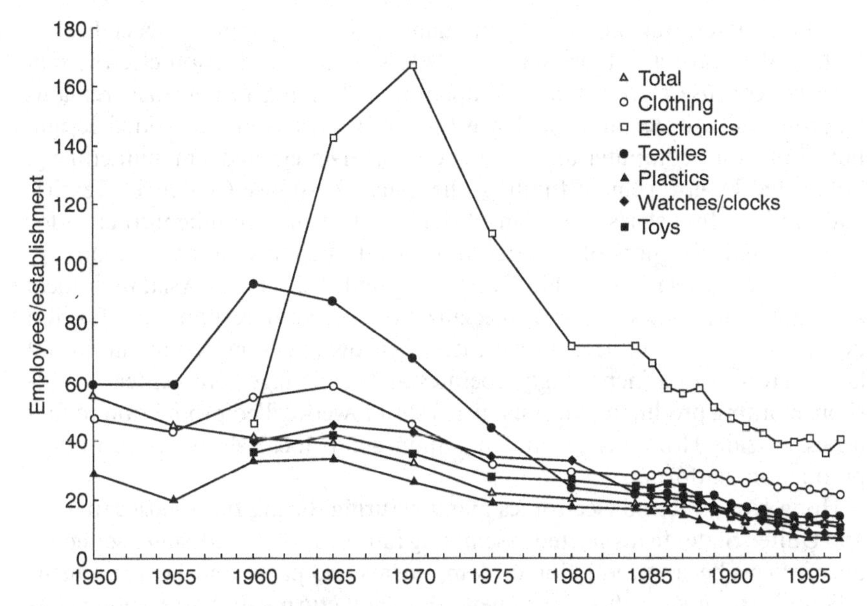

Fig. 7.5. Employees/establishment for total and selected type of manufacturing in Hong Kong, 1950–1997.
Source: Industry Department, *Hong Kong's manufacturing industries*, various years.

coupled with widespread ties to trading firms, allowed factories to respond flexibly to market demand. Industrialists in Hong Kong, therefore, were embedded in local social networks of capital, including other factories and trading firms, and network bridges of the traders provided industrialists unparalleled access to global markets.

Current status of manufacturing

Industrial firms in Hong Kong's leading export sectors in the late 1990s continue to build on their traditional strengths, sophisticated management of labor-intensive production and flexible response to market demands through global networks of trading firms. Their investments in advanced production technologies have not resulted in a substantial shift to capital-intensive manufacturing. Factory size in each industry started dropping by the 1960s, in concert with the overall decline of factory size in Hong Kong (fig. 7.5).[33] Besides the electronics industry, textiles remains one of the more capital-intensive sectors with extensive investment in computer-aided design, computer-

[33] The discussion in this section, unless specifically indicated, draws from text commentary and data in Industry Department, *1996: Hong Kong's manufacturing industries*, and unpublished data.

aided manufacturing, and intelligent manufacturing systems. It specializes in high-quality yarn and fabrics and shifts less-skilled production elsewhere in Asia in response to lower-cost competitors. Clothing, in contrast, remains labor-intensive; its technological investments focus on computer-aided design, but not on computer-aided or computer-integrated manufacturing. Supported by government funding, the Quick Response Center for Textiles and Clothing Industries opened in 1995. Clothing firms are subcontractors for international designers of apparel and for retail chains, and they maintain subcontracting relations within Hong Kong and elsewhere in Asia; their local and global networks support specialization in rapid response to fashion demands. Some local firms have created their own fashion design and brand labels. Hong Kong increasingly operates as the management center for off-shore clothing production in Asia; firms shift lower-skilled production to factories outside Hong Kong, and maintain small local shops as prototype production platforms.

Hong Kong entered electronics manufacturing during the 1960s: Japanese and United States firms started assembling radios; and United States semiconductor firms began production of components and parts, and Japanese firms followed. Hong Kong firms commenced manufacturing and assembling consumer products such as televisions, watches, and clocks. Foreign firms always have been prominent local producers, accounting for one-third to one-half of electronics employment from the 1970s to the 1990s.[34] As total electronics employment plunged (fig. 7.4), the share of electronics exports comprising parts and components rose from 34 percent in 1987 to 74 percent in 1996; Hong Kong firms increasingly export them to subcontractors in mainland China for assembly, or shift entire production units there. The government has boosted support for electronics and other high-technology manufacturing through funds for design and testing centers and universities, but production will not be important in the future. Hong Kong has become the management control center for the electronics industry; besides corporate administration, local functions stress research and development, project management, logistics control, purchasing, and marketing.

Since the late 1980s, employment in watches and clocks, toys, and plastics has also plummeted (fig. 7.4), and these industries have followed similar restructuring processes. Each moved more labor-intensive production to South China and, to a lesser extent, elsewhere in Asia. Watch and clock firms shifted to higher-quality, fashion-style products, whereas most toy manufacturers produce for overseas customers, and local firms carry out little market research or product design. Products of the plastics industry often enter into exports of other goods, and firms specialize in the rapid supply of high-quality components for both large and small orders, responding quickly to changes in

[34] Henderson, *The globalisation of high technology production*, pp. 77–117.

market demand. Local plastics firms emphasize sourcing raw materials, prototyping, engineering support, product design and development, and manufacture of high value-added, precision products, whereas they shift labor-intensive production to mainland China.

The trajectory of Hong Kong's industry points to the imminent demise of labor-intensive export manufacturing, its mainstay since the 1950s, and the relative rise of manufactures typical of an industrial metropolis (commerce-serving, processing, and local market production); remaining industry increasingly focuses on prototype or skilled, small-batch production. Other manufacturing-related activities such as research and development, testing, and design may continue, but these relate more to management control than to production. Industrial firms increasingly use Hong Kong as the high-level administrative center of their manufacturing empires that sprawl into South China, other parts of mainland China, and elsewhere in Asia, and corporate management draws on the social networks of capital of the trade and finance sectors that operate locally and bridge to Asia and the global economy to stay competitive.

Locally owned factories always accounted for the greatest share of total industrial production in Hong Kong, but the government followed an open policy for foreign direct investment in manufacturing. Foreigners could own as much as 100 percent of a firm and freely employ workers, bring in expatriates, and move capital and goods.[35] Because the government did not differentially subsidize foreign factories, they operated under the same conditions as those which were locally owned, making foreign firms subject to rising wages and land values as the Hong Kong economy grew.

Foreign direct investment

Foreigners always owned factories in Hong Kong, but these did not amount to much in the small industrial economy. The pace of foreign direct investment (FDI) in manufacturing accelerated during the 1960s in concert with the growth of locally owned export manufactures; by 1971, foreign factories employed 56,519, about 10 percent of total industrial workers. The trajectory of total FDI employment followed the path of total factory workers, reaching a peak of about 108,000 around 1988; it then fell to 61,483 by 1996 (table 7.5 and fig. 7.4). Electronics firms dominated, accounting for 53 percent of foreign-controlled workers by 1971, and they maintained their employment count during the 1970s. However, their share of Hong Kong's electronics workers fell from 72 percent in 1972 to 32 percent in 1980 because locally owned factories jumped into production. The FDI electronics industry,

[35] Wong, "A comparative study of the industrial policy of Hong Kong and Singapore in the 1980s," p. 267.

however, never lost its position as the largest foreign employer from the early 1970s to the late 1990s. Electronics factories remained the largest in Hong Kong (fig. 7.5), and foreign-owned ones averaged four or more times as many workers as locally owned firms. As of 1996, about 30 percent of foreign electronics firms employed 200 or more workers, whereas less than 1 percent of all factories in Hong Kong were that large.[36]

The restructuring of the FDI manufacturing sector mirrors overall industrial change in Hong Kong. During the 1990s, foreign factories that exported a high share (50–85 percent) of their production had either stable shares of total FDI employment, including electronics and electrical products, or declining shares, including the labor-intensive industries of textiles/clothing and watches/clocks (table 7.5).[37] In contrast, FDI industries that sold much of their output locally, transport equipment and food/beverages, which booked 63 percent and 84 percent, respectively, of their sales locally in 1996, increased their shares of total foreign employment. These trends suggest that foreign factories in Hong Kong will boost the share of total sales they direct to the local market above the 40 percent level that held between 1990 and 1996. Even if export-oriented foreign firms shift to higher-value products, continued decline in total FDI manufacturing employment and the relative shift of foreign employment to locally oriented FDI industries inexorably reduce the impact of foreign export manufacturing on Hong Kong's economy. These FDI firms also make little use of Hong Kong as an export platform to reach the mainland China market; annual sales there remained trendless at 5–6 percent of total sales between 1990 and 1996. Yet, even as the overall impact of FDI factories waned, their share of Hong Kong's manufacturing employment grew by half from 12 percent in 1990 to 19 percent in 1996, and the relative importance of FDI firms in industrial sectors is changing (table 7.5). Foreign firms did not account for the majority of Hong Kong's total employment in any industrial sector in 1990, but FDI firms in electronics, electrical products, and transport equipment dramatically boosted their shares to over half and firms in chemicals and food/beverages raised their moderately high presences by 1996. Each of these industries witnessed jumps in their cumulative value of fixed FDI investment, confirming that foreign firms are targeting those industries (table 7.5). Nevertheless, FDI manufacturing has a minor, declining impact on the Hong Kong economy: between 1990 and 1996, the foreign firms' annual additions of fixed-capital investments averaged only 1.3

[36] Henderson, *The globalisation of high technology production*, tables 5.5–5.6, pp. 88–89, 91; Ho, *Trade, industrial restructuring and development in Hong Kong*, table 7.4, p. 140; Industry Department, *Hong Kong's manufacturing industries* (selected years), and unpublished data; Industry Department, *Survey of external investment in Hong Kong's manufacturing industries* (selected years).

[37] Data on product sales in this paragraph come from Industry Department, *Survey of external investment in Hong Kong's manufacturing industries* (selected years).

Table 7.5. *Foreign direct investment (FDI) by employment and value in Hong Kong, 1990 and 1996*

	FDI employment					
	Percentage of total FDI		Percentage of total in industry		FDI investment HK $, millions	
	1990	1996	1990	1996	1990	1996
Electronics	28.3	27.0	30.0	53.6	5,421	12,393
Electrical products	11.8	9.5	22.6	55.4	1,993	3,886
Textiles & clothing	22.0	13.0	6.2	6.9	1,995	1,646
Watches & clocks	6.8	3.1	22.6	20.7	752	626
Plastic products	1.3	2.4	2.1	11.5	231	1,103
Chemical products	3.4	4.4	37.7	43.8	2,026	3,637
Transport equipment	2.0	11.3	13.5	69.4	952	1,380
Metal products	3.8	3.8	8.3	12.8	1,053	1,108
Food & beverages	5.2	8.4	20.0	24.3	1,226	3,041
Printing & publishing	3.2	4.3	7.7	6.0	1,143	2,352
Other manufactures	12.2	12.8	15.0	16.2	3,359	5,650
Total	100.0	100.0				
Total FDI as percentage of total employment			12.4	18.8		
Total FDI employment	90,262	61,483				
Total FDI investment					20,151	36,822

Sources: Census and Statistics Department, *Hong Kong annual digest of statistics* (selected years); Industry Department, *Hong Kong's manufacturing industries* (selected years); Industry Department, *Survey of external investment in Hong Kong's manufacturing industries* (selected years).

percent of annual gross domestic fixed-capital formation in Hong Kong; and the FDI share of the total Hong Kong labor force dropped from 3.3 percent in 1990 to 2 percent in 1996.[38]

Those foreign firms with the strongest bridges to the social networks of capital of traders and financiers in Hong Kong have the deepest understanding about using Hong Kong as a manufacturing platform. Its leading trade partners since the nineteenth century, Japan, the United States, mainland China, and the United Kingdom, dominate industrial FDI; they retained a

[38] Census and Statistics Department, *Hong Kong annual digest of statistics* (selected years); Industry Department, *Survey of external investment in Hong Kong's manufacturing industries* (selected years).

stable share of about 75 percent of cumulative fixed FDI investment between 1987 and 1996.[39] Japan significantly boosted its share from 25 percent to 40 percent, whereas the United States and the United Kingdom stayed at about 20 percent and 4–8 percent, respectively; mainland China's share, however, declined from a high of 15 percent in 1989 to 7 percent in 1996 because its firms have suitable lower-cost production sites in Guangdong Province. The FDI firms reorganized their production to take into account the decline of manufacturing in Hong Kong and the emergence of industrial satellites in Guangdong. Between 1990 and 1996, the share of FDI firms that engage in some subcontracting stayed at about half, but those firms that subcontract have substantially raised their level of subcontracting.[40] The average share subcontracted jumped from 40 percent to 55 percent, and this increase resulted principally from a sharp rise from 22 percent to 39 percent in the share of firms that subcontract most (75–100 percent) of their production. Firms that subcontract sometimes engage factories in multiple locations, but these factories remain heavily concentrated in either Hong Kong or mainland China. The share of firms that use subcontractors in Hong Kong fell from 61 percent to 41 percent, whereas the share that use subcontractors in mainland China jumped from 54 percent to 75 percent. This rapid change coincided with the fall in the number of factories in Hong Kong and the headlong shift to Guangdong (fig. 7.4), which has served as the industrial workshop of Hong Kong for two decades.

Guangdong Province: industrial workshop of Hong Kong

Instituting reform

Closure of mainland China to foreign investment after 1949 prevented Hong Kong factories from seeking lower labor costs and cheaper land values in rural areas, villages, and cities of Guangdong Province as Hong Kong transformed into an industrial metropolis. The catalyst for change came from reforms, labeled the "four modernizations" (agriculture, industry, science and technology, national defense), that Deng Xiaoping unleashed with the Eleventh Party Congress in December 1978. To develop, Deng believed China must end its isolation and allow capital, information, goods, and technology to enter and help transform the economy. In a brilliant ploy, Beijing's leaders chose Guangdong Province for the starring role and Fujian Province for the secondary part. Each abutted an unresolved geopolitical issue, the reincorporation of Hong Kong in 1997 and reunification with Taiwan, home of many Chinese

[39] Industry Department, *1997 survey of external investment in Hong Kong's manufacturing industries*, table 59, p. 70.

[40] Subcontracting data come from Industry Department, *Survey of external investment in Hong Kong's manufacturing industries* (selected years).

with Fujian roots; success in Guangdong and Fujian would smooth those transitions. Both provinces were sources of most Overseas Chinese in the Nanyang and the rest of the world; Guangdong alone contributed 80 percent of those emigrants. Beijing's leaders saw capital, technology, expertise, and information from Overseas Chinese and from Chinese in Hong Kong and Taiwan as central to the economic development of China. The choice of Guangdong represented an eerie replay of Qing policy in the early nineteenth century: keep the bridge to foreigners at Canton distant from politics in Beijing. This allowed greater experimentation without directly influencing political and economic changes in the rest of China. Both provinces, but especially Guangdong, had sufficiently high economic levels that experiments would not strain the meager budget of the central government. Beijing's leaders who pushed reform retained close ties to Guangdong and knew its cadres were ready for change, and the cadres could access immense human, physical, and economic resources in Hong Kong, conduit to the global economy.[41]

If Beijing's leaders had conceptualized development simply as the creation of the Special Economic Zone (SEZ) of Shenzhen, next to Hong Kong, for export processing factories, benefits would have remained circumscribed to wages, taxes, and export fees for Guangdong and China and to profits for relocated export factories and trading and shipping firms in Hong Kong tied to the SEZ. Fortunately for Hong Kong and Guangdong, these leaders had bigger goals – deep, widespread economic growth and development; therefore, Hong Kong traders, financiers, and industrialists gained a regional hinterland. Beijing's leaders took no chances; they sent Xi Zhongxun and Yang Shangkun, senior government officials, to prepare local cadres for change and to commence reforms, and within two years, Ren Zhongyi and Liang Lingguang, respected, talented politicians and economic managers, took the reins and pushed agricultural reforms much farther than elsewhere in China. Farmers were freed to engage in market production of crops outside their rice quota; consequently, output, especially in the Pearl delta, soared after 1978 as farmers met demands of Hong Kong and the burgeoning urban and industrial populations of the province. Farmers spent their higher incomes on consumer manufactures and housing, and their savings became investment capital for the region. Ren and Liang also pushed reforms of individual enterprises, collectives, and municipal, county, provincial, and state firms, giving them greater freedom to operate and reducing planning and regulations. By 1988, the Guangdong government ended price controls on virtually all consumer goods and most industrial products. From the start of reforms in 1979, Hong Kong and Guangdong became more tightly integrated through transportation and communications improvements. Daily nonstop train service between

[41] Vogel, *One step ahead in China*, pp. 76–87.

Hong Kong and Guangzhou started in 1979, reducing travel time from one day to three hours, and by 1987, four daily nonstop trains left each city; direct-dial phone service linked businesses in the two cities by the mid 1980s. During the decade, hydrofoils, ferries, trucks, and trains increased service between Hong Kong and cities and towns in the province. Guangdong still was not a full market-oriented economy and government remained inefficient, but the base had been set.[42]

Business networks

Hong Kong industrialists, however, initially faced a dilemma because they operated in a prosperous, free-wheeling, market-driven economy, whereas across the border an impoverished society had functioned under socialist ideals of equality, disdain for profits, and strict government controls for thirty years. Reform in Guangdong could not immediately create an investment environment that duplicated the legal guarantees and flexibility of Hong Kong; initially, change had to proceed slowly. As factory owners learned about Guangdong, that news would circulate through Hong Kong's social networks of industrialists, traders, and financiers. Officials and senior staff of Chinese government firms in Hong Kong, including the Bank of China and China Merchants' Steam Navigation Company, always had provided information about China and Guangdong. China's government had founded the venerable China Merchants' in conjunction with leading Chinese merchants in 1873, and its first manager, Tong King-sing from Guangdong, had been the Shanghai comprador for Jardine, Matheson & Company. With its main office in Hong Kong, China Merchants' remained tightly integrated with the great trading houses, and like them, it moved into dock management, marine machinery and engineering, warehousing, tourism, hotels, banking, and investment. In 1979, the Beijing government granted the request of Yuan Geng, the firm's head, to establish an industrial zone in Shekou, a peninsula in Baoan County at the edge of Hong Kong; construction started in 1980 and it soon became part of the Shenzhen SEZ. China Merchants' led the early development of Shekou, and other Hong Kong firms, as became the practice in the zones and elsewhere in Guangdong, participated in construction. This integral involvement of China Merchants', a government firm with access to the corridors of power in Beijing and pivotal member of Hong Kong's business community, must have assured industrialists, traders, and financiers that they could safely invest across the border.[43]

The social networks of the industrialists bridged to a wide array of groups

[42] Vogel, *One step ahead in China*, pp. 60–122; Johnson, "Continuity and transformation in the Pearl River Delta"; Xu, Kwok, Li, and Yan, "Production change in Guangdong."

[43] Hao, *The comprador in nineteenth-century China*, table 22, p. 137; Liu, *Anglo-American steamship rivalry in China, 1862–1874*, pp. 142–52; Vogel, *One step ahead in China*, pp. 61–64, 130–35.

in Hong Kong. Numerous daughters and sons of high-level cadres in Beijing arrived by the mid 1980s to work in Chinese government and "foreign" firms; they provided information about changes in Guangdong and insight into both the security of those investments and support for reforms in Beijing. Industrialists drew on family and village ties in Guangdong, and illegal migration of many young people from there to Hong Kong between 1949 and 1979 augmented this information. Dongguan County, about eighty kilometers north of Hong Kong, was a major source area, and its emigrants became prize reservoirs of information and contacts for industrialists who wanted to build factories and develop relations with firms in that county. Industrialists also acquired information from friends and relatives among the many Overseas Chinese, especially business people, who passed through Hong Kong on their way to visit relatives and ancestral homes. Hong Kong factory owners cautiously invested in new special economic zones that started after 1980, including Shenzhen, which opened quickly and became the leading zone, and Zhuhai, just north of Macau, which started several years later. At first, they forged cooperative agreements with Chinese factories, and then gradually invested their own capital in factories and equipment. Social networks within Hong Kong and across the border in Guangdong gave industrialists sophisticated capacities to evaluate the risks and rewards of investments and to negotiate with Chinese firms and government officials in an environment with an evolving policy of reform, few precedents, and few or no formal legal guarantees. Foreign multinationals from Japan, North America, and Europe, with stylized legal approaches, concerns for guarantees about capital mobility, worries about labor laws, and lack of experience of dealing with the Chinese bureaucracy, could not function effectively in that environment; therefore, few invested in Guangdong during the 1980s.[44]

Pace of investment

Hong Kong industrialists accounted for over two-thirds of the value of agreements with political and enterprise entities in Guangdong from 1979 to 1985, consistent with their greater understanding of potential manufacturing opportunities and sophisticated awareness of risks during those first years of reform. In early surveys, however, they claimed that investments faced numerous difficulties, including government inefficiency, inadequate quality of inputs, low labor productivity, and poor infrastructure. Consistent with those views, their investments totaled only about $100 million (US dollars) annually between 1979 and 1982. During that period, the Guangdong government and Hong Kong firms constructed new economic relationships, developed institutional and physical infrastructures, and built trust; on this foundation, invest-

[44] Vogel, *One step ahead in China*, pp. 62–68, 143–81.

ments almost tripled to $300 million annually between 1983 and 1986.[45] The rate of expansion into Guangdong reached sufficiently large levels that net additions to Hong Kong manufacturing occurred across the border; this comports with the steady total employment in Hong Kong industry from 1980 to 1987 (fig. 7.3). The lag between the surge to Guangdong beginning about 1983 and the start of the collapse in Hong Kong around 1988 probably reflected the time industrialists required to make sure new factories operated effectively before they shuttered old ones.

By 1987, Hong Kong entrepreneurs had financed immense industrial development in Guangdong. As many as 8,700 small- and medium-size (below 200 employees) factories in Hong Kong, about 17 percent of that group, had established outward-processing production in mainland China using inputs mostly supplied by the parent firm. This production is termed outward-processing because the goods are under contractual agreement to be sent from the China factory to Hong Kong. David Wilson, governor of Hong Kong, estimated that factories with Hong Kong owners and those mainland-owned together employed over 1 million factory workers in Guangdong to produce goods for Hong Kong firms, but Xinhua News Agency of mainland China disputed that claim – it said the correct number was 2 million. From the perspective of Hong Kong firms, Guangdong had become an open field for industrial investment, and clothing, textile, and electronics firms, especially large ones, increasingly subcontracted to Guangdong factories. Firms relied on a range of social networks, but kinship ties to Guangdong and business ties with Hong Kong firms that had factories in Guangdong accounted for the majority of subcontractor linkages. Hong Kong subcontractor networks reached throughout the Pearl delta, including Shenzhen SEZ, but these did not dominate. Industrial counties, such as Dongguan, and the city of Guangzhou acquired similar levels of subcontracting, and Hong Kong owners also utilized kinship ties to forge subcontractor relations with firms in villages and rural areas. The sweep of subcontractor relations around the delta reflected the broad nature of industrial development in Guangdong that had emerged by the late 1980s. State firms from elsewhere in mainland China shifted operations to the delta, and village, city, county, and provincial firms either started production or expanded.[46]

[45] Chai, "Industrial co-operation between China and Hong Kong," table 3.19, p. 140 and pp. 114–20; Chen, "The impact of China's four modernizations on Hong Kong's economic development," pp. 96–99; Leung, "Locational characteristics of foreign equity joint venture investment in China, 1979–1985," tables 3, 6, pp. 408, 413; Leung, "Personal contacts, subcontracting linkages, and development in the Hong Kong–Zhujiang Delta region," p. 276; State Statistical Bureau, *China statistical yearbook* (various issues); Thoburn, Chau, and Tang, "Industrial cooperation between Hong Kong and China," table 3, p. 539.

[46] Industry Department, *1995: Hong Kong's manufacturing industries*, table 2.5, p. 25; Johnson, "Continuity and transformation in the Pearl River Delta"; Leung, "Personal contacts, subcontracting linkages, and development in the Hong Kong–Zhujiang Delta region"; Sit, "Hong Kong's industrial out-processing in the Pearl River Delta of China," pp. 565–71; Vogel, *One step ahead in China.*

Nevertheless, most Hong Kong industrialists still had neither initiated their own factories nor forged subcontractor relations to Guangdong Province by the late 1980s, but a notable minority of their competitors had done so. At the same time, the quality, efficiency, and sophistication of Chinese firms in Guangdong rose considerably; the labor supply and productivity grew, in part, because numerous in-migrants arrived and kept shortages from becoming serious; and the industrial infrastructure of factories and transportation improved substantially. Hong Kong industrialists without factories or subcontractors in Guangdong now encountered a new competitive regime. Previously, their competitors in Hong Kong faced the same local costs and participated in the same social networks of capital with trade and financial firms. By the late 1980s, however, factory owners without industrial ties to Guangdong or plants there confronted competitors whose labor and factory-space costs were as much as 75 percent lower. Not surprisingly, Hong Kong industrialists rushed pell-mell to Guangdong after 1987, causing factory employment to decline sharply (fig. 7.3).

By the early 1990s, Guangdong was the workshop of Hong Kong, and its industrialists were investing about $1 billion (US dollars) annually in the province, more than triple the rate of the mid 1980s; this jumped to staggering amounts of $4 billion or more annually by the mid 1990s. Estimates of the number employed in these factories are in the range of three to five million, which is about four times larger than the total in Hong Kong at its peak during the early 1980s. As the corporate headquarters of this vast industrial army, Hong Kong is one of the world's largest centers of senior manufacturing administration. Its industrialists, however, stand at the cusp of change again in the late 1990s, and they are reevaluating the use of Guangdong as principally a convenient production site with low labor and land costs. Its human resources underpin this metamorphosis in their approach; in contrast to the tiny scientific and engineering base of Hong Kong, Guangdong has a large base of 45 universities, 600 scientific institutions, and as many as 500,000 scientific personnel. Exploiting this potential for research and development and upgrading the educational and skill levels of labor to compensate for rising wage levels would allow Hong Kong industrialists to transform their labor-intensive formula for manufacturing. Cooperative efforts between Hong Kong and the cities of Shenzhen and Guangzhou, for example, are directed towards creating this new relation; they are building support for the development of high-technology manufacturing in Guangdong. That signifies a change from the "shop in the front, and the factory in the rear" as a metaphor for the relation between Hong Kong and Guangdong.[47] As Hong Kong's intermediaries

[47] "Guangdong: the factory at the back"; "Overseas investment soars in Shenzhen Economic Zone"; "Shenzhen and Hong Kong enhance cooperation"; "South China's Shenzhen City to promote high-tech industry"; State Statistical Bureau, *China statistical yearbook* (various issues); Zheng, "Promoting economic co-operation between Guangdong and Hong Kong."

of trade and finance provided services to this industrial machine, they accumulated immense amounts of capital, but the underpinnings of the massive growth and structural change of the intermediary sector were much broader. That sector strengthened Hong Kong as the global metropolis for Asia.

8

Global metropolis for Asia

> There is presently some speculation that entrepot trade will experience added growth as a result of China's recent diplomatic initiative; however, it is inconceivable that it will ever occupy its old role in Hong Kong.[1]

Chinese business networks

During the last few decades before World War II, Chinese traders and financiers still operated as an alien minority in the countries of the Nanyang, yet they thrived as participants in colonial exploitation of resources for export and controllers of trade. As colonialism ended during the fifteen-year period after 1945, nationalist passions in newly independent nations led to restrictions and prohibitions on Chinese businesses, and these continued through the 1970s and, in some countries, the 1980s. This inhibited domestic Chinese capital investment and motivated them to transfer capital to safe havens in Singapore and, especially, Hong Kong. Between the late 1940s and early 1950s, capital flight from China represented the earliest large flow to Hong Kong, and continued harassment of Chinese businesses in newly independent nations maintained that flow. As much as $440 million (Hong Kong dollars) entered annually between 1954 and 1963, a rate that equaled the inflow during the previous decade; this capital came from (1954–63 period totals in parentheses): Malaysia and Singapore ($1,200 million); Indonesia ($1,000 million); the Philippines ($800 million); Thailand ($800 million); and South Vietnam, Cambodia, and Laos ($600 million). These sums could have financed 8.4 percent of all imports to Hong Kong during that period; thus, this capital provided powerful support for its trade and industrial expansion.[2]

[1] Riedel, *The industrialization of Hong Kong*, p. 9.

[2] Census and Statistics Department, *Annual review of Hong Kong external trade 1978*; East Asia Analytical Unit, *Overseas Chinese business networks in Asia*, pp. 34–85; Wu and Wu, *Economic development in Southeast Asia*, pp. 29–45, tables 43–44, p. 95, appendix C, pp. 173–79.

Malaysian Chinese

Malaysia illustrates the impact of nationalistic policies on Chinese businesses. After independence in 1957, the new government failed to institute economic reforms and to invest heavily in human capital and rural development, thus leaving the mass of rural Malays impoverished. Chinese businesses in plantation agriculture, trade, and finance expanded, but they remained vulnerable to attacks because Chinese and foreign firms disproportionately controlled the economy. By 1970, accumulated grievances among Malays (*bumiputras*) led the government to drop its *laissez-faire* approach and to institute policies that inserted it into key sectors and that favored *bumiputra* businesses. Government takeovers in the leading primary sectors of plantation agriculture and mining allowed Chinese and foreign firms to withdraw capital from slow-growth businesses rather than face an erosion of capital. This transfer of cash to Chinese and foreign firms, coupled with subsidies to *bumiputra* firms and state enterprises, hugely increased government debt. Government became a pervasive force in the economy that extended from declining primary sectors to construction, trading, insurance, and banking. These efforts raised employment among *bumiputras* and gave them access to capital, but increasing subsidies to inefficient enterprises impeded economic growth; the ruling political party had a higher priority, consolidating control. Smaller Chinese businesses focused on their ethnic market to survive government restrictions, but this removed them as catalysts for Malaysian development; larger firms shunned manufacturing to avoid the possibility of state control or takeover. The government favored state enterprises and *bumiputra* firms in alliances with foreign multinationals; thus, Chinese firms turned to investments in property development, which promised quick returns and large profits. Major firms with bridges to Chinese networks in the Nanyang shifted capital outside Malaysia, primarily to Hong Kong and Singapore, but most relocated only part of their capital and personnel.[3] Chinese businesses in other countries of the Nanyang faced similar harassment to those in Malaysia into the 1980s; only the details and severity differed.

Archetypal intermediary networks

Because Overseas (Nanyang) Chinese occupy pivotal positions in the economies of East and Southeast Asia, their business networks have come under intense scrutiny.[4] Chinese family business organizations are contrasted with

[3] Jesudason, *Ethnicity and the economy*.

[4] From now on the term "Overseas Chinese" will be used; it has become the common term to refer to the Chinese in East and Southeast Asia, outside mainland China. Major works on these business networks include: East Asia Analytical Unit, *Overseas Chinese business networks in Asia*; Hamilton, *Business networks and economic development in East and Southeast Asia*; Kao, "The worldwide web of Chinese business"; Redding, *The spirit of Chinese capitalism*; Whitley, *Business systems in East Asia*; Yeung, *Transnational corporations and business networks*.

professional and bureaucratic forms in North America and Europe, with *kaisha* (large, industrially specialized corporations) in Japan, and with *chaebols* (conglomerates) in Korea. Social, cultural, and political institutions that provide a context and influence organizational structure and behavior contribute to differences among these businesses. The resilience of family businesses among Overseas Chinese, even as such forms decline in the rest of the developed world, tempts analysts to explain that persistence as a product of the social, cultural, and political milieu of the Nanyang. These firms' business networks, according to this logic, rest in family bonds and extend selectively beyond them to clan and dialect groups. Most proponents of this interpretation of Chinese business networks advocate "Chinese exceptionalism."[5] They reify Chinese culture and argue that its Confucian roots provide a template for the behavior of Chinese business people. A paternalistic hierarchy of control governs relations within the family business, and firms interact through the head of the family who constructs a web of reciprocity; these mutual obligations build trust. This behavior does not follow a strictly programmed format that traces directly backwards in time; rather, the dynamic quality of Asian development provides opportunities creatively to modify behavior. Nevertheless, the linkage between behavior and culture in this model of Chinese business organization remains so strong that the model represents "oversocialized" behavior.[6] The exceptionalists argue that Westerners, in contrast to Chinese business people, work in rational, bureaucratic organizations and operate in environments of institutionalized trust supported by legal systems.

Economic sociologists, however, reject this characterization of "Western" business.[7] While accepting that rational, bureaucratic organizations and institutionalized trust exist, researchers argue that economic actors engage in social networks inside and outside those organizations and institutions. They build trust through those networks, and economic calculations are embedded in social behavior that is broader than simple cost/benefit calculations. Thus, economic sociologists explain the resilience of large and small family business organizations in Western economies as an outgrowth of this social behavior, not as a residual form. The networks of business intermediaries cover the gamut from legal and corporate institutionalized relations to informal ones, and neither Chinese nor Western intermediaries are unidimensional.[8] Cultural

[5] For a sample of views of Chinese exceptionalism, see: Mitchell, "Flexible circulation in the Pacific Rim"; Redding, *The spirit of Chinese capitalism*; Redding, "Societal transformation and the contribution of authority relations and cooperation norms in Overseas Chinese business"; Tu, *Confucian traditions in East Asian modernity*; Whitley, *Business systems in East Asia*; Yang, *Gifts, favors and banquets*. For a critique of Chinese exceptionalism, see Hodder, *Merchant princes of the East*.

[6] Granovetter, "Economic action and social structure."

[7] Smelser and Swedberg, *The handbook of economic sociology*; Swedberg, *Explorations in economic sociology*.

[8] Hodder argues that the unidimensional view of Chinese behavior misinterprets it; see Hodder, *Merchant princes of the East*.

practices may vary, but both groups use social networks of capital to gather information, construct cooperative relations, build trust, and enforce sanctions against malefactors. Chinese firms, therefore, act as archetypal intermediaries who face extensive risk: partners to agreements to buy or sell fail to complete transactions, payments are not made, and credits are not repaid. All businesses face these risks, but intermediaries confront greater problems because most of their capital crosses boundaries of local economies and often crosses political borders; they cannot continuously monitor individuals with whom they exchange, and they have limited capacity to enforce legal contracts across political borders. Intermediaries, therefore, always ground exchange in trust as a strategy to reduce risk because legal mechanisms to guarantee the security of transactions raise costs prohibitively.

Binding the Nanyang

As North American and European intermediaries shifted their organizations to more professionalized and bureaucratized models after 1945, Chinese firms may have encountered greater difficulties than under colonialism, when imperial powers protected them as an alien minority. Governments across the Nanyang, except for Hong Kong, Singapore, and Taiwan, harassed Chinese intermediaries and created uncertain geopolitical environments, raising the costs of exchange and making governments parties to attacks on Chinese firms. Therefore, they retained their social networks of families, clans, and dialect groups as means to create trustworthy exchange channels and to enforce negative sanctions on individuals who violated trust, and they strengthened their participation in economic exchange within the Nanyang and between it and the rest of the world; within most countries, their shares in import and export trades, wholesaling, and banking ranged from 25 percent to 95 percent. Chinese firms' prominence in retailing, manufacturing, plantation agriculture, and to a lesser extent, mining, allowed Chinese intermediaries to deploy their networks to link to the consumption and production sides of the economies, just as they had done under colonialism. The shift of capital and family members to Hong Kong, especially, and Singapore, secondarily, enhanced their roles as pivots of the Chinese intermediary networks in the Nanyang. Firms either opened major branch offices or moved their headquarters to those cities, and the retention of family members in each country kept intermediary networks intact. As harassment eased, first in Thailand in the late 1950s, then in the Philippines in the late 1970s, and subsequently in Malaysia and Indonesia during the 1980s, Chinese firms started openly forming multinational alliances.[9]

[9] East Asia Analytical Unit, *Overseas Chinese business networks in Asia*, pp. 34–85; Kunio, *The rise of ersatz capitalism in South-East Asia*, pp. 37–67; Landa, "The political economy of the ethnically homogeneous Chinese middleman group in Southeast Asia"; Redding, *The spirit of Chinese capitalism*, pp. 17–40; Wu and Wu, *Economic development in Southeast Asia*, pp. 46–89, appendix C, pp. 173–79.

Branch networks of Chinese banks headquartered in Hong Kong and Singapore and Chinese banks headquartered elsewhere with branch offices in those metropolises formally created business alliances across the Nanyang. Bangkok Bank, founded in 1844, opened its first foreign branch in Hong Kong in 1954 and subsequently constructed strong local ties to other banks through interlocking directorships and ownership stakes. Multinational capital exchange also occurs within and among large groups; each comprises numerous companies. Some groups such as Li Ka-shing are headquartered in Hong Kong, whereas others locate head offices elsewhere, such as Charoen Pokphand and Sophonpanich of Thailand, Salim and Lippo of Indonesia, Kuok of Malaysia, and Sy of the Philippines. Regardless of the official headquarters, they typically base their chief trade and finance operations in Hong Kong. During the late 1970s, multinational syndicates became more prevalent, and their operations refute the portrayal of Chinese family, clan, and dialect networks as insular; instead, these syndicates form through open, expansive networks with multiple bridges. Investment and banking syndicates usually have lead firms in Hong Kong with members from several places. Taiwan firms have become partners with Hong Kong and Singapore firms in investment projects in Malaysia, Thailand, Indonesia, and the Philippines. With reduced harassment and stabilization of geopolitical conditions, multinational syndicates increasingly include non-Chinese firms from within the region and from Western countries.[10] This cooperation between Chinese and foreign intermediary firms continues a tradition of over 150 years that includes trade and bank exchanges, joint ownership of companies such as steamship lines, and compradors.

Paradoxically, non-Chinese firms supported the growing importance of Chinese business groups in Hong Kong. As late as the mid 1970s, old non-Chinese hongs, such as Jardine's, the Hongkong and Shanghai Bank, Swire's, and Hutchison remained prominent, but Chinese firms began to accumulate sizable capital through industrial and property-development enterprises. Groups such as Li Ka-shing received financing from the Hongkong Bank to expand and take over units of the old hongs, and foreign financial institutions, such as the Canadian Imperial Bank of Commerce, also funded these groups. By the mid 1980s, the Hongkong Bank still remained the most central institution, but Chinese groups reached prominence in local business circles. This collusion across ethnic and national lines that facilitated the rise of these Chinese firms exemplifies the multiple network bridges within Hong Kong. That feature, combined with the numerous bridges that local hub firms forge to social networks across the Nanyang, provide diverse channels that Hong Kong firms exploit for foreign direct investments (FDI). They direct substantial

[10] Brauchli and Biers, "Green lanterns"; Chan, *Li Ka-shing*; East Asia Analytical Unit, *Overseas Chinese business networks in Asia*, pp. 153–75; Hewison, *Bankers and bureaucrats*, pp. 192–204; Weidenbaum and Hughes, *The bamboo network*, pp. 23–59; Wu and Wu, *Economic development in Southeast Asia*, pp. 96–104.

sums to Malaysia, Indonesia, Thailand, the Philippines, and Singapore, and because the social network bridges of Singapore firms reach to nearby countries, Hong Kong firms choose it as their most prominent operational base. Because Hong Kong manufacturers have convenient access to Guangdong Province for low-wage production, their industrial investments elsewhere remain limited.[11]

Reduced harassment and stabilized geopolitical conditions also have allowed Chinese intermediaries to alter their organizations from tightly knit family firms to broader alliances and more professionalized, bureaucratized organizations. For example, First Pacific of Hong Kong, headed by the son of a Manila banker, is heavily funded by the Salim family of Indonesia and includes Americans, Australians, Filipinos, and Hong Kong Chinese in its top executive ranks. Li and Fung, the premier Chinese multinational trading company headquartered in Hong Kong, purchased a unit of a foreign trading company and enlarged its staff to include at least ten nationalities.[12] This transition, nevertheless, does not signify the demise of the family firm; rather, its structure becomes more complex and diffuse. The formation of groups of family firms enhances their capacity to control the exchange of commodity and financial capital because the pooled resources provide greater capital bases and more bridges to other networks. As Chinese firms hire "foreigners" and as some family firms and groups include foreign firms in syndicates, their capacity to exchange capital globally is amplified, and, at the same time, foreign firms acquire greater access to networks of capital in the Nanyang. Because the pivotal Chinese and foreign networks meet at Hong Kong, these Chinese firms concentrate their senior offices there and foreign firms who wish to participate at the most sophisticated level of control of commodity and financial capital in Asia join them. Together, these firms power growth and change in Hong Kong, strengthening it as the global metropolis for Asia.

Hong Kong focuses

Even as observers touted Hong Kong as a newly industrializing country, the composition of the economy shifted dramatically away from local manufacturing production; the share of workers in that sector fell relentlessly from 50 percent to 13 percent between 1976 and 1996, while total employment rose over 70 percent from 1.5 million to 2.6 million (fig. 8.1).[13] The public-sector share of this labor force remained small, reflecting the government's policy to

[11] Chan, *Li Ka-shing*, pp. 79–90, 183–209; Wong, "Business groups in a dynamic environment"; Yeung, *Transnational corporations and business networks*, pp. 80–131.

[12] Clifford, "The new Asian manager"; Fung, "Evolution in the management of family enterprises in Asia."

[13] The data in fig. 8.1 are based on the establishment data; these provide an indicator of the formally organized sectors.

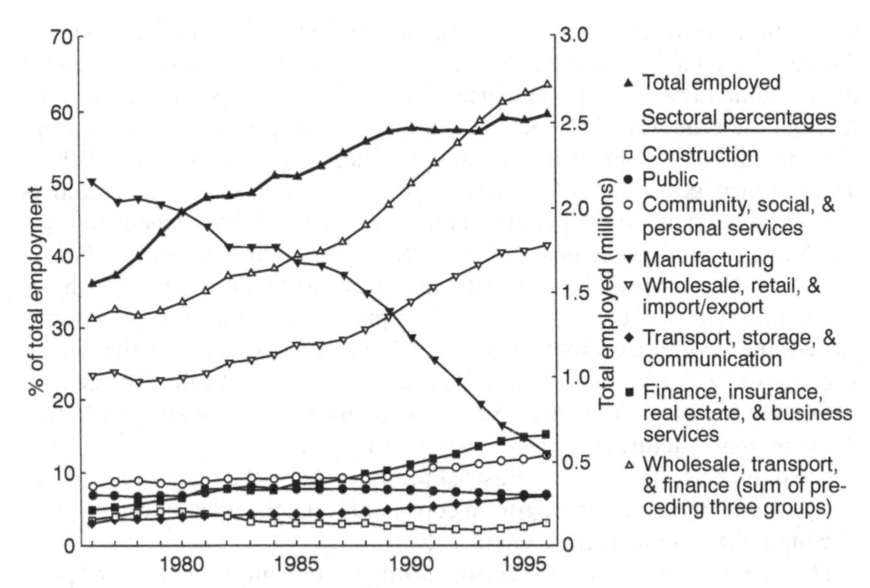

Fig. 8.1. Sectoral change in employment in Hong Kong, 1976–1996.
Source: Census and Statistics Department, *Hong Kong annual digest of statistics*, various years.

provide basic services and not to intervene significantly in the economy and society; it handled interventions such as public housing through contracts to the private sector. Beneath the boom–bust perturbations of construction, the growth of the metropolis created long-term demand for new and refitted infrastructure that kept the construction share of employment close to 3 percent. The steady rise of employment in community, social, and personal services from 8 percent in 1976 to almost 13 percent in 1996 testifies to demand shifts that accompanied increasing per capita income. These sectors, however, do not form the core of the Hong Kong economy; that core became more visible as the swift move out of manufacturing ratcheted through the employment structure.

The sensational shift in employment composition to the wholesale/import/export, transport, and finance groups removed the industrial graft on the economy, but it did not represent a new graft (fig. 8.1).[14] These groups benefited from trading, financing, and shipping the cornucopia of manufactures

[14] These groups include components that meet local demands, such as retailing and commuter transportation, but they are not the engines of the economy. For example, even as per capita income surged, the share of the labor force in retailing rose from only 7.7 percent in 1980 to 8.1 percent in 1996; in contrast, the share employed in import/export jumped from 5.8 percent to 21.0 percent over the same period. Computed from Census and Statistics Department, *Hong Kong annual digest of statistics* (various years).

that poured out of Hong Kong factories for world markets, and as local production shifted to Guangdong Province in the 1980s, the manufacture-related trade and financial services remained vibrant. Other changes also percolated through the economy. Hong Kong intermediaries responded to glimmerings of economic development in East and Southeast Asia during the 1970s as impoverished peasants finally gained some opportunities and as countries tried their hand as export platforms for multinationals from Japan, Europe, and North America. The financial share of employment increased in the late 1970s, stabilized during the mid 1980s, and then surged along with the wholesale/import/export sector; the share of employment in the transport sector moved up slowly throughout the period. The combined share of the wholesale/import/export, transport, and finance groups soared from 31 percent to 64 percent between 1976 and 1996. Their rising total share and the falling share in manufacturing crossed over in the mid 1980s, symbolizing the removal of the graft of local industrial production; by the early 1990s, Hong Kong stood stripped to its long-term core employers, the great intermediaries of commodity and financial capital.

The shares of employment in sectors testify to their importance to the population, but employment does not directly measure contribution to total economic production; sectoral shares of gross domestic product (GDP) provide that indicator (fig. 8.2). Total GDP (current prices) soared eightfold from $128 billion (Hong Kong dollars) in 1980 to $1 trillion in 1995. The share of construction in GDP declined slightly from 7 percent in the early 1980s to 5 percent in the early 1990s, whereas community, social, and personal services rose from 13 percent to 17 percent over that period; both sectors experienced somewhat greater shares of GDP than of employment. The decline in the construction share of GDP confirms that visible signs of skyscraper and housing construction around Hong Kong harbor must be balanced against the huge existing base of infrastructure to avoid overstressing the impact of the property market on the economy. Manufacturing reached its peak share of GDP at just over 30 percent around 1970 (fig. 7.3) and steadily declined after 1980 to 8 percent in 1995 (fig. 8.2). Yet, even at the peak of total manufacturing employment and with industrial shares of GDP between 20 percent and 24 percent during the 1980s, intermediary-related sectors of the economy accounted for the greatest share of GDP. Wholesaling/import/export rose from about 20 percent during the early 1980s to 27 percent by 1995, although this represented a smaller share than it held of employment. Transport and finance had much larger shares of GDP than of employment; transport rose steadily from 1980 to 1995, whereas finance dropped from about 23 percent in the early 1980s to around 16 percent by mid-decade and then reversed course and rose to 25 percent of GDP by the mid 1990s. The combined share that wholesale/import/export, transport, and finance held of GDP stood around

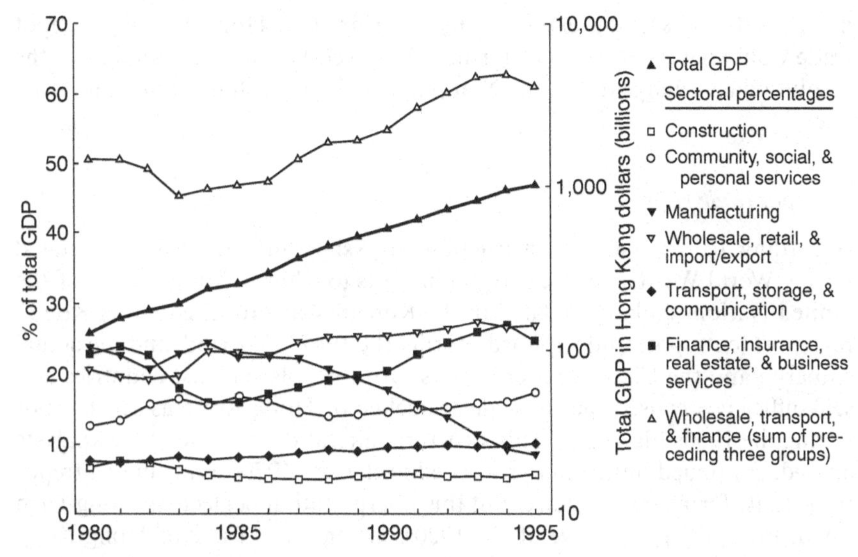

Fig. 8.2. Sectoral change in gross domestic product (GDP) in Hong Kong, 1980–1995.
Source: Census and Statistics Department, *Hong Kong annual digest of statistics*, various years.

50 percent in the early 1980s, well above their 35 percent share of employment, and their share of GDP reached 62 percent by the mid 1990s, about the same as their share of employment. Hong Kong's intermediary sectors, therefore, always dominated the economy.

The trade giant roars

During the nineteenth century, great trading companies in Hong Kong such as Jardine Matheson despaired that the China trade would ever provide much business, and the Nanyang trade offered few opportunities because large plantation and mining corporations internalized major components of long-distance exports of bulk commodities and low-cost, less-specialized Chinese traders dominated trade within and among the impoverished countries. Because the large Hong Kong firms could not efficiently handle small-scale, fragmented trade appropriate to that level of demand, they specialized in trade services; the closure of China after 1949 inserted one more nail in the coffin of Asian trade. Contemporary observers claimed that the growth of export manufacturing in Hong Kong transformed the economy, but Asian impoverishment left entrepot trade (imports of commodities for re-export) moribund. That logic motivated James Riedel's pessimistic prediction in 1974, in the midst of soaring export manufacturing, that Hong Kong's entrepot trade would never return to its former glory; yet, he could not anticipate the revolutionary change that Deng Xiaoping would unleash in China four years later.[15]

Entrepot trade rises

According to analysts, reconstruction of Asian infrastructure devastated during World War II and the surge in imports to China in anticipation of the United Nations embargo related to the Korean War propelled Hong Kong's re-export (entrepot) trade upwards during the 1947–51 period, and the actual embargo and the Chinese government's turn towards socialist countries and self-sufficiency caused the subsequent decline of Hong Kong as an entrepot during the 1950s (fig. 8.3).[16] This structural alteration in trade, these analysts argued, continued until China reopened in the late 1970s, returning entrepot trade to its former significance. But this interpretation neglects the long-term trend. From the perspective of the 1920s, the entrepot trade of Hong Kong rose slowly through the 1950s, and perturbations around that trend line were caused by the depression of the 1930s, World War II, the Korean War, and the closure of China.[17] The fall in re-exports after 1951, therefore, represented a return to the trend line, not structural alteration in the entrepot business of Hong Kong. The closure of China trade hurt importers and exporters, but that trade would not have grown much because China remained impoverished. Hong Kong traders also moderated somewhat the Chinese government's restrictions by surreptitious means such as smuggling and third-party transactions as they had done for decades, and the slow growth of entrepot trade in the Nanyang added other business.

The rising tide of domestic exports that commenced in the late 1940s and shifted to a high-growth trajectory by the late 1950s added a new dimension for Hong Kong traders (fig. 8.3). Previously, they mostly competed with other Asian traders to handle commodities that originated outside Hong Kong, but now their local networks gave them competitive advantages to acquire control

[15] See quote at the heading of the chapter; Riedel, *The industrialization of Hong Kong*, p. 9.

[16] Ho, *Trade, industrial restructuring and development in Hong Kong*, pp. 47–48; Riedel, *The industrialization of Hong Kong*, pp. 7–11, table 1, p. 15; Szczepanik, *The economic growth of Hong Kong*, pp. 45–57.

[17] Because Hong Kong housed little export industrial production before 1940, total exports are a surrogate for re-exports; estimates also are available for the 1948–58 period (Riedel, *The industrialization of Hong Kong*, table 1, p. 15). Combining estimates (before 1959) and actual data (1959 onwards) provides a plausible view of the value of domestic exports and re-exports for the period from 1918 to 1997. Data in figure 8.3 are given in current Hong Kong dollars because estimates of the real value of trade with foreign countries over a long period are difficult to compute accurately, but this does not distort results much because the focus is on broad trends.

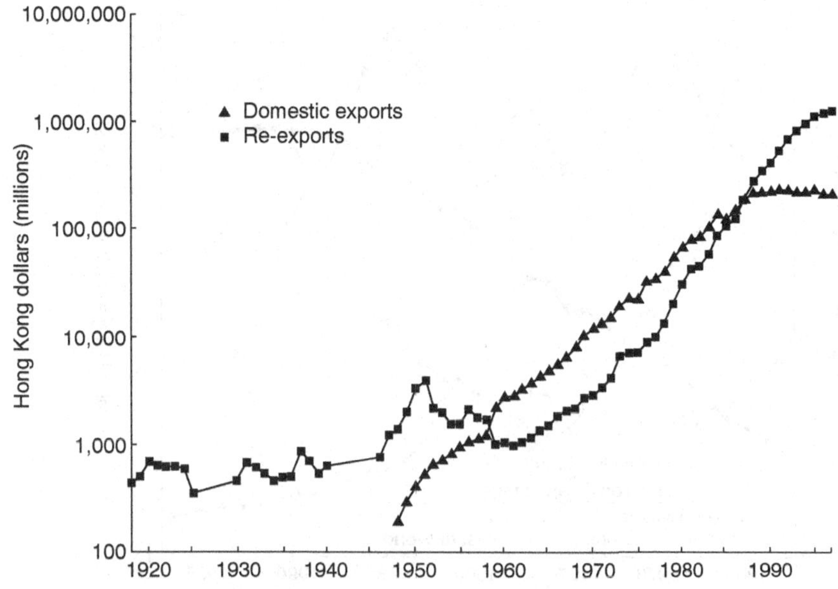

Fig. 8.3. Value of domestic exports and re-exports of Hong Kong, 1918–1997. *Sources:* Census and Statistics Department, *Annual review of Hong Kong external trade*, various years; Riedel, *The industrialization of Hong Kong*, table 1, p. 15; Tom, *The entrepot trade and the monetary standards of Hong Kong, 1842–1941*, appendix 5, p. 108.

over a soaring volume of domestic commodities; from the early 1960s through the 1970s, annual domestic exports were three to four times larger than re-exports. Although Asian peasants remained impoverished, flickers of economic development opened opportunities for Hong Kong's entrepot traders by the mid 1960s, pushing re-export trade above the trend line in place since the 1920s. Together, domestic export, import, and re-export businesses permitted Hong Kong traders to accumulate substantial capital and to develop new specializations in regional trade. The re-export trade increased sharply to a higher growth trajectory after 1978, coincident with the opening of China under the reforms of Deng Xiaoping. Domestic exports never again maintained a growth rate faster than re-exports, and after 1988 the value of domestic exports leveled off as Hong Kong entrepreneurs located new factories in Guangdong Province and shuttered local plants. That same year, the value of re-exports soared past domestic exports, and by 1997, re-exports were almost six times larger; entrepot trade had regained the dominance that it held for a hundred years before 1940. Yet, the re-export trade may have commenced a new, lower-growth trajectory in the mid 1990s, hinting at changes in the physical movement of commodities through Hong Kong.

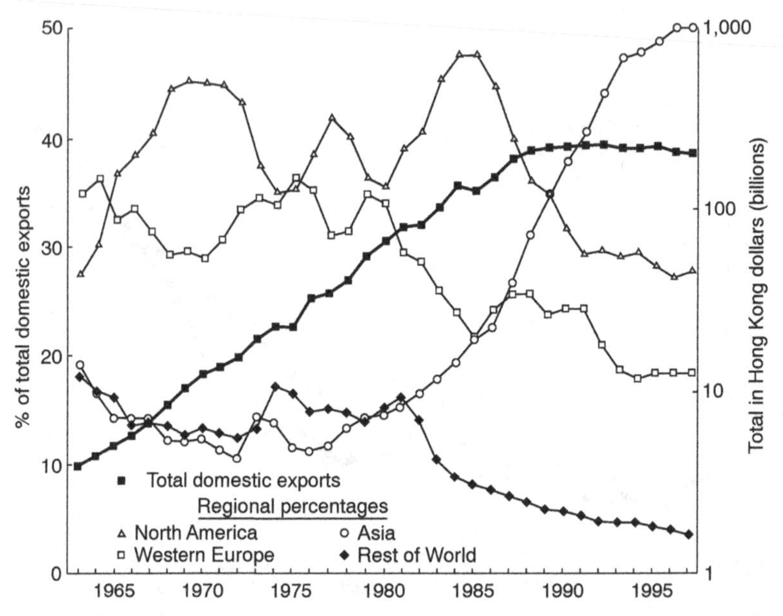

Fig. 8.4. Domestic exports of Hong Kong to world regions, 1963–1997.
Source: Census and Statistics Department, *Annual review of Hong Kong external trade*, various years.

Global networks

The trade linkages embedded in its domestic exports, imports, and re-exports provide a window on the structure and dynamics of Hong Kong as the meeting-place of the Chinese and foreign social networks of capital. Legacies of the nineteenth century cast eerie shadows, demonstrating the resilience of the networks; yet, they adjusted as economic and political conditions changed, sometimes with amazing swiftness, confirming the richness of the local networks and of their bridges emanating from Hong Kong. Its trade networks span the global economy, but limited exchange occurs with low-income markets in the former Soviet Union and Eastern Europe, Central and South America, the Middle East, and Africa, and with prosperous, lightly populated Australasia and Oceania (chiefly Australia and New Zealand). Taken together, this group's (rest of world) share of Hong Kong's domestic exports (fig. 8.4) stayed between 12 percent and 18 percent from the 1960s to the early 1980s and then fell to under 4 percent by the late 1990s; imports (fig. 8.5) from this group dropped steadily from about 10 percent in the early 1960s to under 4 percent by the 1990s; and re-exports (fig. 8.6) to this group stayed between 10 percent and 16 percent from the 1960s to the early 1980s and then fell below 10 percent through the late 1990s. Outside Asia, Hong Kong's trade interme-

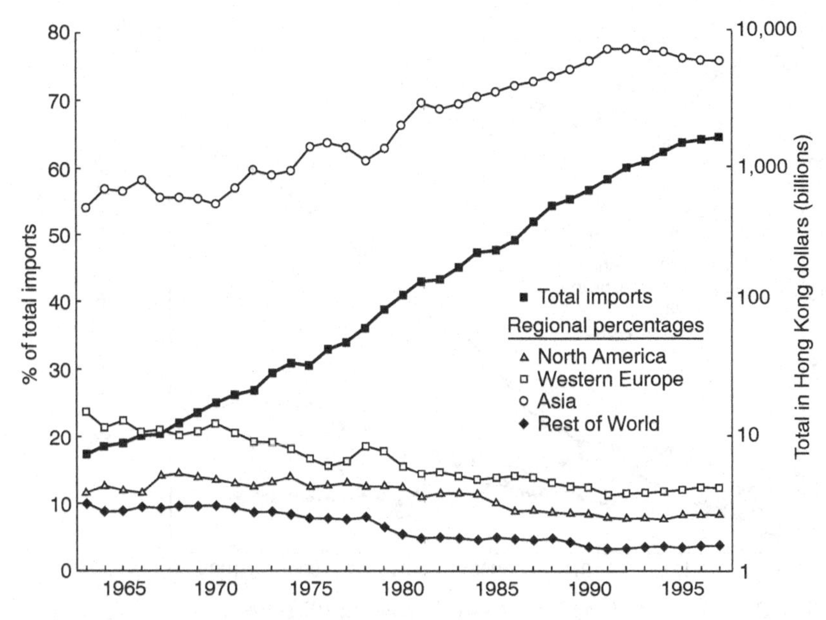

Fig. 8.5. Imports of Hong Kong from world regions, 1963–1997.
Source: Census and Statistics Department, *Annual review of Hong Kong external trade*, various years.

diaries focus, as they have since the nineteenth century, overwhelmingly on the highly populated, affluent regions of North America (chiefly the United States) and Western Europe. From the 1960s to the late 1980s, North America and Western Europe together purchased 60–75 percent of Hong Kong's domestic exports of labor-intensive goods (fig. 8.4). After 1964, North American markets took a greater share; nevertheless, following the mid 1980s, shares of both regions fell as Hong Kong manufacturing shifted to Guangdong Province. Because impoverished Asian peasants could purchase only a small share of Hong Kong's industrial production, Asia's share declined from around 20 percent in the early 1960s to a nadir of about 10 percent in the early 1970s. Its share turned dramatically upwards after 1978 with the start of the reforms of Deng Xiaoping; Asia surpassed Western Europe in 1987, North America in 1990, and the combined total for both regions in 1994. Since that time, Asia has bought close to half of the domestic exports and North America and Western Europe together have taken about 45 percent.

The import and re-export trades present mirror images of each other, except that a portion of the imports are destined for local consumption and production. During the 1960s and 1970s, North America and Western Europe

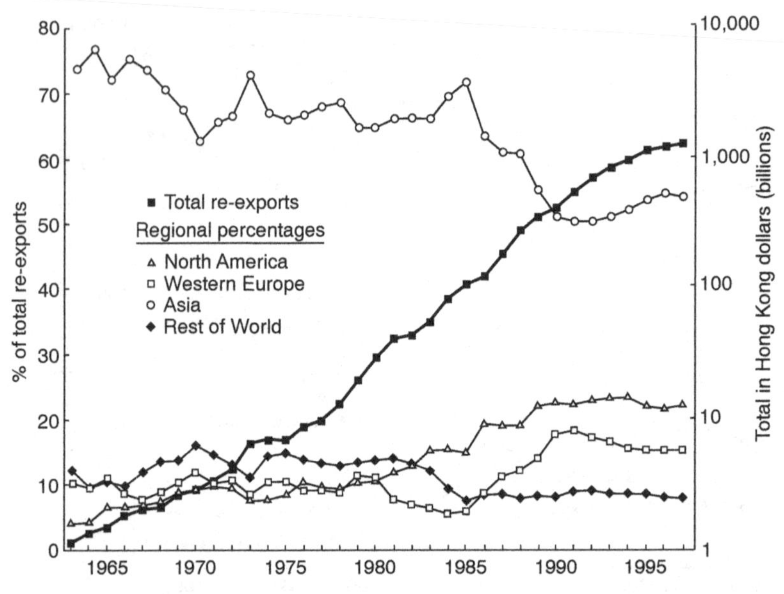

Fig. 8.6. Re-exports of Hong Kong to world regions, 1963–1997.
Source: Census and Statistics Department, *Annual review of Hong Kong external trade*, various years.

supplied about one-third of Hong Kong's imports, and Western Europe consistently supplied a larger share, reflecting the long-term prominence of the old European hongs and their Chinese allies in Hong Kong's trade (fig. 8.5). Asia always ranked as the largest supplier to Hong Kong; its share rose from about 55 percent in the early 1960s to 75 percent around 1990, where it stabilized. Similarly, from the 1960s to the mid 1980s, between 65 percent and 75 percent of the growing re-export trade of Hong Kong stayed within Asia, and countries in North America and Western Europe took another 15–20 percent of the re-exports (fig. 8.6). As Hong Kong's traders sharply raised their entrepot activity after 1978, the share with Asia initially stayed the same, but traders increasingly sent larger shares to North America as that trade expanded faster than shipments to Western Europe during the early 1980s (fig. 8.6). Within the short span from 1985 to 1990, Asia's share of the soaring entrepot trade fell from 70 percent to 50 percent, whereas the combined trade with North America and Western Europe jumped from 20 percent to 40 percent; by the late 1990s, the Asian share has drifted up to just above half, whereas North America and Western Europe have shifted down to a little over one-third. Despite the stunning growth of domestic exports sent outside Asia, the combined import and re-export trades always exceeded it by a large margin, and the majority of that business stayed in Asia, thus continuing the trade pattern from the nineteenth century.

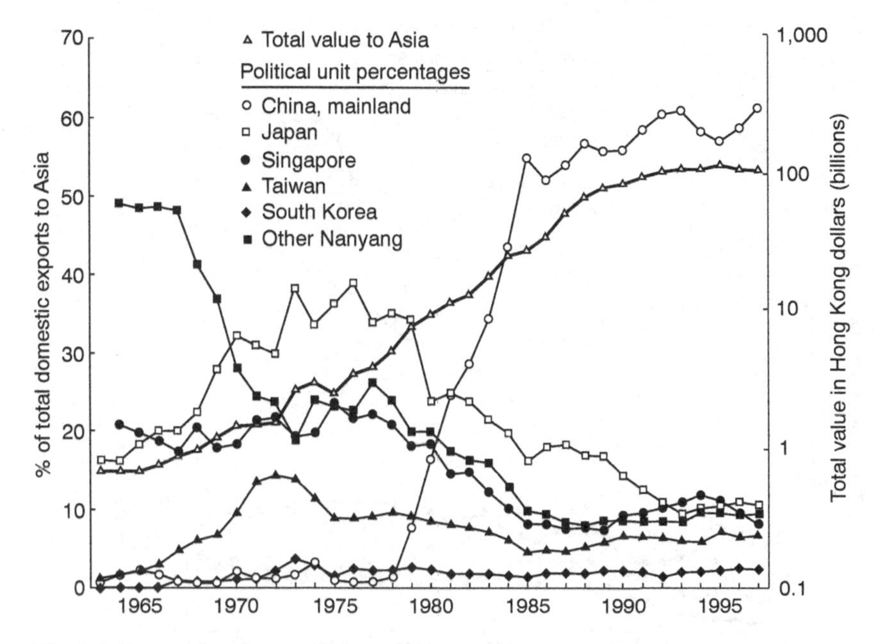

Fig. 8.7. Domestic exports of Hong Kong to Asia, 1963–1997.
Source: Census and Statistics Department, *Annual review of Hong Kong external trade*, various years.

The Asian connection

The transformation of China following the reforms of Deng Xiaoping in 1978 profoundly affected the domestic export (fig. 8.7), import (fig. 8.8), and re-export (fig. 8.9) trades of Hong Kong in Asia, yet its intermediaries' networks continued to span the region even as China's significance soared.[18] Mirroring the developed regions of North America and Western Europe, Japan became the leading Asian destination for industrial exports after the mid 1960s, and it kept that lead until China's share surpassed it in 1982 (fig. 8.7). Singapore typically was the leading export destination after Japan, and by the early 1990s, each took about 10 percent. While exports to Asia stayed low during the 1960s, the other Nanyang countries of Thailand, Malaysia, Indonesia, the Philippines, and Vietnam took almost 50 percent of those exports, but their share fell to about 25 percent during the 1970s and leveled at around 9 percent after the mid 1980s. The absolute amount of exports to them failed to rise much until the late 1970s, testimony to their limited economic transformation and their barriers to imports of labor-intensive goods from Hong Kong firms that

[18] The selected Asian economies and the "other Nanyang" group (Thailand, Malaysia, Indonesia, Philippines, Vietnam) in figures 8.7–8.9: took 90 percent to 95 percent of domestic exports, supplied 92 percent to 98 percent of imports; and took 90 percent to 95 percent of re-exports.

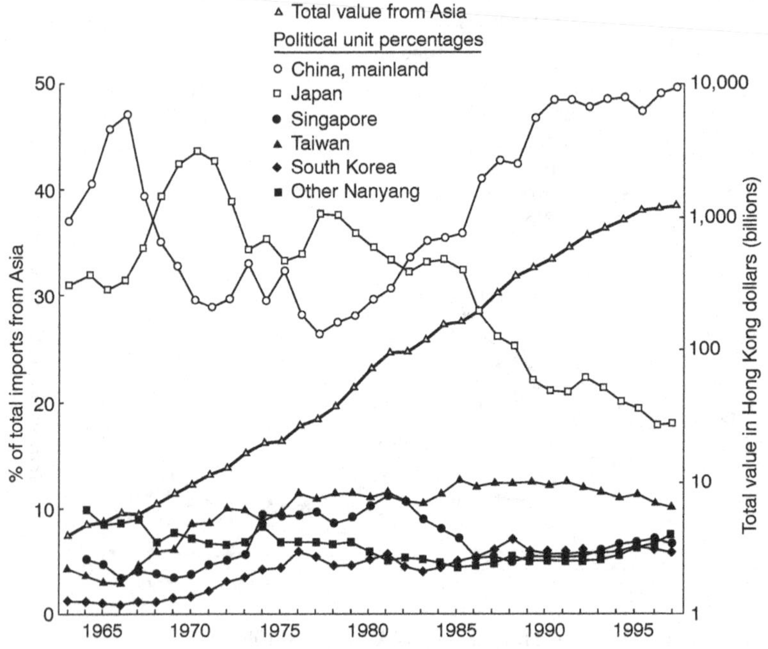

Fig. 8.8. Imports of Hong Kong from Asia, 1963–1997.
Source: Census and Statistics Department, *Annual review of Hong Kong external trade*, various years.

might compete with their factories. Close ties between Taiwan and Hong Kong, partly based on manufacturing firms that have operations in both places, lifted Taiwan's rank to just below those of Japan and Singapore; however, Korea's restrictions on industrial imports place it near the bottom in Asia among Hong Kong's trade partners. After the reforms of Deng Xiaoping, the share of exports sent to China vaulted vertically from 8 percent in 1979 to 55 percent in 1985, and by the late 1990s, it drifted up to about 60 percent.

Even before Deng Xiaoping's reforms, China often supplied 30 percent or more of Hong Kong's imports, far greater than any other Asian economy, except Japan (fig. 8.8). That country supplied more than China from the late 1960s to 1982, when China surpassed it permanently; imports from Japan comprised shipments of inputs from domestic Japanese factories to their Hong Kong units and of goods that Japanese distributors and Hong Kong traders resold. The import trade with the other Nanyang economies (Thailand, Malaysia, Indonesia, the Philippines, and Vietnam) languished during the 1960s and 1970s because those poor economies had little, except food and raw materials, to offer Hong Kong traders. Surging imports from Taiwan starting in the late 1960s, followed by those from Korea during the 1970s, boosted shares of those rising industrial economies as suppliers for consumption in Hong Kong and for goods to re-export to Asia. From the early 1960s to the late

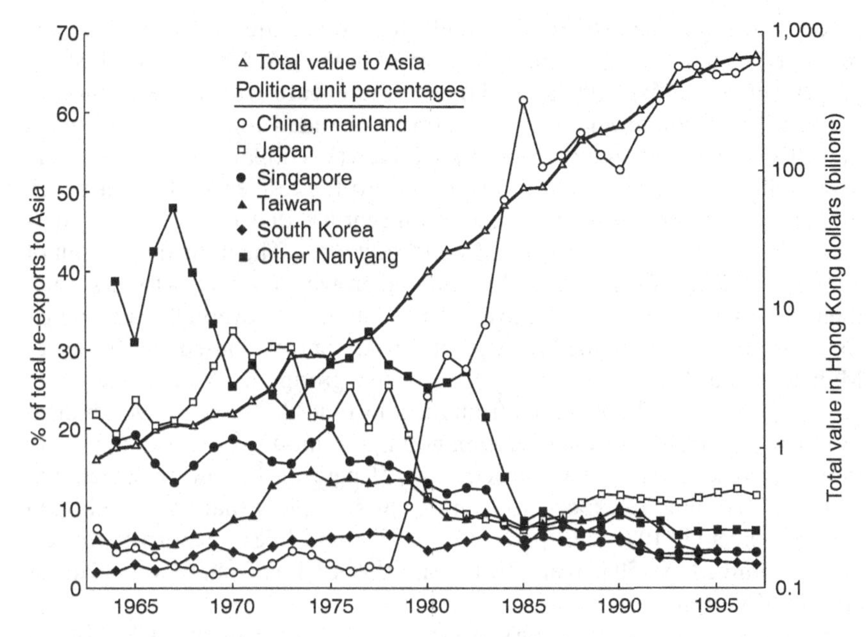

Fig. 8.9. Re-exports of Hong Kong to Asia, 1963–1997.
Source: Census and Statistics Department, *Annual review of Hong Kong external trade*, various years.

1990s, Singapore maintained its traditional status as supplier of imports that its traders collected from Malaysia and Indonesia, and multinationals with regional headquarters in Singapore forwarded manufactures to Hong Kong for redistribution in Asia and elsewhere. China's share of Hong Kong's imports rose steadily after 1978 and surged in the mid 1980s as Hong Kong factories relocated to Guangdong Province and mainland firms boosted their subcontract work there; Hong Kong traders also expanded their business as re-exporters of goods imported from other factories in China. Its prominence as the source of imports after the mid 1980s, however, obscures the momentous transformation of the import trade; simultaneously, imports from other Asian economies soared. Some came from the industrial powers of Taiwan and Korea, and imports from Japan also continued to grow. But equally as significant, imports from other Nanyang economies surged over tenfold from the mid 1980s to the late 1990s as they became multinational export platforms; nevertheless, their collective share remained below 8 percent.

As of the mid 1960s, Hong Kong's re-export trade followed familiar lines across the Nanyang as large shares went to the entrepot of Singapore and to other economies (Thailand, Malaysia, Indonesia, the Philippines, and Vietnam); their collective shares stayed between 40 percent and 50 percent through the 1970s (fig. 8.9). Japan purchased a substantial share (20–30 percent) of the re-exports from the mid 1960s to the late 1970s, and its share recovered from a low point around 7 percent in the mid 1980s to reach about 12 percent by the late 1990s, the largest Asian share with the exception of China's. The absolute value of Japan's trade expanded continuously from the mid 1960s, and it amounted to about $10 billion (US dollars) by 1997, up from $700 million in the mid 1980s. This testifies to the resilience of the network bridges, first created in the mid nineteenth century, that Hong Kong traders maintain between the Nanyang and Northeast Asia. The industrialization of Taiwan and Korea during the 1970s powered growing re-exports to them, and their shares remained above most other economies. Although Singapore's share dropped, it continued as a pivotal trade partner based on its ties to Malaysia and Indonesia. Re-exports to the group of "other Nanyang" economies jumped about sixfold from the mid 1980s to the late 1990s, indicative of their appetite for consumption goods and inputs for export manufactures; nevertheless, their share stayed below 10 percent. Following the reforms of Deng Xiaoping, re-exports to China grew so rapidly that its share leaped from 10 percent in 1979 to 61 percent by 1985, even while re-exports to the rest of Asia climbed by 150 percent; by the late 1990s, China's share was steady at around 65 percent.

A large share of Hong Kong's growing trade with mainland China comprises "outward-processing goods"; firms in China process them, but the goods are under contractual agreement to pass through Hong Kong. Those goods accounted for 79 percent of the domestic exports to China in 1990, and this declined slightly to 73 percent by 1996; nonetheless, rapid shrinkage of local manufacturing in Hong Kong points to the demise of this trade. The share of imports involved in outward processing rose steadily from 62 percent in 1990 to 80 percent in 1996, whereas the re-export share peaked around 50 percent in 1990, followed by a decline to 43 percent by 1996. The astounding volumes and high shares of the import and re-export trades included in outward processing underscore the strength of the "shop in the front, and the factory in the rear" relation between Hong Kong and Guangdong Province. Yet, this neglects an equally consequential transformation, the expansion of trade between Hong Kong and China that does not involve outward processing. Re-exports of those goods destined for China soared over fourfold between 1990 and 1996 from $7.1 billion (US dollars) to $30.8 billion, and re-exports of Hong Kong that originated in China rose 40 percent between 1991 and 1996 from $12.1 billion to $16.9 billion. To place that trade into perspective based on 1996 data, re-exports destined for China were 93 percent of the value of Hong Kong's re-exports to North America and they were 29 percent larger than the value shipped to Western Europe.[19] Therefore, Hong Kong's

[19] Computed from Census and Statistics Department, *Hong Kong annual digest of statistics* (various years).

China trade that is pure entrepot business, without direct manufacturing relations, has become equivalent to its trade with affluent regions of the world; the potential of the vast Chinese market that lured foreign firms in the nineteenth century finally has been realized.

Since the middle of that century, locally headquartered and foreign banks with branch offices have financed Hong Kong's trade. Although foreign banks finance their nation's export trade to Hong Kong from their corporate headquarters, they require offices in Hong Kong to assist their domestic exporters and importers based there. These offices also provide opportunities to finance Hong Kong's exporters and importers, Asian production of firms from their home nations, and infrastructure developments in Asia.

Emporium of finance

International banks flock to Hong Kong

Rapid, large-scale industrialization of Hong Kong from the 1950s to the mid 1960s attracted foreign banks, propelling the number of licensed banks from about thirty-five to around seventy-five, but then the total stabilized because intraregional Asian trade languished and its poor countries could not afford infrastructure projects (fig. 8.10).[20] Small consumer and industrial loans and credit for domestic exporters to North America and Europe offered opportunities in Hong Kong; however, fierce local competitors had better access to social networks to capture those businesses. In contrast, foreign banks retained competitive advantages to finance their national importers in Hong Kong who dealt with their home countries and to finance their multinationals engaged in resource extraction and processing activities in Asia, but those demands remained modest. Expansion of trade within Asia that passed through Hong Kong and the spread of export industrialization across the region contributed to the almost doubling of the number of banks between 1977 and 1984. The opening of China under the reforms of Deng Xiaoping probably tantalized banks; nonetheless, the halting pace of investment before 1985, much of it dominated by Hong Kong Chinese, could not sustain much foreign bank business. The continued rise in the number of banks to the present, though at a slower rate, mirrors the economic growth of Asia, especially China.[21]

[20] Full-service banks (licensed banks) that accepted deposits of any size and loaned money for consumer and business purposes ranked among the most important banks.

[21] Jao, *Banking and currency in Hong Kong*, pp. 32–71; Tai, "Commercial banking." Various moratoriums on bank licenses between 1966 and 1984 affected the timing of bank openings, but these disruptions probably did not influence the overall trend much because banks would not establish a fully operational bank unless economic conditions warranted. If demand for bank services remained high and unfulfilled, international banks had leverage to influence the colonial government to loosen restrictions. See Tai, "Commercial banking," pp. 1–2.

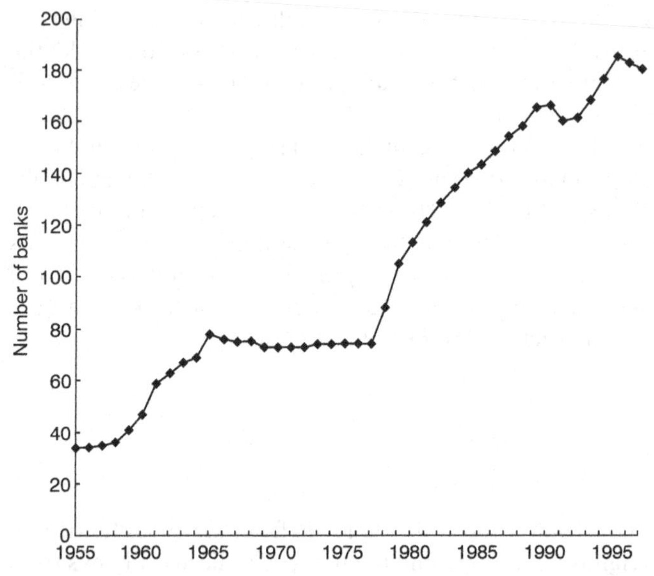

Fig. 8.10. Number of licensed banks in Hong Kong, 1955–1997.
Sources: Hong Kong Government Information Services, *Hong Kong*, various years; Census and Statistics Department, *Hong Kong annual digest of statistics*, various years; Hong Kong Monetary Authority, *Monthly statistical bulletin*.

Resolution of a paradox

Analysts concluded that the growing number of international financial institutions in Hong Kong since the late 1970s had transformed it into the regional financial center for Asia. Their choice of Hong Kong as the pivot to service the growing economies, according to the prevailing view, follows a clear logic: Hong Kong occupies a central location in Asia and provides close proximity to China; its time-zone position allows firms to participate in 24-hour global trading; it has excellent transportation and communication infrastructures; and it has receptive governmental policies that include low taxes and minimal regulations.[22] Because other Asian business centers had similar advantages, this persuasive argument raises puzzling issues. Although Japan had more regulations, Tokyo housed huge financial institutions active in Asia, and it had Hong Kong's other advantages, including proximity, time-zone position, and superb infrastructure. Singapore was distant from China, but, like Tokyo, it had the other advantages that Hong Kong offered. Bank expansion in Hong Kong came at the start of the reforms of Deng Xiaoping, before evidence existed that those reforms represented more than the onset of another convul-

[22] For examples of this widely held view, see: Jao, "The rise of Hong Kong as a financial center"; Jao, "The development of Hong Kong's financial sector 1967–92"; Nontapunthawat, "Hong Kong as a financial centre in the Asian Pacific region in the 21st century."

Table 8.1. *World's largest 500 banks with licensed bank in Hong Kong, 1986–1997*

	Number of banks			Percentage of group		
World ranking	1986	1991	1997	1986	1991	1997
1–20	19	19	20	95.0	95.0	100.0
21–50	27	21	24	90.0	70.0	80.0
51–100	27	32	29	54.0	64.0	58.0
101–200	14	31	37	14.0	31.0	37.0
201–500	13	16	20	4.3	5.3	6.7
Totals	100	119	130	20.0	23.8	26.0

Sources: Office of the Commissioner of Banking, *Annual report of the Commissioner of Banking for 1989*, table 1.7, p. 61; Hong Kong Monetary Authority, *Annual report 1995*, table 3, pp. 92–93; Hong Kong Monetary Authority, *Annual report 1997*, table 4, pp. 122–23.

sion. Furthermore, Hong Kong had a striking disadvantage: the looming return to China in 1997 cast a pall of uncertainty over its viability as a financial center. Rather than its transformation into the regional financial center for Asia being inevitable, a paradox exists: Hong Kong soared in the face of extraordinary uncertainty, and viable alternative centers existed, such as Tokyo and Singapore, with greater long-term political security. This can be resolved: Hong Kong has functioned as the regional financial center for Asia since at least the 1870s. Its foreign and Chinese financial firms always have been the most specialized, highly capitalized in Asia (outside Japan), and they have the best access to information about the political economy of the region. Foreign banks that entered Asia after the late 1970s set aside the political uncertainty of 1997; to participate in the surging financial business in the region, they had to join Hong Kong's social networks of capital. Their arrival in large numbers, therefore, augmented Hong Kong's long-term status as the great Asian regional financial center; financial scale increased and more-specialized activities emerged as its institutions reacted to competition from other financial centers in Asia.

World's largest banks

A substantial share of the world's largest banks with headquarters outside Hong Kong choose it for a regional headquarters at the highest level of banking, licensed banks, and many of them have facilities in lesser financial centers in Asia (table 8.1). Virtually all of the top fifty and almost three-fourths of the top hundred world banks operate from Hong Kong, and they maintained that presence during more than a decade of uncertainty preceding its return to China's control, testimony to Hong Kong's centrality for intermediaries with the greatest capital and highest levels of specialization in financial services. Banks at lower levels of the top 500 banks increased their overall presence in Hong Kong, confirming that international banks are elaborating a hierarchy of financial services. As smaller banks arrive to meet growing demands for financial services in Asia, larger ones use their stronger capital positions to react to this competition by offering more-specialized services.

Enthusiastic promoters of globalization portray the circulation of capital as a seamless exchange among financial centers, implying that Hong Kong should have foreign banks from countries throughout the world. It houses major banks from almost thirty economies (table 8.2), but their origins mirror the commodity trade of Hong Kong with Western Europe, North America, and Asia and multinational sources of investment for Asia; most highly developed nations in Western Europe and North America have licensed banks in Hong Kong.[23] No licensed banks come from the less-developed world outside Asia (except Bahrain and Iran); nevertheless, their poverty does not fully explain this absence because several Asian economies, such as India, Indonesia, and the Philippines, are also poor, yet have banks in Hong Kong. Regional exchanges of commodity and financial capital that date back over a century integrate poor and prosperous Asian economies, whereas Hong Kong retains feeble ties to poor countries outside Asia. Intensified Asian exchanges of capital encouraged regional banks to raise their share of total foreign banks in Hong Kong from 49 percent in 1986 to 56 percent in 1997. Japanese banks are the premier circulators of capital through Hong Kong, and mainland China is in second place, far above other Asian economies; by 1997, China surpassed all developed economies, except Japan, underscoring that Hong Kong's financiers are pivotal intermediaries of capital with the mainland.

Circulator of financial capital

The amount of capital that licensed banks in Hong Kong control increased 324-fold from 1970 to 1997, reaching $7.7 trillion (Hong Kong dollars), equivalent to $1 trillion (US dollars) (fig. 8.11).[24] It grew at a 21 percent compound annual rate under the regime of domestic export expansion between 1970 and 1978 and accelerated to an annual rate of 34 percent between 1978 and 1987 during the first decade of reforms unleashed by Deng Xiaoping and as a result of greater trade within Asia; from 1987 to 1997, the annual growth rate slowed

[23] The drop in the number of United States banks probably results from bank mergers.

[24] Fig. 8.11 records liabilities and these are exactly equivalent to assets under financial accounting principles.

Table 8.2. *Number of licensed banks in Hong Kong by political unit of beneficial ownership, 1986–1997*

Political unit	1986	1991	1997
Asia and Pacific	84	89	108
Hong Kong	20	15	16
Mainland China	15	15	18
Australia	4	4	4
India	4	4	4
Indonesia	2	3	3
Japan	25	33	44
Korea, Republic of	3	3	3
Malaysia	2	2	3
New Zealand	0	1	0
Pakistan	1	1	1
Philippines	2	2	2
Singapore	5	5	5
Taiwan	0	0	4
Thailand	1	1	1
Europe	37	48	50
Austria	0	2	2
Belgium/Luxemburg	3	3	4
Denmark	0	2	2
France	8	8	8
Germany	8	8	10
Ireland, Republic of	0	1	0
Italy	4	7	6
Netherlands	3	3	3
Spain	1	3	3
Sweden	0	3	2
Switzerland	3	3	3
United Kingdom	7	5	7
North America	28	24	20
Canada	6	6	6
United States	22	18	14
Middle East	2	2	2
Bahrain	1	1	1
Iran	1	1	1
Total licensed banks	151	163	180

Sources: Office of the Commissioner of Banking, *Annual report of the Commissioner of Banking for 1989*, table 1.5, pp. 57–58; Hong Kong Monetary Authority, *Annual report 1995*, table 4, pp. 94–95; Hong Kong Monetary Authority, *Annual report 1997*, table 3, pp. 120–21.

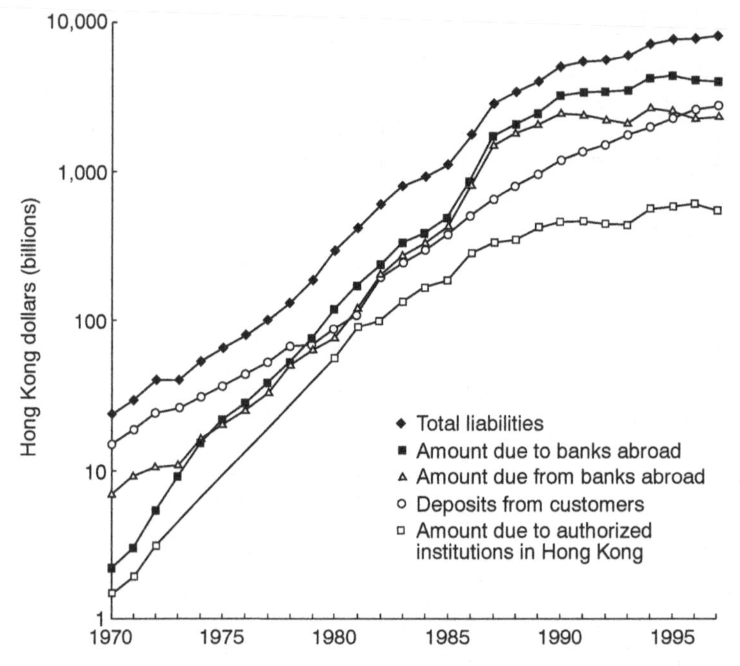

Fig. 8.11. Balance-sheets of licensed banks in Hong Kong, 1970–1997.
Sources: Hong Kong Government Information Services, *Hong Kong*, various years; Census and Statistics Department, *Hong Kong annual digest of statistics*, various years; Hong Kong Monetary Authority, *Monthly statistical bulletin*.

to 10 percent. Previous rates were not sustainable as the cumulative investment base enlarged and the onset of economic problems in Japan and in countries of the Nanyang forced retrenchment by the mid 1990s. The increasing wealth of businesses and consumers in Hong Kong provides a massive amount of capital that banks circulate; since 1980, deposits from customers consistently accounted for 25–30 percent of the total liabilities of banks. Those deposits exceeded deposits received from other banks in Hong Kong (fig. 8.11, "amount due to authorized institutions in Hong Kong") and rose far more rapidly after 1980. During the 1970s, customer deposits supplied much of the capital available for lending, but surging flows of capital from banks outside Hong Kong ("amount due to banks abroad") deposited in local banks roared past customer deposits in 1979; that capital from external sources comprised the majority of the liabilities of banks from 1987 to 1997, reaching a peak share of 63 percent in 1990. Nevertheless, rising per capita income and accumulating capital in the business sector continued to swell customer deposits at a rapid rate, pushing their share to over one-third of liabilities by 1997, the highest level since the 1970s; this underscores the importance of Hong Kong as a generator of capital.

Hong Kong banks also deposit capital in external banks to cover business activities and to gain returns on capital in the form of interest. The difference between the deposits those external banks make in Hong Kong ("amount due to banks abroad") and the deposits that its banks make externally ("amount due from banks abroad") indicates the net transfer of capital to Hong Kong. During the early 1970s, Hong Kong generated surplus capital for use abroad, but after 1974 the inflow of deposits ("amount due to banks abroad") always exceeded outflows ("amount due from banks abroad"), although the two came relatively close to balancing from the late 1970s to the mid 1980s (fig. 8.11). This shift from exporter to importer of capital during the 1970s suggests that the start of export industrialization in Asia and the growth of regional trade may have motivated this net transfer of capital to Hong Kong for redistribution, and after 1978, the reforms of Deng Xiaoping and the start of investment in China, especially Guangdong Province, impelled global banks to augment their net inflow of capital. Since 1993, that net inflow has ranged between $150 billion (US dollars) and $220 billion, a massive amount for redistribution to Asia. The domestic travails of Japanese banks and the financial crisis in Asia that reached serious proportions in 1997 contributed to the leveling off in total liabilities of banks and in deposits that foreign banks send to Hong Kong ("amount due to banks abroad") during the 1990s. Japanese banks accounted for half or more of all liabilities of foreign banks during the early to mid 1990s, but between 1995 and 1997 they reduced their deposits in Hong Kong at a 10 percent compound annual rate, a contraction of $66 billion (US dollars) over that period. Although the United Kingdom and Singapore banks retained stable deposit levels and both the United States and mainland Chinese banks increased their deposits at annual rates of 14 percent, the plunge in Japanese capital swamped the net deposit increases from the United States ($4 billion) and mainland China ($9 billion).[25]

While global banks started balancing their portfolios towards net transfers of capital to Hong Kong during the 1970s, their total loans and advances ("loans") rose steadily (fig. 8.12). The finance of non-trade business in Hong Kong and the rest of Asia always dominates loans because the export and import trade mostly relies on credit, not direct loans, and loans for other activities cover a wide range of the economy including government, manufacturing, construction, and consumers; thus, loans for non-trade uses inside and outside Hong Kong are not shown separately on fig. 8.12. Prior to the mid 1980s, the majority of loans financed business inside Hong Kong; loans for use outside jumped from a small base, yet the small amounts emphasize the meager economic development in Asia through the early 1980s. Loan volume

[25] Hong Kong Monetary Authority, *Annual report 1997*, table 11, p. 132; Hong Kong Monetary Authority, *Monthly statistical bulletin*, no. 45 (May 1998), table 2.9, pp. 31–35; Sapsford, "Bankers skeptical Tokyo has power to force reform."

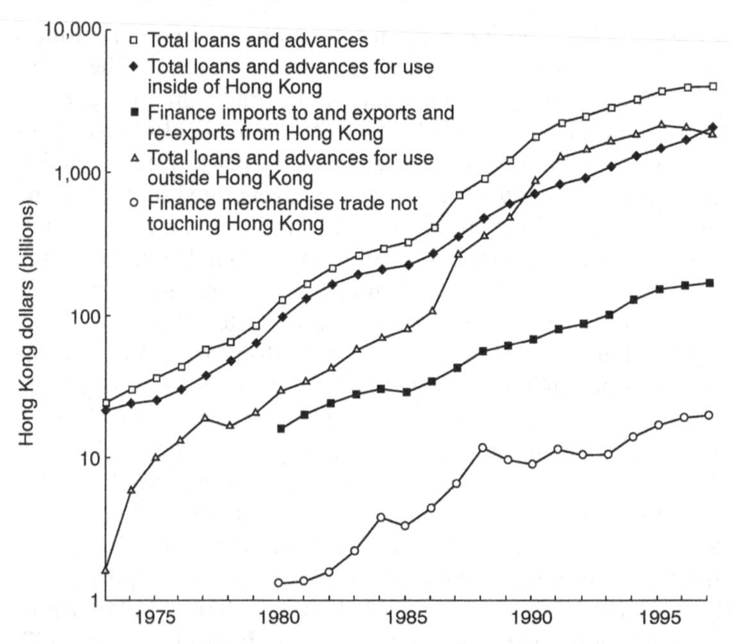

Fig. 8.12. Loans and advances of licensed banks in Hong Kong, 1973–1997.
Sources: Hong Kong Government Information Services, *Hong Kong*, various years; Census and Statistics Department, *Hong Kong annual digest of statistics*, various years; Hong Kong Monetary Authority, *Monthly statistical bulletin.*

grew both inside and outside Hong Kong after the start of the reforms of Deng Xiaoping in 1978, and loans for use outside rose faster than for use inside after 1985. Economic growth in the Nanyang countries and the pell-mell rush of Hong Kong industrialists to Guangdong Province prompted a surge of loans for outside use in 1990, vaulting them past the loans for use inside. This greater emphasis of global banks on financing business and government projects throughout Asia continued in the 1990s, but accumulating problems in the region caused banks, especially Japanese ones, to contract their lending after 1995. Although loans to finance trade represented a small share of loans, their shifts provide another window on the relation between Hong Kong intermediaries and Asian development (fig. 8.12). Loans for trade touching Hong Kong (imports, exports, re-exports) expanded at a steady, robust rate, consistent with the prominence that local firms retained over trade within Asia and with developed countries, and they always exceeded loans for trade not touching Hong Kong by substantial margins. Those latter loans jumped during the 1980s, but they experienced slow growth after 1988. Continued expansion of export industries in other Asian countries (excluding China) probably will restrain the expansion of loans for trade not touching Hong Kong because traders in metropolises such as Bangkok, Jakarta, and

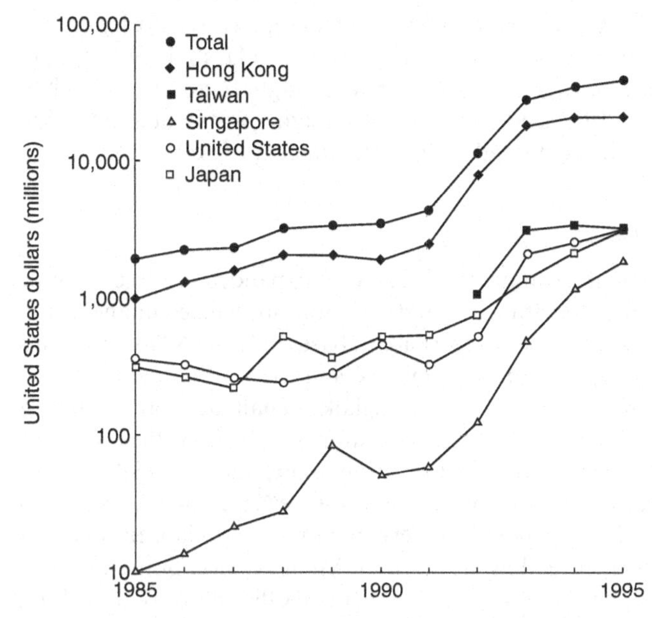

Fig. 8.13. Annual foreign direct investment utilized in China by origin, 1985–1995. *Source:* State Statistical Bureau, *China statistical yearbook*, various years.

Kuala Lumpur have greater access to information about those export industries and tighter integration with domestic networks of capital. Those firms may operate as branches of Hong Kong firms, but in any case, global banks with branches in those metropolises probably will handle more of that trade finance.

Because Hong Kong is the meeting-place of the Chinese and foreign social networks of capital, it offers the optimal base to scan business opportunities in China; therefore, Hong Kong provides much of the annual foreign direct investment (FDI) utilized in China, typically accounting for 50–70 percent (fig. 8.13). Capital from Hong Kong doubled from about \$1 billion (US dollars) annually in 1985 to about \$2 billion by 1990, but then it surged to rates of \$20 billion annually by the mid 1990s. Japan and the United States alternate between second and third rank, but eased restrictions on formal relations between Taiwan and China led to an outpouring of official capital investment by Taiwan firms in 1992 that immediately vaulted it to second rank. The bridges of Singapore's intermediaries to networks of capital in mainland China always have been much weaker than Hong Kong's; even after rapid expansion of Singapore's FDI, Hong Kong's annual investments remained eleven times larger as of 1995. The centers of Overseas Chinese capital in Hong Kong, Taiwan, and Singapore collectively provide most external funds for the economic development of China, averaging around 70 percent in the mid 1990s, thus realizing the vision of Deng Xiaoping. The majority of this FDI heads to the provinces of Guangdong (adjacent to Hong Kong), Fujian (adjacent to Taiwan), and Jiangsu (adjacent to Shanghai), and to Shanghai, an eerie reproduction of the leading centers of foreign investment and trading in China during the nineteenth and early twentieth centuries.[26]

Financial specialization

Following 1970, capital accumulation in Asia expanded with economic growth and development, creating demands for more specialized financial services and encouraging global banks to establish bases in Hong Kong. As countries started their rise out of poverty, financiers in the leading metropolises, such as Kuala Lumpur (Malaysia), Bangkok (Thailand), and Jakarta (Indonesia), exploited their greater access to information about the economy, and in many cases, protection from competition, to expand financial services. Hong Kong financial institutions reacted to these newer competitors and to their long-term ones in Singapore and developed more-specialized financial services to maintain their leading position in Asian finance. Some of these institutions, especially the largest, operate a regional office network; the Hong Kong office houses the most-specialized units that control the greatest amount of the firm's capital, as well as less-specialized operations for local and nearby regional activities, whereas regional offices in other metropolises house less-specialized operations. Large global banks headquartered elsewhere often replicate this office organization of firms headquartered in Hong Kong. New specialized institutions also emerge to provide financial services. The Hong Kong government promotes this highly competitive financial regime through policies that allow foreign firms easy entry, minimize regulatory costs, and encourage sound, well-run businesses; thus, the government rejects the option of allowing locally based financial institutions to "appeal to force" to protect their business.[27] This approach maintains a fiercely competitive environment that encourages financial institutions to react continuously to competition through marshaling large capital resources and specializing. On the other hand, the enormous, growing agglomeration of firms provides superb local social networks of communication about capital opportunities in Asia and a setting conducive to cooperative ventures.

Even the state-controlled banks of China in Hong Kong, most of which originated before 1949, vigorously expanded branch-office networks in Hong

[26] China also receives capital through loans from international capital markets, multilateral agencies, and foreign governments. For data on FDI by province, see State Statistical Bureau, *China statistical yearbook* (various years).

[27] Policies of the Hong Kong government towards financial institutions are listed in numerous documents. For recent statements, see: Hong Kong Monetary Authority, *Annual report 1997*; Hong Kong Monetary Authority, *Money and banking in Hong Kong*.

Kong to acquire access to local capital and to participate better in this competitive financial market. From 1969 to 1981, their branch numbers surged fourfold from 47 to 193, and they utilized their greater capital to offer more-specialized financial services and to expand beyond financing directly related to China. Several also raised their access to capital to sufficient levels by 1980 that they formed two joint-venture merchant banks to offer specialized services, including participation in syndicated loans, that required large amounts of capital. These efforts solidified financial bonds between Hong Kong and China through offering sophisticated financial services, and they extended the reach of these banks in Asia as they serviced Hong Kong firms with business outside China. By the 1990s, mainland Chinese banks had more deposits than foreign-owned banks from any other nation, and they ranked second, after Japanese banks, in assets, total loans to customers, and loans for use inside and outside Hong Kong.[28]

Other sophisticated financial intermediaries emerged in Hong Kong. The first stock exchange dated from 1891, but from 1969 to 1973 the growth in stock activity encouraged the formation of three more exchanges; all four merged into the Stock Exchange of Hong Kong in 1986. The gold market also started early; the Chinese Gold and Silver Exchange, which engages in local purchase and sale of gold and physical exchange of it, was formed in 1910. After the government lifted restrictions on gold imports, an extension of the international gold market developed in 1974; Hong Kong gained professional traders who buy and sell gold on the spot market in London and effect payments in New York, and since 1980 gold contracts have traded on the Hong Kong Futures Exchange. By that time, Hong Kong had become the fourth biggest gold market after London, New York, and Zurich. The capacity of its gold dealers and traders to dominate the gold market of Asia from the start rested on the Chinese business networks that could circumvent restrictions within the region. Hong Kong firms served as intermediaries in exchanges among international gold suppliers in South Africa, Canada, and Australia, gold markets in Western Europe, and gold purchasers in Asia.[29] The large, growing number of firms and employment in most sectors of stocks, futures, and bullion from 1984 to 1997 reflects Hong Kong's increased prominence in these specialty financial services (table 8.3). International banks that flocked to Hong Kong starting in the late 1970s (fig. 8.10) sharply boosted the foreign exchange market. Initially they specialized in their national currencies in trades among themselves, but as the market expanded they entered a broader array of currencies. By the mid 1990s, Hong Kong had become the fifth largest

[28] Hong Kong Monetary Authority, *Annual report* (various years); Jao, "Hong Kong's role in financing China's modernization," pp. 30–38.

[29] Freris, *The financial markets of Hong Kong*, pp. 97–98, 157–60; Schenk, "The Hong Kong gold market and the Southeast Asian gold trade in the 1950s."

Table 8.3. *Specialized financial institutions in Hong Kong, 1984–1997*

	Number of establishments		Number employed	
Type	1984	1997	1984	1997
Stock, bullion, and commodity exchanges	6	3	190	651
Stock and share companies	777	1,071	5,412	13,260
Futures and gold bullion brokers and dealers	161	204	1,578	1,475
Investment and holding companies	1,275	3,879	14,244	28,498

Sources: Census and Statistics Department, *Employment & vacancies statistics (detailed tables) in transport, storage & communication, financing, insurance, real estate & business services, community, social & personal services, 1984*, table 1, pp. 1.3–1.5; Census and Statistics Department, *Employment & vacancies statistics (detailed tables), series A (services sectors), 1997*, table 1.2, pp. 9–13.

global market for foreign exchange trading, including specialty products such as spot and forward options and futures.[30]

Efforts of the Hong Kong government to minimize regulations and taxes on financial transactions during the late 1970s paved the way for Hong Kong's dramatic expansion as a regional investment management center. Growing capital accumulation within Asia and opportunities for investment created a market for unit-investment trusts, including mutual funds, that permit individuals and organizations to buy units in pools of assets that may include bonds, stocks, and currencies. The number of authorized trusts reached over a hundred as of 1983, over half established under local law, with foreign financial firms introducing the remainder; the number of these funds soared to 903 by 1994.[31] Hong Kong managers did not run all of these funds; many had headquarters elsewhere but had authorization to sell in Hong Kong. Nevertheless, the tripling of the number of investment and holding companies and the doubling of their employees from 1984 to 1997 attests to the growing stature of Hong Kong as a regional investment management center for Asia (table 8.3). That growth, coupled with swelling demand for capital investment in transportation, communication, power plants, factories, and offices, motivated leading investment banks, especially from the United States, to enlarge operations considerably in Hong Kong during the early 1990s as underwriters

[30] Freris, *The financial markets of Hong Kong*, pp. 180–81; Ho, "The money market and the foreign exchange market," pp. 79–82; Hong Kong Monetary Authority, *Annual report 1995*, p. 17.

[31] Freris, *The financial markets of Hong Kong*, pp. 160–62; Hong Kong Monetary Authority, *Money and banking in Hong Kong*, p. 40; Scott, "Unit trusts and insurance companies," pp. 123–29.

of stocks and bonds and providers of venture capital to Asian firms. These banks tap the capital of regional investment funds run from Hong Kong and mobilize capital from elsewhere in the world.[32] Regional investment firms have little choice; they must base senior managers in Hong Kong to access the social networks of capital that bind Chinese and foreign firms because those networks provide critical information about investment opportunities and risk assessment. The networks also attract international corporations in all types of businesses that need a regional management center for Asia.

Corporate management center

The global headquarters of firms internalize the highest levels of intermediary decision-making regarding the exchange of commodities, financial capital, and services within and among firms. Yet, they cannot manage international operations solely from the global headquarters if they require access to complex information about a world region's trustworthy firms and about sources of demand and supply of commodities and financial capital; they must participate in the region's social networks of capital. A regional headquarters supervises offices or subsidiaries in other countries without the need for sustained supervision from the global headquarters; therefore, the regional headquarters represents the top of a management hierarchy within the world region. Typically, the office structure comprises a regional corporate headquarters in one metropolis and regional offices in other cities. Some firms use one regional office to manage operations in the world region, rather than create a hierarchical management structure under a regional headquarters.[33]

The expansion of multinational regional headquarters and offices since World War II seemingly represents a new wave in the global economy, but from the perspective of Hong Kong, this rests on an old tradition. At its inception in the 1840s, Hong Kong was the regional headquarters for great trading firms, such as Jardine, Matheson & Company, and by the 1860s, Hongkong and Shanghai Bank made Hong Kong the global and regional headquarters that supervised offices in other countries. During the late nineteenth and early twentieth centuries, firms from Europe and the United States housed regional headquarters in Hong Kong, Shanghai, and Singapore and others relied only on regional offices. As of the early 1990s, multinationals with global headquarters outside Asia made Hong Kong their preferred choice for an Asian regional headquarters; it accounted for 51 percent of them, whereas Tokyo had 29 percent and Singapore had 20 percent. The number of firms with global headquarters outside Hong Kong and which choose it as a regional headquarters or office center has soared since the 1970s, indicating that these

32 Sender, "Guns for hire."

33 For concepts of corporate headquarters location, see chapter 2, and for an application to Hong Kong, see Wilson, "Hong Kong as regional headquarters."

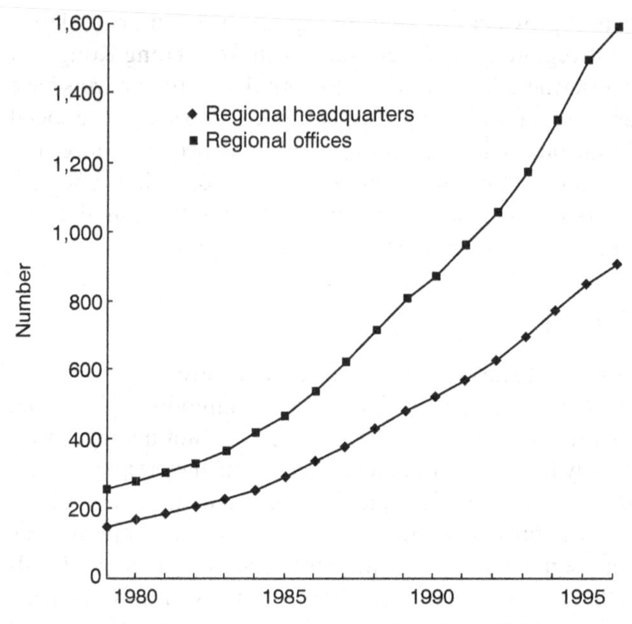

Fig. 8.14. Number of regional headquarters and regional offices in Hong Kong, 1979–1996.
Source: Industry Department, *Survey of regional representation by overseas companies in Hong Kong*, various years.

firms increasingly use Hong Kong as the platform for managing their Asian business (fig. 8.14).[34] The steady rise in the number of these management units implies that the broad economic growth of Asia and China promoted this expansion, not a specific geopolitical event.

The business lines of regional headquarters and offices follow similar patterns (table 8.4). Among manufacturing firms headquartered outside Hong Kong, about 120 use it as a regional headquarters to supervise offices and industrial operations in Asia and about eighty use it for a regional office, testimony to its significance as a management center for industrial firms, even where they do not engage in local manufacturing. Although Hong Kong Chinese dominate manufacturing in Guangdong Province, firms from Taiwan and Singapore also own factories there, but they account for a tiny share of the regional headquarters and offices in Hong Kong. Most come from Japan,

[34] Wilson, "Hong Kong as regional headquarters," p. 107. The data in fig. 8.14 come from a survey of 10,229 overseas companies with operations in Hong Kong; the survey had a response rate of 42 percent. The total number in the figure at any point is based only on those responses. The actual numbers of regional headquarters and offices remain indeterminate, but they amount to substantially more than the reported numbers. See Industry Department, *1997 survey of regional representation by overseas companies in Hong Kong*, pp. 5–7.

the United States, United Kingdom, and other Western European countries, confirming that Hong Kong is a base for industrial global management that does not simply rest on ethnic Chinese manufacturing firms.[35] Nonetheless, these global firms access the Chinese networks in Hong Kong to gain insights about optimal manufacturing sites and to form cooperative ventures and subcontracting relations for Asian production. The regional headquarters and offices that focused on manufacturing supervision constituted about 7 percent of all "offices"; however, over 40 percent of the parents of global firms operated predominantly as industrial corporations (table 8.4). The clue to the puzzle rests in the shares of "offices" that operated in the wholesale/retail and import/export areas; they were about 50 percent of the total, but only 30 percent of their parent firms had those business lines. Global manufacturing corporations, even if they have only limited production in Asia, use Hong Kong as their regional headquarters or offices for sales and distribution to Asia.[36]

Most regional headquarters and offices in Hong Kong operate in trade and financial services (table 8.4), consistent with the focus of locally headquartered firms. Offices that manage transportation and communication businesses continue the status of Hong Kong as a regional management center in those lines from the mid nineteenth century. Formerly, this included the great steamship companies; now firms manage airlines and telecommunications, as well as shipping. Hong Kong ranks fifth globally as an airline hub, just below Tokyo and above seventh-ranked Singapore, and it is near the top of the hierarchy in telecommunications. Among Asian cities, Hong Kong is the leader in global telecommunications, operating as the fiber optic hub of the region, and numerous communications satellites are managed by local firms; within Hong Kong, all-digital, high-volume lines provide linkages among firms.[37] It also serves as an Asian management base for firms that provide business services, such as law, accounting, real estate, construction, architecture, and engineering (table 8.4). They buttress the finance, trade, and manufacturing firms in the region and provide services to other clients across Asia.

Major industrial economies head the list of the home bases of firms with

[35] If a global manufacturing firm only had factories in Hong Kong, the office would not have received the designation of regional headquarters or regional office. The conclusion that ethnic Chinese manufacturing firms from Taiwan and Singapore have few regional headquarters and offices for manufacturing supervision rests on an examination of lists of office operations. See Industry Department, *1997 survey of regional representation by overseas companies in Hong Kong*, pp. 29–120.

[36] This conclusion is confirmed by a perusal of the list of offices in wholesale/retail and import/export; it includes numerous large industrial corporations. See Industry Department, *1997 survey of regional representation by overseas companies in Hong Kong*, pp. 29–120.

[37] Enright, Scott, and Dodwell, *The Hong Kong advantage*, pp. 90–92, 157–58; Langdale, "The geography of international business telecommunications"; Lovelock, "Information highways and the trade in telecommunications services"; Smith and Timberlake, "Conceptualising and mapping the structure of the world system's city system," table 1, p. 298.

Table 8.4. *Regional headquarters and offices in Hong Kong and their overseas parent firms by major business line, 1997*

	Percentage distribution			
	Regional headquarters		Regional offices	
Business line	Regional headqts.	Overseas parent	Regional offices	Overseas parent
Trade & finance				
Wholesale/retail & import/export	48.5	28.8	51.1	34.1
Finance & banking	11.4	14.9	13.0	14.9
Insurance	1.4	1.7	2.4	3.4
Manufacturing				
All	13.0	41.6	5.0	41.7
Transport & communication				
Transport & related services	9.7	8.2	5.0	6.1
Communication services	1.3	2.5	1.3	1.6
Business services				
Construction, architectural & civil engineering	4.7	5.3	4.5	6.2
Real estate	3.8	3.0	1.2	2.5
Other business services	18.3	10.4	20.7	9.0
Other				
Restaurants & hotels	1.0	2.3	0.6	1.6
Diversified	1.3	7.3	0.1	1.5
Other	1.2	4.4	1.5	2.2
Number of companies	1,067	1,205	1,709	2,004

Note: Percentages based on original counts of companies. The total number of companies is higher than the survey total of 924 regional headquarters and 1,606 regional offices because some companies were in more than one line of business.
Source: Industry Department, *1997 survey of regional representation by overseas companies in Hong Kong*, tables 2.3, 3.3, pp. 12, 17.

regional headquarters or offices in Hong Kong, and this is consistent with the importance of their global corporations as suppliers of goods and capital to Asia, providers of services (transportation, communication, legal, and accounting), and as buyers of Asian products (table 8.5). The number of global corporations from the United States that choose Hong Kong as their Asian management base far surpasses every industrial economy outside Asia. Numerous United Kingdom firms continue to build on their long-term business in Asia and the advantages of former colonial control over Hong Kong. Corporations headquartered in other Asian economies do not locate regional headquarters and offices in Hong Kong solely to carry on business with local firms; those "offices" would not fit the survey category, which refers to supervision of activities in other political units. Hong Kong offers advantages as a regional management center that firms cannot duplicate elsewhere in Asia, even from their home bases; it is the pivot of the Chinese and foreign social networks of capital.

Japanese firms lead among regional offices in Hong Kong, consonant with their widespread manufacturing, trading, and financing in East and Southeast Asia (table 8.5). These offices sometimes report to a regional headquarters, and in the case of Hong Kong, most probably report directly to the corporate headquarters in Japan. The second-rank position of Japan among regional headquarters, which supervise offices or subsidiaries in other economies without the need for sustained supervision from the global headquarters, nevertheless seems paradoxical. Tokyo shares the top of the global hierarchy with New York and London; thus, its corporate headquarters have the sophisticated range of financial, producer, transportation, and telecommunications services to manage far-flung operations.[38] Corporate managers in Hong Kong face a grueling one-day flight to the United States, the United Kingdom, or other countries in Europe, and time-zone differences make contemporaneous back-and-forth telecommunications about decisions difficult; international corporations headquartered in those nations, therefore, must house regional headquarters in Hong Kong. Managers of Japanese corporations in Hong Kong, however, can reach their Tokyo corporate headquarters within four hours by airplane and operate in an adjacent time zone.

The numerous regional headquarters of Japanese corporations, therefore, signify that they cede autonomy and authority to officers in Hong Kong to manage complex operations, especially across Southeast Asia, because Tokyo and other Japanese cities miss two key ingredients: they are inaccessible to Chinese capital networks, and they are unattractive to foreign firms from outside Asia who engage in activities external to Japan. Japanese corporations with operations in China can house regional offices in mainland cities, especially Shanghai, that report directly to corporate headquarters, but Hong Kong also serves as a place to meet regional headquarters and offices from China firms. Mainland China ranked between third and fourth on these units, and this emphasizes the fact that Chinese firms with international operations look to Hong Kong as their entrée to Asian markets. No Chinese city offers comparable contacts; even Shanghai, the metropolis of Central and North China, remains removed from the networks of capital in the Nanyang, continuing the structure of business networks existing since the nineteenth century. Mainland Chinese firms are rapidly expanding their management presence in

[38] For the evidence on Tokyo as a premier global metropolis, see Sassen, *The global city*.

Table 8.5. *Regional headquarters and offices in Hong Kong by political unit of parent, 1997*

	Number of companies	
Political unit of parent	Regional headquarters	Regional offices
North America		
United States	219	262
Canada	7	18
Europe	250	372
United Kingdom	86	130
Germany	53	77
France	35	65
Switzerland	30	22
Netherlands	27	47
Sweden	12	16
Italy	7	15
Asia and Pacific	303	718
Japan	121	378
Mainland China	117	128
Taiwan	28	49
Korea, Republic of	14	81
Singapore	9	44
Malaysia	6	5
Indonesia	2	7
Thailand	1	4
Australia	5	22
Caribbean offshore centers	33	12
Others	127	229
Total	939	1,611

Note: Some companies are joint ventures; therefore, the total of regional headquarters is greater than the 924 identified in the survey and total of regional offices is greater than the 1,606 identified in the survey. Some political unit values are incomplete owing to non-disclosure.
Source: Industry Department, *1997 survey of regional representation by overseas companies in Hong Kong*, tables 2.2, 3.2, pp. 11, 16, appendices I and II, pp. 29–120.

Table 8.6. *Region of responsibility for regional headquarters of overseas companies in Hong Kong, 1997*

Region of responsibility	Number of companies	Percentage of total
Hong Kong and mainland China	364	39.4
Southeast Asia (excluding mainland China)	54	5.8
Southeast Asia (including mainland China)	184	19.9
East Asia	112	12.1
Asia Pacific	210	22.7
Total	924	99.9

Note: Definitions of regions:
Southeast Asia (excluding mainland China) Hong Kong, Taiwan, Philippines, Indonesia, Thailand, Malaysia, Singapore, and Vietnam;
East Asia Southeast Asia (including mainland China) plus Japan and Korea;
Asia Pacific East Asia plus Australia and New Zealand.
Source: Industry Department, *1997 survey of regional representation by overseas companies in Hong Kong*, table 2.4, p. 13.

Hong Kong beyond the numerous offices for local business. From relatively trivial numbers in 1991, their totals jumped to sixty-seven regional headquarters and forty-two regional offices by 1993; and those numbers, respectively, had doubled and tripled by 1997 (table 8.5).[39]

The sweep of control of regional headquarters in Hong Kong across all combinations of territorial units testifies to its status as the regional management center for Asia (table 8.6). The ranking of territories replicates the spheres of influence of firms in Hong Kong since the mid nineteenth century: mainland China business stands at the top, accounting for 39 percent of the firms; and Southeast Asia (the Nanyang), including the small set of firms that focus on the Nanyang and the larger set that also manage business in China, accounts for about one-fourth of them. Over one-third of the regional headquarters have responsibility for most of Asia (East Asia plus Asia Pacific in table 8.6), ratification of Hong Kong as the global metropolis for Asia. The details of territorial control remain secondary to the evidence that global firms see Hong Kong as the premier meeting-place of the Chinese and foreign social networks of capital for Asia. These regional headquarters and offices of foreign firms augment the local corporate headquarters of firms to create a vast agglomeration of intermediaries that control and coordinate the

[39] Industry Department, *Survey of regional representation by overseas companies in Hong Kong* (various years).

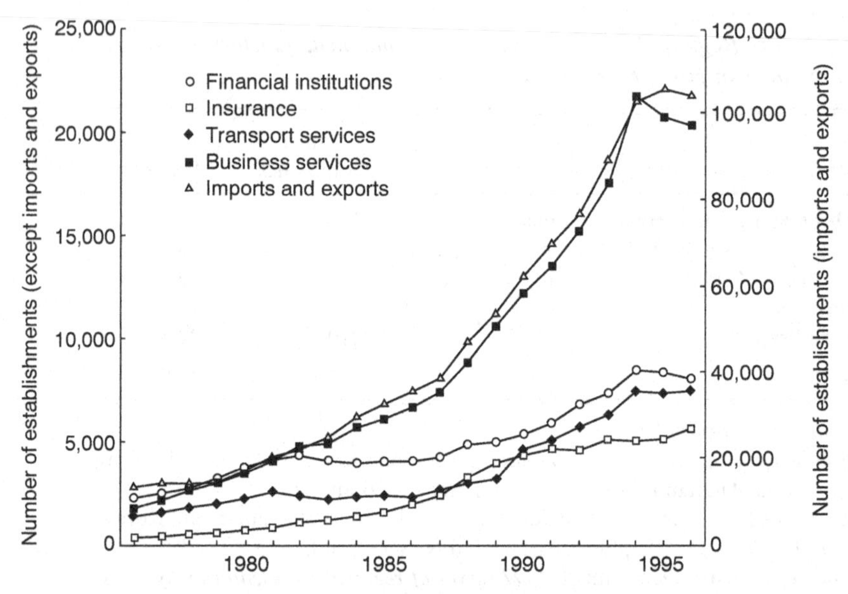

Fig. 8.15. Number of intermediary businesses and their support services in Hong Kong, 1976–1996.
Source: Census and Statistics Department, *Hong Kong annual digest of statistics*, various years.

exchange of capital. Growing numbers of firms, both financial and non-financial, and the increased size and complexity of their operations across diverse cultural and linguistic boundaries spawn a demand for specialized services that support those operations. Large corporations may internalize some services, but even they cannot afford to internalize the full gamut or the most specialized ones; small firms must buy most or all of these services from external vendors. These intermediary and producer services define Hong Kong as the global metropolis for Asia.

Intermediary and producer service center for Asia

The mid 1970s offers a baseline from which to track the expansion of intermediary and producer services in Hong Kong. These sectors expanded as rural development initiatives and industrial growth based on import substitution and export manufactures generated rising per capita incomes in many economies (fig. 7.2), but the stimulative impact of Deng Xiaoping's reforms in 1978 would not commence for several years. From the mid 1970s to the mid 1990s, both sectors soared to astounding heights (figs. 8.15–8.16). The number of intermediary businesses in finance, including commercial and investment banks, brokerage firms, and money management firms, rose 252 percent to 8,004, whereas those in insurance soared 1,821 percent to 5,551; import and

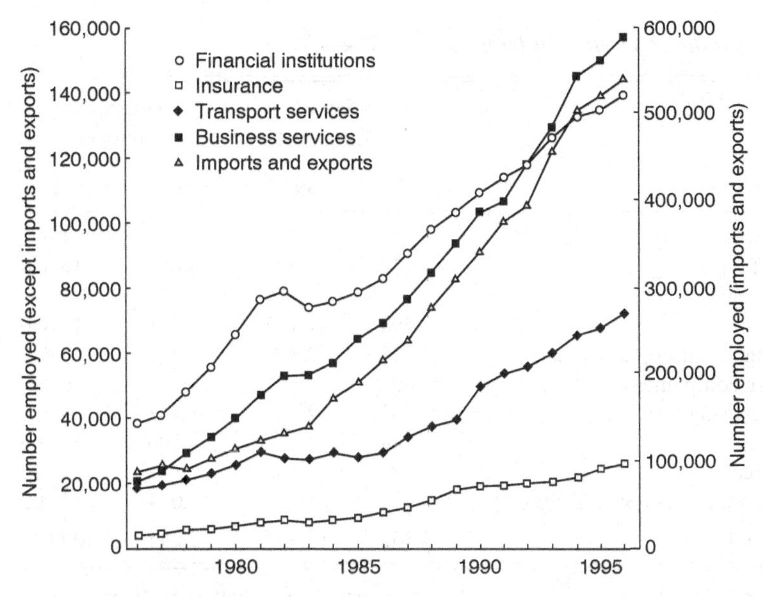

Fig. 8.16. Employment in intermediary businesses and their support services in Hong Kong, 1976–1996.
Source: Census and Statistics Department, *Hong Kong annual digest of statistics*, various years.

export firms jumped 642 percent to 103,961. These intermediary operations expanded their employment by huge amounts: finance rose 264 percent to 138,238; insurance climbed 517 percent to 25,495; and import and export jumped 524 percent to 538,049. Although some of these firms and workers serviced the local economy, their growth far surpassed the 66 percent increase in total employment, indicating that most of the expansion related to business outside Hong Kong.

The intermediary firms of Hong Kong generated enormous demands for producer services from the mid 1970s to the mid 1990s (figs. 8.15–8.16). The number of firms that provided transport services, such as freight forwarding, but excluding transport companies (shipping lines and airlines), rose 409 percent to 7,456, and firms in the broad category of business services, excluding rental of machinery and equipment, soared 1,063 percent to 20,187; transport services employment increased 282 percent to 72,030 and business services workers rose 688 percent to 156,571. The vibrant producer services sector exhibited considerable richness as firms multiplied in both traditional and non-traditional areas from 1984 to 1997 (table 8.7). Sea cargo forwarding and shipbrokerage date from the start of Hong Kong in the 1840s, yet those sectors expanded even faster than a twentieth-century specialty, air cargo forwarding. Numerous specialties that serve small firms and large corporations, including law, accounting, advertising, public relations, market research, and

Table 8.7. *Producer services in Hong Kong, 1984–1997*

	Number of establishments		Number of employees	
Selected services	1984	1997	1984	1997
Air cargo forwarding	269	793	4,840	11,690
Sea cargo forwarding	332	2,076	3,650	16,354
Shipbrokerage	29	126	177	661
Legal	498	1,162	7,271	17,440
Accounting, auditing, & bookkeeping	725	2,880	7,541	17,722
Advertising companies & agencies	608	821	4,158	6,908
Other advertising services	502	2,558	1,743	8,618
Public relations	29	109	206	805
Market research	58	66	563	717
Business management & consultancy	312	3,277	2,094	19,909
Selected total	3,362	13,868	32,243	100,824

Sources: Census and Statistics Department, *Employment & vacancies statistics (detailed tables) in transport, storage & communication, financing, insurance, real estate & business services, community, social & personal services, 1984*, table 1, pp. 1.1–1.5; Census and Statistics Department, *Employment & vacancies statistics (detailed tables), series A (services sectors), 1997*, tables 1.1–1.2, pp. 5–13.

business management and consultancy, were between two and eight times larger by 1997, compared to 1984.

The industrial sector of the economy that had developed from the late 1940s to the early 1980s declined precipitously from the late 1980s to reveal the firms of Hong Kong in their traditional status as the intermediaries of commodity and financial capital for Asia. The cast of firms diversified beyond banks, trading companies, and trade (producer) service firms that had typified Hong Kong during the mid nineteenth century. Its largest, most-capitalized intermediaries in trade and finance developed elaborate specializations to control and coordinate greater flows of commodity and finance capital, and less-capitalized, less-specialized intermediaries proliferated to serve growing subregions of Asia. Global non-financial corporations joined trade and finance firms to utilize their services and acquire the rich information embedded in Hong Kong's social networks of capital, thus expanding the scale of Hong Kong as the regional corporate management center for Asia. Collectively, these intermediaries of capital created a huge demand for producer service firms that support their businesses. That is the global metropolis for Asia that China inherited in 1997 when it resumed control as the sovereign power.

9

Hong Kong, China

The Government of the Hong Kong Special Administrative Region shall provide an appropriate economic and legal environment for the maintenance of the status of Hong Kong as an international financial centre.[1]

1997: past, present, and future converge

Rescission

Although Hong Kong housed one of the world's largest agglomerations of sophisticated trade, financial, and corporate management intermediaries and their producer services, its future rested on geopolitical events of the nineteenth century. Under the Treaty of Nanking (1842) that ended the Opium War, the British acquired Hong Kong in perpetuity, but no informed observer doubted that China eventually would demand that the British leave. They had taken Hong Kong, according to the Chinese, in an unequal treaty when China was weak, and its government did not recognize British rights to continued control. Paradoxically, the impetus for Hong Kong's return to China rested in the ticking clock of an obscure treaty from 1898 when Britain acquired the New Territories, an area north of the Kowloon Peninsula, from China; the lease terminated on July 1, 1997.[2]

After 1949, the leaders of China's Communist Party kept Hong Kong secure as an international trade and financial center under British rule. That pragmatic decision gave China access to information and foreign exchange during its struggle to develop under socialist principles and the restrictions on economic ties to foreign countries. China would have gained little economically from taking control, and the political risk of confrontation with foreign powers must have outweighed considerations of pride. The motivation to

[1] Article 109. *The Basic Law of the Hong Kong Special Administrative Region of the People's Republic of China.* [2] Endacott, *A history of Hong Kong*, pp. 21–23, 260–69.

broach the subject of 1997 seems to have originated within the financial sector of Hong Kong by the early 1970s as concerns about property leases rose; with less than two decades to go, investment returns over the lives of leases started to look unpromising. The British government ultimately recognized that China would not agree to a continued British presence. Following protracted negotiations, the two governments signed the Sino-British Joint Declaration on the Future of Hong Kong in September 1984, and China proclaimed the Basic Law, the document that formed the constitution for governing Hong Kong, in April 1990. China's consistent stance during the negotiations is reflected in the Basic Law; it insisted on absolute sovereignty and aimed to ensure that Hong Kong remained an international financial center. The Basic Law established a governance structure with a locally elected chief executive, and it stated a commitment to a capitalist system for Hong Kong that included free capital mobility and provided for Hong Kong's own laws and regulations to run the economy. It would operate as a Special Administrative Region for fifty years, with expiration in 2047, and China would prohibit Hong Kong from becoming a base for political subversion or challenges to Beijing. Hong Kong's return to Chinese rule on July 1, 1997, was unparalleled; no global metropolis of its stature had ever been transferred peacefully between major powers.[3] This return stimulated immense commentary about Hong Kong's future, and outside mainland China most observers predicted negative consequences.

Skeptical talk and counter-action

The principle of "one country, two systems" struck many skeptics as absurd. They interpreted evidence of political uncertainty over the details of life in Hong Kong, economic problems and corruption in China, assertions of Beijing leaders that challenges to their control were unacceptable, and stock market and real estate gyrations in Hong Kong as proof that the mighty business center would decline after 1997. Confusion over the details of governance and uncertainties about ultimate intentions stemmed from the incongruity of combining a rich, capitalist city-state with an impoverished nation ruled by a Communist Party hierarchy. The citation that 500,000 Hong Kong Chinese emigrated between 1985 and 1995 seemingly provided unequivocal evidence of its demise.[4] On the eve of the formal return to China, a poll of Asian executives counted 31 percent as expecting that Hong Kong businesses would shift their headquarters outside the city after the handover, but this share fell to 14 percent when the universe was limited to those based in Hong Kong. From

[3] "China resumes control of Hong Kong, concluding 156 years of British rule"; Overholt, *The rise of China*, pp. 249–65; Welsh, *A borrowed place*, pp. 502–36.

[4] Gilley, "Darkness dawns"; Kraar, "The death of Hong Kong"; Szulc, "A looming Greek tragedy in Hong Kong."

1991 to 1995, foreign companies with regional headquarters or offices in Hong Kong also voiced concern about its viability, yet their fears diminished dramatically by 1997.[5]

This more benign view of China's rule from locally headquartered and foreign companies in Hong Kong on the eve of 1997 matches the substantial evidence on their actual decisions. From the Joint Declaration of 1984 to 1997, while political and economic uncertainty swirled, the financial, trade, regional corporate management, and producer services sectors of Hong Kong rose to unprecedented heights.[6] That expansion coincided with one of the swiftest, most successful economic restructurings of a global business center ever seen. From the start of the plunge in manufacturing in 1987 until 1996, factory employment fell by 63 percent, yet total employment rose 10 percent. This loss in manufacturing plus the net gain in total employment meant that Hong Kong generated a total of 773,750 new jobs by 1996, equal to 33 percent of the workforce in 1987.[7] While foreign firms bemoaned the unfavorable political climate and the deterioration in it, the share that planned to maintain their regional headquarters and offices in Hong Kong stayed around 95 percent from 1992 to 1997, resounding ratification of its continued prominence as a global metropolis. These firms remained sensitive to the political climate, but they ranked that issue between fourth and eighth on a list of relevant factors affecting their decisions; instead, banking and financial facilities and infrastructure were the most important factors, and on those grounds they remained quite pleased.[8] The number of foreign residents that obtained visas to work in Hong Kong soared after the mid 1980s, and the much discussed exodus of Hong Kong Chinese, which rose significantly after 1985, was counter-balanced by a rising tide of new permanent residents (fig. 9.1). In most years, the number of arrivals exceeded those leaving, and Hong Kong gained 122,403 net migrants between 1985 and 1997.

The China question

This evidence on Hong Kong's vibrancy, nevertheless, fails to refute fully the gloomy future that skeptics paint. Local democracy never existed prior to the early 1990s; the British government appointed the Governor who had full authority to run Hong Kong, and the rule of law came indirectly through the implied support of Britain's parliament. The members of the Legislative Council included those selected by the Governor, government officials, and individuals elected by professional bodies and district boards. Demonstrations

[5] Asia Studies Limited of Hong Kong, "Asian executives poll"; Industry Department, *Survey of regional representation by overseas companies in Hong Kong*, tables 4.1–4.3, surveys of 1991–97.
[6] See evidence in chapter 8.
[7] Census and Statistics Department, *Hong Kong annual digest of statistics* (various years).
[8] Industry Department, *Survey of regional representation by overseas companies in Hong Kong*, tables 4.1–4.3, surveys of 1991–97.

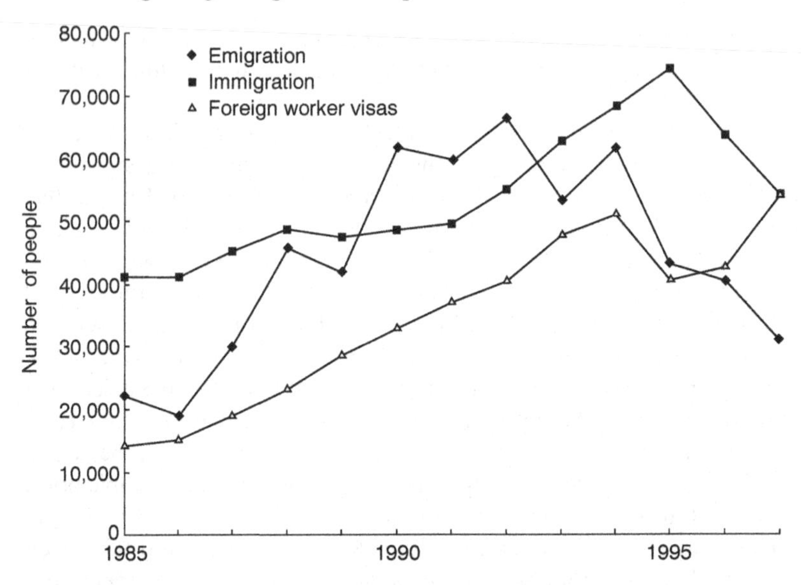

Fig. 9.1. Emigration from and immigration to Hong Kong and foreign worker visas, 1985–1997.
Sources: Immigration Department, Government of the Hong Kong Special Administrative Region, unpublished tabulations; Security Branch, Government of the Hong Kong Special Administrative Region, unpublished tabulations.

in Hong Kong in response to Beijing's crackdown on protesters in Tiananmen Square in June of 1989 provided vent for many citizens' fears about their future under Chinese rule after 1997. Chris Patten, the new Governor who arrived in 1992, became a rallying point for politicians who wanted to create a more democratic government, but he dealt with Beijing through confrontation and unilateral decisions, a sharp departure from previous British governors who operated in a consultative mode with China. It viewed Patten's efforts, supported by political activists, to institutionalize democratic elements prior to 1997 as a challenge to China's sovereignty over Hong Kong; on that score, Beijing would not negotiate. In the Basic Law and its own statements, China's government saw the institutionalization of democracy as a gradual process, always under the authority of the central government.[9]

The British colonial government dealt firmly with demonstrations and strikes; similarly, Beijing expects Hong Kong's government to maintain control. The absence of violent demonstrations after 1989, when political tensions rose during Patten's efforts to alter the electoral processes, suggests that even as Hong Kong's citizens express support for more open democratic pro-

[9] Cheng and Lo, *From colony to SAR*; Cohen and Zhao, *Hong Kong under Chinese rule*.

cedures, they place greater weight on stability than on confrontation with China. To avoid provocation from a military presence, Beijing maintains only a small garrison of elite, disciplined soldiers of the People's Liberation Army in Hong Kong, and they have little contact with citizens; that is the task of the regular police force. These measures, nevertheless, cannot guarantee that political conflict will remain quiescent. The business community supports a legal system that guarantees contracts and property rights and consistently rates a secure business environment and stability more highly than freedom of speech and the press; this could place them in conflict with citizens who want to promote their interests through mobilization and the electoral process. Resolution of these contentious issues, while being overseen by Beijing, challenges the political skills of the Hong Kong government.[10]

Because Hong Kong is irrevocably bonded to China's future, skeptics also claim that China's difficulties or even collapse will harm Hong Kong. Barring political and social collapse, military resurgence, regional breakdowns, democratic revolution, and Maoist revival – events that have dramatic cachet – the more likely scenarios include: succession fights, struggles over lines of authority among governmental levels from the center in Beijing down to the villages, and social and economic strains instigated by rapid economic growth and development.[11] Emphasis on internal stresses, however, dismisses the broad political, cultural, and ethnic unity at the core of China covering several thousand years. Initial successes in the climb out of poverty support a consensus among Chinese leaders and citizens that economic development has the highest priority. This means shifting the economy towards greater market orientation by freeing commodity prices from regulations, reducing subsidies for state enterprises, creating capital markets, and opening competition through encouraging private firms. The leaders' views differ on the speed of change that promotes optimal development and the degree that the state should retain a hold on the economy; however, agreement does exist on the direction of change. The immense task of raising 1.3 billion people out of poverty and building a political and legal structure that manages the task carefully, given disparities in the speed of development, inevitably produces a rocky path with power struggles.[12]

This consensus that emerged with the success of Deng Xiaoping's reforms has created a China that is not "one country, two systems"; rather, it is "one country, many systems." Heavy state regulation of all sectors of the economy has declined, bringing into the foreground regional economic complexities

[10] Cohen and Zhao, *Hong Kong under Chinese rule*; Gilley, "Man on the spot"; Welsh, *A borrowed place*.

[11] Baum, "China after Deng"; Goldstone, "The coming Chinese collapse"; Huang, "Why China will not collapse"; Lieberthal, *Governing China*.

[12] Chen, "China fills cabinet with younger technocrats"; Chen, "China to test waters of political reform"; Fairbank, *China*; Forney, "Private party"; Overholt, "China after Deng."

that existed in the nineteenth century. At that time, the weak Qing government could not implement strong economic policies, and many regions, especially around the treaty ports, forged their own external links. Initial reform efforts after 1978 that permitted Guangdong and Fujian provinces to experiment with less regulation, decontrol prices, manage their foreign trade, create special economic zones, and acquire fiscal independence, set China on a path of greater regional differentiation, and as other provincial governments implement different economic policies, paths diverge more.[13] The special administrative region of Hong Kong, therefore, looks less like an aberration. This skepticism about Hong Kong's future also detracts attention from consideration of another scenario, normal change in a global metropolis.

Normal change expected in Hong Kong

Metropolis and region

Many observers portray Hong Kong as a nation-state, but that appellation misconstrues its essence. The intermediaries of capital make Hong Kong the global metropolis for Asia, the international metropolis for China, and the regional metropolis for South China. As in the cases of London and New York, steady erosion of the physical manufacturing base results from high land and labor costs. Flirtations with firms that dangle the offer of jobs in return for monetary concessions, including tax breaks and subsidized training, land, and buildings, may slow that erosion of production, but they will not stop it. Manufacturing that remains in Hong Kong will produce goods that require immediate access to local consumers, make specialty products in small workshops, operate as demonstration plants for production slated elsewhere, or serve as pilot plants for research and development facilities. As a pivot of information, Hong Kong can prosper as a center of research and development, design, testing, and other sophisticated information processing services related to manufacturing. Those efforts are more likely to succeed if combined with upgrading the capacities of universities and colleges to offer science and technology education and courses of study to meet the needs of the intermediary and producer service sectors, the dominant parts of the economy. Firms started by students and faculty from local universities and colleges are more likely to stay in Hong Kong, and the government can facilitate that through supporting partnerships of investors and organizations that help new firms through the early growth stages. Because predictions about the future of the rapidly changing high-technology industry are often erroneous, Hong Kong must avoid offering subsidies to create a center of such manufac-

[13] Chen and Ho, "Southern China growth triangle"; Linge and Forbes, *China's spatial economy*; Vogel, *One step ahead in China*, pp. 76–122.

turing through targeting industries or attracting foreign firms to science parks. Regardless of their success, local high-technology firms will not retain much manufacturing in Hong Kong anyway.[14]

Hong Kong manufacturing firms and foreign firms with factories in South China or elsewhere in Asia use Hong Kong as a corporate management base, not a production base. The vast production enterprise of Hong Kong firms in nearby Guangdong Province is continuing, although rising wage levels and land costs inevitably create pressures to reorganize manufacturing. Low-wage manufacturing shifts into the interior of China or elsewhere in Asia, while Hong Kong and foreign firms establish higher value-added, more-sophisticated manufacturing in Guangdong. Cooperative relations between Hong Kong and the cities of Guangzhou and Shenzhen aim to encourage the development of high-technology manufacturing in Guangdong. Since 1997, senior politicians in Hong Kong and Guangdong have supported greater integration by means of a number of developments: high-speed train service with smooth visa entry linking Hong Kong with Shenzhen and Guangzhou; agency cooperation promoting the region as a tourist destination; the building in Shenzhen of low-cost housing for Hong Kong residents; and the construction by Hong Kong companies of more infrastructure (highways, utilities, housing) in Guangdong. Advocates of a manufacturing (industrial) policy for Hong Kong downplay this transformation of regional relations and mistake Hong Kong for a nation-state; on the contrary, it is a global metropolis for intermediaries of capital.[15]

Changes in commodity shipments

From the start in the 1840s, the small scale of individual trades encouraged firms to aggregate commodities at Hong Kong before long-distance shipment and to break down shipments there for redistribution within Asia. These commodity movements characterized Hong Kong as a center of physical movement, and the industrial expansion that started in the late 1940s provided increasing supplies of local goods that powered the surge in shipments. When intra-Asian trade accelerated in the 1970s, Hong Kong also gained import trade from Europe and North America for re-export to Asia; this trade, including shipments to China and exports of Hong Kong's factories in China starting in the 1980s, augmented shipments through the port. From shares

[14] "Hong Kong should move into hi-tech"; Tong, "Hong Kong plans to push for high technology science park project seen as way to boost manufacturing sector."

[15] Berger and Lester, *Made by Hong Kong*; "China: closer ties expected with Hong Kong"; "China: Shenzhen to focus investment on housing for local, Hong Kong residents"; "Good scientific cooperation between Guangzhou, Hong Kong"; "Hong Kong, Macao, Guangdong explore ways to develop tourism"; "Hong Kong–Shenzhen express train services to open"; Kwok and So, *The Hong Kong–Guangdong link*; "Shenzhen and Hong Kong enhance cooperation in financial sector"; Sito, "Hongs speed up Guangdong plan for investment."

below about 10 percent of the imports and exports of China in the early 1960s, the port of Hong Kong raised its shares to about half by the early 1990s.[16]

These shipments have reached such a large scale that they could pass directly between foreign and Chinese ports to avoid hierarchical aggregation and disaggregation at Hong Kong, but the port infrastructure of China remains poor. The improved ports in the Pearl River delta, such as at Shenzhen, capture exports from, and imports to, Guangdong Province because these ports offer cheap land and labor and, as inland transport improves, factories in Guangdong can avoid congestion on the roads to and from Hong Kong; yet, Hong Kong is building a new river container terminal to service the Pearl delta ports to undercut those competitors. Similar port developments at places such as Shantou or Xiamen (Foochow) may reduce the relative significance of Hong Kong for bulk shipments, and Shanghai probably will reclaim its position as the commodity transshipment center of China based on its trade ties with Central and North China. Domestic and foreign direct investments in Shanghai also provide growing output for transshipment, and the local government has concentrated that development in Pudong, a new zone that includes a large port. Nevertheless, Hong Kong has capitalized on its trade ties, superb deep-water facilities, and highly efficient operations run by private-sector firms to become the world's largest container port. Other competitors in Asia, including Singapore, Kaohsiung (Taiwan), Pusan (South Korea), and Japanese ports (Tokyo-Yokohama, Kobe-Osaka), do not have equivalent access to the vast, growing Chinese economy and to the entrepot trade of Asia. On top of that, Hong Kong has become the world's largest international air cargo center, and its new airport, Chek Lap Kok, has almost double the capacity of the old airport.[17] Even if Hong Kong declines relatively or absolutely as a commodity shipment center, this would not signify the decline of control and coordination that its importers and exporters exert over the commodity trade of Asia. They do not require physical contact with shipments; rather, they rely on their access to information and their social networks of capital to make exchanges.

Office center

These importers and exporters, along with the other intermediaries of commodity and financial capital and the producer services firms linked to them, generate a prodigious, expanding demand for office space. Those skyscrapers give Hong Kong its signature landscape surrounding the harbor between

[16] Chen and Ho, "Southern China growth triangle," tables 4–5, pp. 45–46.

[17] Enright, Scott, and Dodwell, *The Hong Kong advantage*, pp. 122–29; "Hong Kong – international shipping center"; Shi, Lin, and Liang, "Shanghai as a regional hub"; Rimmer, "Transport and communications in the Pacific economic zone during the early 21st century"; Yeung, "Infrastructure development in the southern China growth triangle."

Hong Kong island and Kowloon, but the intensification of office construction poses the problem of high-cost space. Surveys of foreign firms taken from 1991 to 1997 consistently demonstrate that about 50 percent of them rate the cost of office space as unfavorable and 75 percent of those with regional headquarters say high rental cost is a problem for them.[18] This presents a paradox; at the same time as these overseas firms targeted the cost of office space as a serious problem, they dramatically expanded their presence in Hong Kong, and locally headquartered firms did likewise. Observers anticipate that thousands of private and government firms from China will augment this demand for offices as they seek a window on international trade and finance.[19] One resolution to the paradox rests in the obvious fact that firms use space intensively; most employees, including senior executives, occupy tiny offices.

The other resolution follows time-honored principles. The most highly capitalized and specialized intermediaries and their sophisticated producer services firms require frequent face-to-face contact to carry on complex negotiations and consultations; at the same time, however, high-cost office space represents only a small share of their expenses. This assures their continued agglomeration in Hong Kong, especially in prime office clusters in Central, Admiralty, Wan Chai, and Causeway Bay, on Hong Kong island, and Tsim Sha Tsui in Kowloon. Savings on office space from relocation to another metropolis would be negated by an even greater loss of business from being outside the networks of capital in Hong Kong. If their back-offices, which process routine information, expand, firms shift these components to lower-cost sites in Hong Kong, such as vacant industrial buildings that are renovated as back-office space or demolished and replaced with new office buildings. With improvements in infrastructure (telecommunications and transport) in Guangdong Province, large multinationals have opened back-offices in Shenzhen and Guangzhou to acquire low-cost space and cheap labor.[20] Intermediary and producer services firms, therefore, continually identify the high cost of office space as a problem, while they resolve it through intensification and decentralization. As the global metropolis for Asia, Hong Kong experiences these normal changes in manufacturing, commodity shipment, and office space; yet, skeptics still claim that metropolitan competitors in Asia will undercut it.

[18] Industry Department, *Survey of regional representation by overseas companies in Hong Kong* (various years).

[19] This point comes from confidential interviews in Hong Kong. Also, see: Chen, "Early birds"; Smith, "Chinese regions chase Hong Kong cash."

[20] "Hong Kong service sector attracted to southern Chinese province"; Sito, "Bank to centralise back-office functions." Examination of comparative cost figures for Asian metropolises demonstrates that Hong Kong does not stand out as an exceptionally costly place compared to the most widely promoted competitors, Tokyo and Singapore. Part of the confusion over cost figures for Hong Kong comes from an overemphasis on the cost of space, rather than taking into account the total cost structure, including corporate tax rates; those are substantially lower in Hong Kong. See Enright, Scott, and Dodwell, *The Hong Kong advantage*, pp. 337–50.

Alternatives to Hong Kong

Promotional rhetoric of local and national politicians portrays competition among metropolises, but metropolises differ according to intermediary levels of capitalization and specialization. Because Hong Kong always has housed the headquarters of firms with the greatest reach across Asia, other firms often choose it for regional headquarters or offices. In turn, the largest of the locally headquartered firms establish branch offices elsewhere in Asia to enhance their control and coordination of capital. The rise of Asian peasants out of poverty enhances opportunities to exchange capital, thus underpinning the emergence of more capitalized and specialized intermediaries in national (and regional) metropolises of each nation. Some are indigenous intermediaries, but others are "foreign" firms that join agglomerations to participate in local social networks of capital. This process accounts for the rising trade and financial significance of Bangkok, Jakarta, Kuala Lumpur, Ho Chi Minh City, and Manila, and political decisions such as deregulation can boost their competitiveness; but these metropolises do not diminish the importance of Hong Kong. Economic growth and development within each country offers opportunities for the most highly capitalized intermediaries in Hong Kong to accumulate capital and to react to competition from intermediaries in other metropolises through devising new, more-specialized services. Some firms internalize this competition by raising the amount of capital committed to their regional headquarters and offices and offering more-specialized services from them. Those less-specialized services that firms in Hong Kong provided to other nations are eliminated as intermediaries in other metropolises subsume them. Intermediaries in Bangkok, Jakarta, Kuala Lumpur, Ho Chi Minh City, and Manila have little chance of overtaking those in Hong Kong as leaders in Asia, but observers frequently present the Tokyo, Singapore, and Shanghai agglomerations as challengers to the dominance of those in Hong Kong.[21]

Tokyo: outside the networks

According to numerous measures of global scale, including number of the largest banks, total bank assets, number of the largest securities firms, total stock market turnover, and number of the largest multinational headquarters, Tokyo stands at the top of the global hierarchy along with London and New York.[22] Foreign intermediaries enter Tokyo to participate in the huge domestic economy of Japan. In the early 1990s, about 90 foreign banks operated in

[21] Barnathan, "Asia's supercities"; Enright, Scott, and Dodwell, *The Hong Kong advantage*, pp. 241–79.

[22] Sassen, *The global city*, tables 7.1–7.5, pp. 170–79; Suzuki, "The future role of Japan as a financial centre in the Asian-Pacific area."

Tokyo, but this number contrasts with over 160 foreign licensed banks in Hong Kong at that time (fig. 8.10). Few foreign intermediaries manage their Asian operations, except those in Northeast Asia, from Tokyo, and their regional headquarters or office typically reports directly to corporate headquarters in, for example, New York or London, and has little or no management control over other Asian offices.[23] The juxtaposition of Tokyo as the agglomeration of huge Japanese intermediaries, many with substantial global operations, with its unimportance as a base for foreign intermediaries to operate in Asia seems paradoxical. Analysts frequently explain this by claiming that heavy government regulation inhibits foreign firms, and that limited facility in English, the international language of business, in Tokyo and the rest of Japan poses problems for foreign regional headquarters; both explanations make valid points, but they are insufficient.[24] Although regulations hinder foreign firms in the Japanese market, they are not so restrictive that they prevent firms from using Tokyo as their regional management center for Asia, and foreign firms could communicate among themselves in English even if contact with Japanese speakers is difficult.

A complete explanation for the paradox mirrors the reasoning for the heavy presence of regional headquarters and offices of Japanese corporations in Hong Kong. Tokyo and other Japanese metropolises remain outside the Chinese social networks of capital in Asia, whereas Hong Kong is the pivot of those networks. Overseas Chinese firms, like other foreign firms, only place offices in Japan to operate in the domestic market. The chief administrative offices of these Chinese firms remain in Hong Kong, and secondarily, Singapore and other metropolises of the Nanyang. Large Japanese trading companies such as Mitsubishi, Mitsui, and Sumitomo are exceptions to the relative isolation of Japanese intermediaries from the Chinese networks of capital. Besides their large operations in Hong Kong, they pursue alliances with Overseas Chinese firms in Hong Kong, Singapore, Bangkok, and Jakarta to invest in infrastructure projects in the Nanyang. Japanese financial firms, however, tend to isolate themselves from those Overseas Chinese networks. Offices of Tokyo firms in the metropolises of the Nanyang hire few senior local people; thus, they fail to bridge social networks through family members of Overseas Chinese. The Bank of Japan epitomized this isolation when its senior officials predicted that Hong Kong would serve primarily as the financial center for China and the growth triangle comprising Guangdong Province, Fujian Province, and Taiwan, thus not recognizing that Overseas Chinese firms in Hong Kong also form the bridge to networks of capital in the Nanyang.[25]

[23] Nicolle, "Hong Kong in a regional perspective," pp. 82–83; author's interviews with business executives in Hong Kong.

[24] McGurn, "Money talks"; Nicolle, "Hong Kong in a regional perspective," pp. 82–83.

[25] Rowley, "Retreat to China"; Sender, "The sun never sets"; Sender, "Local hero."

The failure of Japanese financial institutions to make Tokyo the hub of the capital networks in Asia demonstrates that size alone does not confer power; instead, power comes from position in the network.[26] Tokyo's financial institutions occupy weak positions in Asian capital networks because they are removed from the Chinese networks. In the late 1990s, the highly regulated Japanese financial markets were opened more to foreign competition because the capital bases of Japan's firms became dangerously eroded under mountains of bad domestic and foreign loans, mostly in Asia. The influx of large financial firms, especially from the United States, may restructure Tokyo's financial role in Asia because offices of these foreign firms will primarily operate in Japan, whereas their Hong Kong offices will supervise Asia-wide finance (outside Japan) and offices in the metropolises of each country will handle that business.[27] This will sharply reduce the presence of Japanese financial institutions in Hong Kong and the rest of Asia because foreign firms will transfer capital from Tokyo to their other offices.

Singapore: bridge to the Nanyang

In contrast to Tokyo, Singapore is a beehive of Overseas Chinese businesses that participate in the Chinese social networks of capital in the Nanyang, and, as a former British colony, it is also a meeting-place of the Chinese and foreign networks. Singapore's exports of local manufactures surged twentyfold in real value from 1969 to 1990, and the number of industrial workers rose from 143,100 in 1970 to 447,436 in 1990; yet, Hong Kong, whose manufacturing had declined by 1990, still employed over 60 percent more workers than Singapore, most in locally owned firms (fig. 7.3). Foreign multinationals funded Singapore's industrial expansion; their share of value of output and value-added rose from about 45 percent in 1968 to 75 percent in 1990. Nevertheless, Singapore firms, similar to those in Hong Kong, are relocating or building new factories elsewhere to achieve lower-cost production and to serve markets in those countries, but Singapore firms look mostly to Malaysia, Indonesia, and Thailand, whereas Hong Kong firms focus on China.[28] In conjunction with the rise of intra-Asian trade, the growth of export industrialization in East and Southeast Asia, and Deng Xiaoping's reforms in China, the financial and service sectors of Singapore, similarly to those in Hong Kong, soared, and total employment rose from 23,071 in 1970 to 167,222 in 1990. Chinese banks, such as Industrial and Commercial Bank, Far East Bank, Development Bank of Singapore, and United Overseas Bank, actively control and coordinate capital flows with Malaysia and Indonesia. Malaysian banks

[26] See chapter 2.

[27] Pitman, "Merrill Lynch heads stampede into battle with Tokyo's titans"; Tett, "'Big bang' adds to pressure on Japanese banks."

[28] Fong, "Staying global and going regional"; Huff, *The economic growth of Singapore*, fig. 11.2, p. 304, tables 11.3, 11.11, A.11, pp. 305, 319, 410–11.

are active in Singapore, but greater restrictions on finance in Indonesia hinder capital flows.[29]

Singapore officials present it as a competitive alternative to Hong Kong for the title of greatest trade and financial center of Asia.[30] As of the early 1990s, their aggregate shares of gross domestic product (GDP) in intermediary sectors (finance and commerce) and in those that directly serve them (transport, communications, and business services) tracked closely at 58 percent for Singapore and 55 percent for Hong Kong (table 9.1). Manufacturing in Singapore contributed much more to GDP than in Hong Kong, reflecting the greater production of large multinationals in Singapore. Although individual sectoral employment shares differed, both metropolises fit the label of trade, finance, and corporate management centers; nevertheless, the dramatic growth of Singapore's GDP still left its economy about half the size of Hong Kong's. They employed similar numbers in the physical movement of commodities (transportation) and information (communications), but these do not measure the degree of decision-making over the control and coordination of the exchange of capital. Hong Kong employed 2.5 times as many as Singapore in commerce, and in 1990, the 341,583 importers and exporters in Hong Kong even outnumbered all the commerce employees in Singapore, including importers and exporters, retailers, and wholesalers.[31] Its trade sector always has focused on Malaysia, Indonesia, and Thailand whereas Hong Kong traders range over the Nanyang, China, and Northeast Asia. Foreign multinationals, which dominate manufacturing in Singapore and nearby Malaysia, internalize many of their buying, marketing, and selling functions. This eliminates Chinese importers and exporters from large areas of the industrial economy of Singapore, whereas in Hong Kong those same firms are pivotal global distributors of manufactures produced in Hong Kong and Guangdong Province. Hong Kong importers and exporters participate in more-sophisticated trade networks, accumulate greater capital, and develop higher levels of specialization in commodity trading. Most foreign multinationals market and distribute their goods in Asia from Hong Kong because they need access to these social networks of importers and exporters.

Compared to Singapore, Hong Kong also is a much greater center of financial intermediation, with almost twice as many employed in finance and business services, close to 50 percent more foreign banks, and a stock market about twice as large; they only reach equivalence in foreign exchange trading, a specialty that Singapore developed early (table 9.1). The extensive local funds in the government-controlled Central Provident Fund and Post Office Savings Bank reduce the amount of capital controlled by Overseas Chinese Banks

[29] Huff, *The economic growth of Singapore*, tables 11.7–11.8, 11.20, A.11, pp. 315–16, 335, 410–11; Regnier, *Singapore*, pp. 118–38.

[30] Porter, "Monetary body strives for spot on world stage."

[31] Census and Statistics Department, *Hong Kong annual digest of statistics, 1994*, table 3.11, p. 34.

Table 9.1. *Selected comparisons of Singapore and Hong Kong*

	Gross domestic product (1990)		Employment (1990)			
	Percentage of total		Number		Percentage of total	
Industrial sectors	Singapore	Hong Kong	Singapore	Hong Kong	Singapore	Hong Kong
Manufacturing	29.0	17.6	447,436	715,597	29.1	26.0
Commerce	17.6	25.2	337,519	829,448	22.0	30.2
Transport & communications	14.2	9.5	146,553	132,792	9.5	4.8
Finance & business services	26.2	20.2	167,222	276,621	10.9	10.1
Selected totals	87.0	72.5	1,098,730	1,954,458	71.5	71.1
Other	13.0	27.5	438,281	793,642	28.5	28.9
Totals	100.0	100.0	1,537,011	2,748,100	100.0	100.0
Total GDP in current Hong Kong $ (billions)	285.5	559.4				
			Singapore	Hong Kong		
Banking (1994)						
Foreign banks (number)			119	171		
Representative offices (number)			50	156		
Foreign assets (US $, billions)			366	615		
Equity markets (1994)						
Market capitalization (US $, billions)			135	267		
Annual turnover (US $, billions)			77	126		
International bond issues by financial institutions (number in 1994)			0	10		
Foreign exchange turnover (percentage of world total in 1992)			6	6		

Sources: Huff, *The economic growth of Singapore*, tables 11.2, A.11–A.12, pp. 303, 410–12; Census and Statistics Department, *Hong Kong annual digest of statistics, 1994*, tables 3.1, 3.11, 7.8, 9.11, pp. 29, 34, 117, 137; Hong Kong Monetary Authority, *Money and banking in Hong Kong*, tables 1–4, pp. 35–38.

headquartered in Singapore and foreign banks with local offices, hindering their capacity to specialize in financial services, whereas locally generated capital in Hong Kong provides opportunities for local and foreign banks to specialize.[32] The dominance of foreign manufacturing multinationals in Singapore and nearby Malaysia also hinders local and foreign banks from accumulating capital and specializing because the multinationals obtain most of their sophisticated financial services from large banks in their home countries, whereas numerous Chinese manufacturing firms headquartered in Hong Kong provide a huge demand for financial services.

Singapore financiers retain their long-term focus on Malaysia, Indonesia, and Thailand, and they serve as conduits between these countries and markets in Japan, Western Europe, and North America. They react to competition from financiers in nearby countries of the Nanyang, although Indonesian financial firms are weak. The Indonesian government maintains restrictions that constrain opportunities for Singapore financiers, and their gains from flight capital do not compensate for limited chances to exchange capital in Indonesia. To fund projects in China, Singapore financiers continue their old practice of operating through Hong Kong offices or in cooperation with its institutions. Singapore's modest success in attracting mutual fund firms still leaves it much less important than Hong Kong, but when these firms follow routine investment strategies, Singapore offers an alternative; similarly, some financial institutions have moved back-office operations there. These activities do not require access to the sophisticated social networks of capital in Hong Kong that bridge to Asia-wide Chinese and foreign networks. Singapore also attracts regional headquarters, but these focus mostly on Malaysia, Indonesia, and Thailand. Hong Kong's management sweep, however, spans Asia and includes China (table 8.6); as of 1998, it had almost 1,000 regional headquarters, whereas Singapore had only 120.[33]

Shanghai: gateway to Central and North China

The widely proclaimed competition between Shanghai and Hong Kong for the title of financial titan of China receives as much attention as Hong Kong's battle with Singapore for the top position in Asia. According to Sha Lin, deputy mayor, "Shanghai will catch up with Hong Kong," and other local advocates claimed that Shanghai would soon take the lead. Prior to 1997, Beijing authorities asserted that they would not boost Shanghai over Hong Kong; instead, they planned to maintain its international

[32] Singapore's government finally introduced some reforms that allow professionals outside the government to manage about one-third of the money; see Hiebert, "Luring new wealth."

[33] Industry Department, *1997 survey of regional representation by overseas companies in Hong Kong*, table 2.1, p. 11; Regnier, *Singapore*, pp. 124–35; Silverman, "Now the good news"; Solomon, "Salim assailed for unit's shift to Singapore"; "Three MNCs awarded regional headquarters status."

stature.[34] Regardless of Beijing's precise balance of support for Shanghai or Hong Kong, the metropolis of Central and North China remains a weak semblance of its past glory as the trade and finance center for China's heartland; most exports of Central China exit through Hong Kong rather than Shanghai. Beijing and Shanghai authorities, therefore, are devoting huge resources to rebuild Shanghai's infrastructure, such as new buildings and renovations in the Bund, heart of the old financial district, and the development of the Lujiazui financial and trade zone across the Huangpu River from the Bund in Pudong; extensive foreign capital also underwrites these investments. Authorities hope to rectify other weaknesses, including the lack of qualified personnel to work in sophisticated firms, an inadequate stock exchange, and restrictions on financial and trade activities. As of the late 1990s, however, Shanghai remained a tiny replica of Hong Kong; total assets of foreign banks in Shanghai amounted to only $18 billion (US dollars) compared to $678 billion in Hong Kong.[35]

To pose a putative battle between Shanghai and Hong Kong misconstrues the issues. If Beijing and Shanghai authorities continue to improve the interior transportation and communications infrastructure to Central and North China, upgrade the physical infrastructure of Shanghai, and loosen restrictions on the trade and financial activities of domestic and foreign intermediaries, then Shanghai will return to its former position as the gateway to Central and North China. Nevertheless, the existing, much improved transportation and communication linkages between Central China and Hong Kong will restrain the full return of Shanghai to its former stature. Chinese intermediaries in the city, including private and government firms, maintain the social networks of capital within China that made Shanghai the pivot of much of China. To operate competitively, foreign traders, financiers, and multinationals with production and sales in the heart of China must establish their regional headquarters in Shanghai to link to those networks; Hong Kong never served as the regional headquarters for that part of China. Thus, multinationals such as Kodak (United States) and BASF (Germany) are using Shanghai for the regional headquarters of their materials processing plants in China, and American International Group, an insurance multinational, also has placed its regional headquarters in Shanghai and secondary offices in Guangzhou and Beijing.[36]

[34] Kahn, "A tale of 3 cities"; Peng, "Shanghai briefing"; Richburg, "Shanghai or Hong Kong" (quote).

[35] Harding, "Arriving in a trickle"; Hong Kong Monetary Authority, *Annual report 1997*, tables 1, 11, pp. 118, 132; Miller and Chan, "Foreign pioneers hope for busy life in quiet Pudong"; Shi, Lin, and Liang, "Shanghai as a regional hub"; Sung, "'Dragon head' of China's economy?"; Yeh, "Pudong."

[36] Guoli, "Strategic center for China operations established by BASF"; "Huge investment put by Kodak in business expansion in China"; Miller, "US insurance giant in line as Mainland set to open prime market."

Continued economic growth and development of China will catapult Shanghai to the upper tier of the world's metropolises because its trade, finance, and corporate management firms dominate exchange networks in Central and North China, an economy with a population of about 800 million or more. Foreign intermediary firms dealing with those regions will require a major operation in Shanghai. Nevertheless, Hong Kong will continue as the pivot of intermediaries who control and coordinate exchange with South China, an economy that includes perhaps 400 million or more people, and with Central China in competition with Shanghai intermediaries; Hong Kong's regional economy, therefore, surpasses most nations. In contrast to Shanghai, however, Hong Kong intermediaries represent the highest-capitalized, most-specialized bridges between the global economy outside Asia and the economies of the Nanyang, China, and Northeast Asia. Mainland Chinese and foreign intermediaries, therefore, that need access to these wider global capital markets will base their premier offices in Hong Kong, not Shanghai.[37]

Fallacy of technological intermediation

Skeptics' claims that intermediaries in other metropolises will usurp the dominance of those in Hong Kong rests on another misperception, the "fallacy of technological intermediation."[38] The argument is that modern telecommunications make most face-to-face contact unnecessary; thus, traders, financiers, and corporate managers can operate from anywhere to engage in non-local exchange and manage their dispersed assets. As proof of that, skeptics refer to global trading systems that operate twenty-four hours a day, with investors and traders exchanging capital seamlessly over networks of computers. This logic predicts that intermediaries in Hong Kong will relocate elsewhere in Asia at the slightest sign of problems with Chinese governance, but this conflates technological means with decision-making about the exchange of capital. Technologies permit global trading and improvements in telecommunications allow firms to routinize transactions that formerly involved non-routine decision-making, yet these advances also create opportunities for new, more-specialized, complex decision-making about the exchange of capital. Those decisions, such as capital investment in widely separated regions, joint-venture investment, and private equity placements, require trust built on social bonds and access to sophisticated networks of capital. Even if technology does not undercut Hong Kong, skeptics believe that the uncertainty of Chinese sovereignty produces a risky future, but they dismiss too readily the grave uncertainty facing most countries of Asia.

[37] Chen, "Early birds"; Smith, "Chinese regions chase Hong Kong cash."
[38] See chapter 2 for further discussion.

Risks in Asia

Rapid economic growth and development taxes the capacity of social, economic, and political institutions in impoverished countries to distribute benefits to the majority of the population and to target investments that sustain expansion. Because the recently industrializing countries of Malaysia, Thailand, Indonesia, and the Philippines invested minimally in rural development prior to their export industrial surges, the unproductive peasant masses provide little demand for goods and services. The political and economic elite captures most of the gains from investment in low-wage export manufacturing and from investment in expensive public infrastructure (airports, highways, and government buildings) and real estate (housing, office buildings, and retail stores) in the national metropolis, but these contribute few productivity gains for the broader economy. The military in Thailand and Indonesia intrude into government and private decisions, and Indonesia faces social, political, and economic problems because rural development lagged and former President Suharto's family and associates captured disproportionate economic gains. The Philippines achieved a semblance of democracy, but the powerful landed oligarchy retards advances of impoverished peasants. Bankruptcy looms over South Korean *chaebols* as a consequence of overinvestment in industrial capacity, and this undermines Korean banks who provided the loans. Problems in the rest of Asia exacerbate economic stagnation in Japan because its firms invested heavily in the region, and their losses inflate the already huge portfolios of worthless loans of Japanese banks, especially in Tokyo, who financed that investment. Weak financial regulations and excessive bank lending to speculative real estate projects and industrial capacity also left mountains of bad debts on the balance-sheets of banks in other Asian countries. Because a large share of these loans are denominated in US dollars and owed to foreigners, the collapse of currencies considerably enlarged debt levels in those countries.[39]

Negotiation through these heightened financial risks requires participation in the most sophisticated social networks of capital. Japanese financial institutions headquartered in Tokyo confront a dual problem: the head office is outside the Chinese networks and removed from the pivotal meeting-place in Hong Kong of the Chinese and foreign networks; and they remain saddled with worthless domestic and foreign loans. Tokyo's weakened institutions face intensified competition in Japan from foreign firms as the government deregulates financial markets. Because these foreign firms mostly concentrate on the

[39] Brauchli, "Speak no evil"; Goad, "Slew of dismal economic news in Asia positions more countries for recession"; Hamilton, "Japan's new premier launches image offensive as economic problems pile up almost daily"; Overholt, "Indonesia after Suharto"; Overholt, "Thailand"; Overholt, "Korea"; Pesek, "Dis-oriented markets"; Tasker, "The Thai, Indonesian and Burmese armies have long been entrenched in politics and business."

Japanese market, Tokyo will not advance relatively as a global metropolis for Asia. Singapore confronts increasing risk from the economic problems in Asia, and it does not have a powerful national state to support it. Malaysia, Indonesia, and Thailand, the focus of Singapore's intermediaries of capital, face severe economic problems and need deregulation to make their economies efficient. The problems in Indonesia, Singapore's uncomfortably close neighbor, undermines the city's tranquillity. Though little discussed, Singapore's ethnic/racial make-up contains the seeds for conflict if regional tensions rise. Since the 1950s, the population composition has remained around 75 percent Chinese, 7 percent Indians, and 14 percent Malays, plus small shares of other groups, whereas Hong Kong remains overwhelmingly unified at about 95 percent Chinese.[40]

Risk management center for Asia

Hong Kong's intermediaries of capital also confront economic difficulties in Asia: the profitable franchise of the Hongkong and Shanghai Banking Corporation as the premier bank in Asia is declining, and prominent firms such as Peregrine Investments are collapsing. Reduced demand for intermediary services in Asia translates into declining property values in Hong Kong as businesses retrench, threatening large portfolios of loans to developers who built offices, retail structures, and housing for the trade, financial, and corporate management sectors.[41] These difficulties that confront Hong Kong's intermediaries threaten their absolute level of business, yet the problems jeopardize neither their relative stature as the most highly specialized and capitalized firms in Asia nor their dominance based on participation in the pivotal meeting-place of the Chinese and foreign social networks of capital. They maintain deep recognition of their history that spans over 150 years of coping with political and economic uncertainty, including the Taiping rebellion, the Opium Wars, imperialism, two world wars, depression, and revolutions in many countries after 1945, and they self-consciously identify Hong Kong as the risk management center for Asia.[42] Nevertheless, critics maintain that this experience is irrelevant, because Britain no longer protects Hong Kong firms from the corruption and favoritism to the politically connected that is endemic to China, and this critique crystallizes as the claim that *guanxi* (connections) will undermine Hong Kong.[43]

[40] Hiebert, "Singapore"; Huff, *The economic growth of Singapore*, table 10.5, p. 292; Landers, "Men and boys"; McDermott, "US banks say loan exposure in Asia is limited"; Spindle and Sapsford, "Japan's investors fear a game with new rules."

[41] Guyot, "HSBC's growth is fettered by Asian turmoil"; Kohli, "Foreign banks slash exposure to Hong Kong"; Yiu, "Independent Peregrine crash probe considered."

[42] Based on interviews with business executives in Hong Kong.

[43] "Guanxi maximizing"; Weidenbaum and Hughes, *The bamboo network*.

Guanxi

According to this scenario, *guanxi* weaken the competitive capacity of Hong Kong intermediaries and many Overseas Chinese, local Hong Kong, and foreign firms will depart for other metropolises in Asia; however, even if corruption and favoritism increase, its intermediaries dealt with conditions that were at least as bad in the past. This scenario also assumes that Chinese control produces a *guanxi* effect (corruption and favoritism) that is worse than in other Asian metropolises, but government activities in South Korea, the Philippines, Malaysia, Indonesia, and Thailand illustrate the fallacy of that assumption; even Singapore, reputed to have minimal corruption, has a one-party state that makes unilateral decisions. In Japan, bonds among businesses, political parties, and government bureaucracies produce a competitive environment tilted against foreign firms. Overseas Chinese and foreign firms face a dilemma; those who move their senior offices out of Hong Kong or refuse to locate there sever themselves from full participation in its social networks of capital and that means missing opportunities to exchange capital within Asia and between Asia and the rest of the world. Because these firms exhibit mistrust of the Chinese government, it may limit their access to China, and if it continues rapid economic growth, these firms will not fully profit from the expansion of one of the world's largest economies in the twenty-first century. These firms confront a further dilemma; the alternatives to Hong Kong are weaker: Tokyo remains outside the Chinese networks of capital, Singapore serves primarily as the regional base for Indonesia, Malaysia, and Thailand, and other Asian metropolises represent even poorer options. Taipei cannot serve as an alternative because Beijing considers Taiwan an integral part of China, and the networks of capital in other metropolises such as Bangkok, Jakarta, or Manila are mostly intranational, with external links chiefly to nearby trade partners. The most specialized and capitalized Chinese and foreign firms, therefore, are unlikely to reconstitute their meeting-place of social networks of capital elsewhere.

The behavior of the intermediaries of capital contradicts the view that *guanxi* will seriously undermine Hong Kong. Foreign firms invest heavily in the heart of the *guanxi* environment, mainland China. Prior to the mid 1990s, risk-averse multinationals, such as Procter & Gamble, Motorola, Nestlé, Sony, and Unilever, left investment in China to Hong Kong and Taiwanese firms whose sophisticated understanding was gained from their social networks of capital, but these multinationals now pour huge investments into China. The preeminent symbol of British colonialism, the Hongkong and Shanghai Bank, reopened the branch office in Beijing in 1995 that it had closed in 1955. Swire Pacific, another symbol, sold a stake in its Cathay Pacific Airways at a discount to Citic Pacific and China National Aviation, China-backed firms. Outsiders berated Swire's for collapsing under pressure from China, but the firm claimed the sale made strategic sense. Its aggressive moves to invest in industrial enterprises in China demonstrates that the firm saw opportunities elsewhere; if Swire's feared *guanxi* effects in Hong Kong, its expansion in China seems bizarre.[44]

Red chips

For the pessimists, the move of mainland-owned Chinese companies, known as "red chips," to Hong Kong directly infiltrates *guanxi* and undermines the free-market environment with its legal protections. The stance that red chips, with their direct or indirect ties to government, introduce a new business relation to Hong Kong rests on a false comparison. Hong Kong never has been a bastion of free-market business, unfettered by the hand of politics; business and government have intertwined since the 1840s. Taipans of the great British firms such as Jardine's and Swire's assiduously cultivated officials, they served as advisors to, and members of, the government, including service on the executive and legislative councils, and private banks issued the currency under a government monopoly. Business and social clubs served as venues for sharing information and advice, and the Hong Kong Jockey Club, founded in 1884, epitomizes these bonds. The taipans ran the Jockey Club from the start and the leadership of the business elite continues; now prominent Chinese business executives are taking over. It holds the government monopoly on gambling, but this private institution is accountable only to itself and serves as a private charity, contributing as much as 7 percent of the government's total revenue.[45]

The intense public focus on the arrival of red chips started several years before 1997, but Chinese firms with government ties have operated in Hong Kong since the nineteenth and early twentieth centuries; these include the venerable China Merchants' Steam Navigation Company and the Bank of China. After Deng Xiaoping's reforms in 1978, the pace of arrival accelerated; China Overseas Holdings, owned by China's Ministry of Construction, and China Liaoning, owned by Liaoning Province, entered Hong Kong in 1979. China Holdings constructed numerous high-rise office towers, about 5 percent of the total housing, and the terminal at Chek Lap Kok, the new airport. Shanghai Industrial Investment Holdings, owned by the Shanghai municipal government, entered Hong Kong during the early 1980s. By the late 1990s, over 1,800 mainland companies had officially registered to do business in Hong Kong. Even before 1997, these red chips, including banks, conglomerates, and holding companies, accounted for as much as 25 percent of the local bank deposits, trade, and cargo handling, and supplied most of the meat

[44] These points are discussed in Meyer, "Expert managers of uncertainty." For evidence, see: Baum, "Chafing at the bit"; Brauchli, "China stays a magnet for overseas money"; "HK Bank open its door in Beijing"; Kalathil, "Swire Pacific Ltd. goes courting in China."

[45] Kwan, *The making of Hong Kong society*, pp. 19–62; Silverman, "A run for their money."

and produce consumed locally. Citic Pacific, owned by China International Trust & Investment, an agency of the Chinese government, holds major stakes in telecommunications (Hong Kong Telecommunications), electricity (China Light & Power Company), and aviation (Cathay Pacific Airlines). The major rise in construction awards to red chips, such as China Overseas Holdings, merely replicates earlier awards that British hongs captured.[46]

During the months preceding the return of Hong Kong to the control of China in 1997, the enthusiasm of investors in Hong Kong and mainland China for red-chip firms reached extreme levels; soaring stock prices were portrayed as a "delirium" and "scarlet fever." Yet, investors also bid the shares of other Hong Kong firms higher, indicating that they believed "blue chips" would not suffer irreparable harm from red chips. Investors viewed the *guanxi* of red chips as an indicator of their future success. Chinese government entities placed assets with these firms, typically at a discount to their perceived value, and investors believed an implicit promise existed that more assets would be transferred at such discounts in the future. Investment advisors in Hong Kong predicted that the speculative bubble in the stock prices of red chips would burst, and that happened during the collapse of Asian currencies and stock markets in late 1997. Skeptics viewed this speculative excess, the collapse of stock prices, and seemingly blatant valuation of firms according to their *guanxi* appeal as evidence that the arrival of red chips undermined Hong Kong. Yet, red chips strengthen it when they shift their headquarters there or buy stakes in local firms because they add to Hong Kong's sophisticated business skills, boost demand for specialized services, and contribute new bridges to social networks of capital in China; and foreign firms recognize those contributions.[47]

Because increasing numbers of red chips boost demand for services, they bolster specialization and capitalization of firms and encourage new, less-specialized and less-capitalized intermediaries to emerge. The plans of Shanghai Industrial Holdings to issue convertible bonds brought together a sophisticated consortium led by Barclays de Zoete Wedd, Hongkong and Shanghai Investment Bank Asia, and Morgan Stanley & Company. When Beijing Enterprises Holdings sold stock, the foreign firm, Morgan Stanley Asia, and the local firm, Peregrine Capital, garnered huge underwriting profits, and the foreign bank, Standard Chartered, gained large proceeds from short-term holdings of check deposits for the stock. These heightened prospects also encourage mainland Chinese banks to expand their presence in Hong Kong.

[46] Brauchli, "Hong Kong is sold on Beijing's merchant"; Chen, "Early birds"; "China Liaoning firm secures world market foothold"; Guyot, "Citic Pacific reaffirms link to China as net doubles"; Tong, "Red chips are down, but Hong Kong hopeful"; Welsh, *A borrowed place*.

[47] Glain and Stein, "Worries remain after Asian markets stage a rebound"; Guyot, "Scarlet fever"; Guyot, "Red-chip investments lose their allure"; Kahn and Guyot, "Red-chip fever hits Hong Kong market"; Stein, "China's push into Hong Kong stocks raises concerns over insider trading"; Webb, "Chinese shares show 'guanxi' appeal."

The Bank of China and its investment banking arm, China Development Finance Company (CDFC), compete with foreign banks. The inexperience of CDFC staff does not permit it to lead many large underwritings, but its parent's power confers a competitive advantage as a partner or secondary underwriter. Foreign firms claim political connections of the Bank of China bestow that clout, but that influence is also common in North America and Europe. Less-specialized, less-capitalized investment banks with red-chip pedigrees, such as Fotic Capital, Guangdong Securities, and China Everbright Securities, also compete for business. They typically serve as lower-level underwriters for major stock and bond offerings, but they sometimes participate at senior levels. When Guangdong Brewery Holdings came public, the foreign firm, Jardine Fleming, served as the lead underwriter, but Guangdong Securities, drawing upon deep contacts in the province, served as co-lead underwriter.[48] Critics dismiss these red-chip investment banks as relying on *guanxi* for their success, but they bring valuable bridges to the networks of capital within China, the same advantages that any domestic intermediary has *vis-à-vis* a foreign competitor. In an eerie replay of the nineteenth century, these red-chip banks also have much lower costs than foreign firms.

The concentration of red-chip headquarters in Hong Kong bolsters it as the information hub about those firms. Traders and analysts who follow their stocks live in Hong Kong, and their superior capacity to evaluate local red chips and Chinese firms on the mainland means that Chinese companies obtain the best prices for their stocks in Hong Kong. Intermediaries of capital based elsewhere, therefore, shift their Asian headquarters to Hong Kong to participate in its social networks of capital. National Mutual Funds Management (Asia), a unit of Axa, the French insurance giant, for example, expanded in Hong Kong, and funds from other units of the firm shifted to that office for investment. According to Ophelia Tong Tze-ming, managing director and chief financial officer, "We . . . feel the best way to play China is through Hong Kong through the red chips."[49] Paradoxically, what the skeptics claim is one of Hong Kong's greatest weaknesses, the arrival of red chips and the *guanxi* they bring, contribute to its greatest strength; they augment it as the hub of the social networks of capital that bridge China and global markets.

Inordinate focus on the *guanxi* of red chips detracts from the larger question: do their *guanxi* confer disproportionate advantages that undermine foreign and Overseas Chinese firms in Hong Kong? The deluge of red chips moving to Hong Kong suggests that heightened competition results, rather

[48] Guyot, "Bank of China arm scores in Hong Kong"; Guyot, "'Red chip' IPOs whip Hong Kong into fever pitch"; Guyot, "China firm to cut Hong Kong holding, casting shadow over other 'red chips'"; Kalathil, "Hong Kong is feeling clout of red-chip banks."

[49] French, "National Mutual sees surge in Asian assets," p. 14 (quote); Gopalan, "Analysts say red chip frenzy repeat unlikely"; Smith, "Chinese companies hope dual listings in Hong Kong will bolster their shares."

than the introduction of protected firms with cozy government relations. Besides the huge monopolistic firms of the central government, red chips include enterprises from many of the largest cities of China, including Beijing, Shanghai, Tianjin, and Guangzhou, and from innumerable provinces, counties, and small cities. Many of these firms engage in similar businesses because political fragmentation produced government firms with local monopolies. Their *guanxi* extend to their local political units, but when they expand their market areas they compete with other government firms who have *guanxi*; Hong Kong, therefore, becomes the headquarters of innumerable firms competing head-to-head in China. Red chips with pure government pedigree, including those of the People's Liberation Army, also face competition from new private firms in mainland China who move to Hong Kong, and their senior executives often claim government connections. Yet, foreign and Overseas Chinese (including Hong Kong Chinese) firms are not idle; they vigorously form joint ventures with government and private mainland firms. China International Capital Corporation is a joint-venture investment bank of the China Construction Bank and Morgan Stanley Dean Witter, and Shanghai Industrial Investment Company and Salomon Brothers formed a joint-venture asset management firm. Foreign and Overseas Chinese firms also attack *guanxi* directly; they hire mainland Chinese business people, many with foreign degrees, who bring their *guanxi* to the firms. And in reverse, red chips hire senior executives of foreign firms; Tianjin Development Holdings, a unit of the Tianjin municipal government, hired a former Merrill Lynch investment banking director as its executive deputy general manager. Hong Kong, therefore, is the battleground of a vast array of private, government, and joint-venture firms, all claiming *guanxi*, and these firms create complex networks. In that environment, *guanxi* become a euphemism for social networks of capital rather than a term that indicates special government access. If China continues to reduce government ownership of businesses, that government link of *guanxi* will gradually decline.[50]

Deeply rooted social networks of capital

Skeptics who question the future of Hong Kong dismiss too readily the deep roots of the networks centered there. Since the mid nineteenth century, Hong Kong has been the premier meeting-place of the Chinese and foreign social networks of capital in Asia, and these networks continuously elaborate, strengthen, and widen. Whereas in the nineteenth century the Hong Kong hub counted intermediary firms in the hundreds, the total in the late twentieth century surpasses 100,000 (fig. 8.15). These numbers, nevertheless, obscure the

[50] Chan, "Shanghai Industrial dismisses talk of Seapower Securities bid"; Chan, "Tianjin lures ex-Merrill director"; "CICC chairman outlines success"; Sender, "Tarnished lustre"; Smith, "Chinese regions chase Hong Kong cash"; Smith, "Municipal-run firms helped build China."

more significant point; it houses the greatest agglomeration of the most-specialized, highly capitalized firms that reach throughout the Nanyang, China, and Northeast Asia, and they bridge these areas both within Asia and to the global economy. Overseas Chinese and government and private firms from within China provide the widest and deepest penetration of its regions, and they typically house their senior offices for China in Hong Kong. Some skeptics who question the future of Hong Kong focus on the legacy of family and ethnic ties of Overseas Chinese with Guangdong and Fujian Provinces.[51] Yet, the Overseas Chinese have wider ties to mainland firms that bond firms in coastal metropolises to those in the interior, and the skeptics too readily dismiss the degree that red-chip and private mainland firms with senior offices in Hong Kong strengthen its networks.

Emigration: the false alarm

Skeptics ignored these deeply rooted networks when they argued that rising numbers of emigrants from Hong Kong after 1985 constituted a "brain drain" with drastic consequences, and they conflated Hong Kong with a national economy, rather than viewing it as a global metropolis that soon would shift from colonial outsider to being controlled by its sovereign power. The number of emigrants stayed in a band from 18,000 to 22,000 annually from 1980 to 1986, even while economic and political volatility roiled Hong Kong, including the uncertainty preceding the Joint Declaration of 1984 between China and Britain on the return of Hong Kong to Chinese control; but the skeptics dismissed this evidence. The 1997 date for return still remained a decade away; therefore, critics of China readily pointed to the ticking clock to assert that the surging number of emigrants meant that Hong Kong faced a dim future (fig. 9.1). These emigrants came disproportionately from the highly educated, upper-level white-collar groups, and this emigrant elite, China's critics argued, constituted part of the "functional core" of Hong Kong; their absence would undermine its economic and political viability.[52] Evidence for the negative impact of emigration, however, becomes less persuasive when scrutinized closely. The mounting exodus coincided with increased emigration from all Asian countries as regional economic growth and development encouraged gifted people to seek opportunities in new venues. Although emigrants from Hong Kong formed a talented group, the concept of a "functional core" presumes a static labor force. But Hong Kong is a global metropolis that offers opportunities to the "best and brightest" in the world; both immigration to

[51] Weidenbaum and Hughes claim that the spread of China's economic reforms inland diminishes the role of the "bamboo network" of Overseas Chinese; large Western, Japanese, and Korean multinationals will replace them. See Weidenbaum and Hughes, *The bamboo network*, p. 223.

[52] Skeldon, "Emigration and the future of Hong Kong"; Skeldon, "Hong Kong in an international migration system," tables 2.2–2.4, pp. 30–33.

Hong Kong and the count of people taking out employment visas swelled after 1985 (fig. 9.1). The people who acquired work visas joined the huge expatriate community employed in the trade, financial, corporate management, and business service sectors. Surveys also revealed that between one-fifth and one-third of the emigrants returned within about five years under the passports of their adopted countries. The vibrancy of Hong Kong overrode the impact of the fear of the return to China in 1997; instead of a decline in the number employed in upper-level white-collar occupations, the number jumped from 304,000 in 1988 to at least 443,000 in 1996, and may have reached 858,000 by that date.[53]

Hong Kong's expansion as the global metropolis for Asia, therefore, overwhelmed negative impacts of rising emigration after 1985; moreover, this emigration also augmented its stature. The main destinations of the *émigrés* – the United States, Canada, and Australia – were also leading choices during the nineteenth century, thus deepening the networks; and the dispersal elsewhere formed new bridges to social networks of capital. Because highly educated professional groups adroitly bridge networks, their substantial representation among recent emigrants strengthens those bridges. The transfer of capital, for example, between Hong Kong and Vancouver, British Columbia, for real estate development followed paths through those networks, and, within Asia, emigrants followed long-term trade and financial networks. Singapore's officials portray their city as a competitor to Hong Kong, but in a perverse blow, the number of emigrants from Singapore to Hong Kong exceeds the reverse flow.[54]

China supports Hong Kong

Ultimately, Hong Kong's future rests on China's commitment to maintain it as the global metropolis for Asia. A deliberate policy to undermine Hong Kong through capital controls, restrictions on expatriate workers, and military repression would result in a wholesale exodus of Overseas Chinese and foreign firms. Such an approach would signal an end to China's engagement with the global economy that it has followed since the start of Deng Xiaoping's reforms

[53] Skeldon, "Emigration and the future of Hong Kong"; Skeldon, "Emigration from Hong Kong, 1945–1994"; Wu, "Hong Kong emigrants return home." The data on employment in Hong Kong come from: Census and Statistics Department, *Hong Kong annual digest of statistics, 1989*, table 3.3, p. 32; Census and Statistics Department, *Hong Kong annual digest of statistics, 1997*, table 2.4, p. 15. The census changed the occupational classification after 1992. In 1989, the categories for upper-level white-collar workers consisted of professional, technical and related workers and administrative and managerial workers, whereas in 1996 the categories consisted of managers and administrators plus professionals (for the lower bound); associate professionals are added to this for the upper bound.

[54] Mitchell, "Flexible circulation in the Pacific Rim"; Sinn, "Emigration from Hong Kong before 1941"; Skeldon, "Migration from Hong Kong"; Skeldon, "Turning points in labor migration"; Skeldon, "Singapore as a potential destination for Hong Kong emigrants before 1997."

in 1978. That policy change and the accompanying implosion of the Chinese economy would damage every Asian country; intermediaries of capital would leave, not only Hong Kong but also each metropolis. In contrast to that dire scenario, China offers extraordinary assurances to support Hong Kong. The government of China starts with the premise that "the Hong Kong Special Administrative Region is an inalienable part of the People's Republic of China" (Basic Law, Article 1). Yet, in a gesture that is inconceivable for a sovereign power, China "authorizes the Hong Kong Special Administrative Region to exercise a high degree of autonomy" (Article 2). It explicitly excludes the socialist system from Hong Kong for fifty years (Article 5) and protects the right of private property (Article 6). And to leave no doubt about its intentions, Beijing enshrined its promise to protect Hong Kong in Article 109, which committed the local government to maintain Hong Kong as an international financial center.[55] Skeptics might claim that China could abrogate these guarantees, but the likelihood of such an action remains remote, unless some challenge to Beijing emerges from within Hong Kong. Even when the regime of Mao Zedong implemented extreme policies from 1949 to 1976, China never threatened Hong Kong's status; thus, to argue that China now will deliberately implement policies that damage Hong Kong stretches credulity. Problems of governance are possible, and should internal turmoil in China arise, Hong Kong will be affected, but its business sector will not necessarily suffer irreparable harm.[56]

China's commitment to Hong Kong, however, goes far beyond written statements. Since at least the mid 1980s, China has remained the largest holder of Hong Kong dollars, accounting for 20–30 percent of those dollar assets and deposits in the banks of Hong Kong.[57] This gives China an immense interest in Hong Kong's stability because that determines the value of the currency. Chen Yuan, deputy governor of the People's Bank of China, its central bank, repeatedly affirmed the commitment of China to maintain the Hong Kong dollar as an international currency that circulates in China as a foreign currency. The Chinese Ministry of Finance plans to issue its future debt in Hong Kong and United States dollars through the debt market in Hong Kong. Because government units in China will issue enormous, increasing volumes of debt, those moves will provide powerful undergirding to the expansion of Hong Kong as a financial center; officials of the Chinese Ministry of Finance see their debt plans as directly supporting that expansion. The Bank of China relocated the headquarters of its international unit from London to Hong

[55] *The Basic Law of the Hong Kong Special Administrative Region of the People's Republic of China.*

[56] Chung and Lo, "Beijing's relations with the Hong Kong Special Administrative Region"; Overholt, "Hong Kong's financial stability through 1997."

[57] Office of the Commissioner of Banking, *Annual report of the Commissioner of Banking for 1989*, tables 2.5A–2.5B, pp. 74–75; Hong Kong Monetary Authority, *Annual report 1997*, tables 11–12, pp. 132–33.

Kong and consolidated its international investment banking operations there; the bank explicitly stated that this demonstrated its support for Hong Kong as an international financial center.[58] The growing tendency of Chinese firms to list their stocks on the Hong Kong Stock Exchange increases the capitalization and liquidity of that market and supports greater specialization of financial intermediaries and China business analysts. As the number of those listings rises, the relative importance of local real estate development firms on the Stock Exchange declines; this transforms it into a national stock market for one of the world's largest economies, enhancing the Exchange's potential to attract listings from all over Asia.

The rush of red chips and private mainland firms to Hong Kong to establish their international, and in some cases, domestic headquarters confirms that they view the city as their gateway to the global economy, and they have invested extensive tangible capital in offices, retailing, housing, utilities, and trade infrastructure.[59] China also has a large, growing commitment of human capital to Hong Kong, including numerous daughters and sons of China's political and economic elite, who work in mainland firms and in the foreign "capitalist" bastions of trade, finance, corporate management, and producer services. They come directly from the top universities in China with bachelor's or master's degrees or doctorates, and others arrive with undergraduate degrees, MBAs, or doctorates from leading universities in the United States, Canada, the United Kingdom, and Australia.[60] The social networks of capital, therefore, not only bond mainland Chinese firms with Overseas Chinese and other foreign firms in Hong Kong, but these networks also extend directly to the corridors of power in Beijing, giving the central government input about the nuances of managing capitalist Hong Kong. China's commitment to maintain Hong Kong as its international window to global capital has momentous implications. It enhances Hong Kong's intermediaries as bridges into China and to the rest of the global economy because other intermediaries who aim to participate in the economic growth and development of China, comprising about one-fourth of the world's population, must base their key decision-makers in Hong Kong.

Asia's economic travails impact Hong Kong: falls in the value of real estate erode the capital of local property firms, tourism falls, stock prices gyrate and plunge, and the gross domestic product of the economy declines. In response

[58] "Bank of China launches investment banking arm in Hong Kong"; Yuan, "Monetary developments in China and monetary relations with Hong Kong"; Yuan, "Financial relations between Hong Kong and the Mainland"; Peng, "HK, China explore bond issue."

[59] Lau, Hawkins, and Chan, "Monetary and exchange rate management with international capital mobility," p. 109; No and Chan, "Hope for progress on other projects"; Shih, "The role of the People's Republic of China's Hong Kong-based conglomerates as agents of development"; Sito, "CITIC outbids all, wins Tamar site for $3.35b."

[60] Based on interviews with business executives in Hong Kong. Also, see Kahn, "Hong Kong draws in Chinese diaspora."

to economic fears, the government intervenes in the stock market, eroding investors' confidence in the commitment of the Hong Kong government to open markets.[61] These events have parallels in national economies such as those of the United Kingdom and the United States; nevertheless, Hong Kong is not a country, it is a global metropolis similar to London and New York. Property, retail sales, tourism, or the stock market do not fundamentally shape its outlook. Instead, Hong Kong's future is determined by the capacity of its intermediaries of commodity and financial capital to maintain dominance of exchange within Asia and between it and the global economy and to increase their capitalization and specialization as economic exchange recovers. That capacity emerges from their participation in the premier meeting-place of the foreign and Chinese social networks of capital. No other Asian metropolis has comparable networks; thus, Hong Kong is the risk management center of Asia.

Critics strenuously argued that the return of Hong Kong to China would damage it as an international business center. Yet, paradoxically, what they saw as Hong Kong's greatest weakness is its greatest strength; China's leaders repeatedly express unequivocal support for Hong Kong through words and actions.[62] China has responded to the humiliation of the Opium Wars after which Britain forcibly took Hong Kong when China was weak. It champions Hong Kong as regional metropolis for a hinterland of 400 million or more people, as international metropolis for China, and as global metropolis for Asia. That positions Hong Kong to approach the stature of London and New York sometime in the twenty-first century, and to serve as the global window for one of the world's largest economies.

[61] Guyot, "Hong Kong dollar captive of real estate"; Guyot, "Fears rise in Hong Kong over credit"; Lucas, "Hong Kong dilutes its free-market image as it grapples with financial crisis."

[62] Cheung, "CSRC chief meets with red chips"; Tong, "Red chips are down, but Hong Kong hopeful"; Wong and Lee-Young, "Hong Kong looks to Mainland for economic, political support."

Bibliography

Adas, Michael, *The Burma Delta: economic development and social change on an Asian rice frontier, 1852–1941* (Madison, WI, 1974).

Allen, G. C., and Donnithorne, Audrey G., *Western enterprise in Far Eastern economic development: China and Japan* (New York, 1954).

Allen, Robert, "Agriculture during the Industrial Revolution," in *The economic history of Britain since 1700*, vol. I: *1700–1860*, edited by Roderick Floud and Donald McCloskey (2nd edn.; Cambridge, 1994), pp. 96–122.

Asia Studies Limited of Hong Kong, "Asian executives poll," *Far Eastern Economic Review* (July 3, 1997), p. 30.

Balassa, Bela, *The newly industrializing countries in the world economy* (New York, 1981).

Banister, T. Roger, *A history of the external trade of China, 1834–1881* (Shanghai, 1931).

"Bank of China launches investment banking arm in Hong Kong," *Xinhua News Agency* (August 28, 1998).

Bard, Solomon, *Traders of Hong Kong: some foreign merchant houses, 1841–1899* (Hong Kong, 1993).

Barnathan, Joyce, "Asia's supercities: Hong Kong is still no. 1 – but rivals are nipping at its heels," *Business Week*, international edition (April 24, 1995), pp. 26–29.

Barr, Pat, "Jardines in Japan," in *The thistle and the jade*, edited by Maggie Keswick (London, 1982), pp. 153–65.

Barrett, Richard E., and Whyte, Martin King, "Dependency theory and Taiwan: analysis of a deviant case," *American Journal of Sociology*, 87 (March 1982), pp. 1064–89.

Baster, A. S. J., *The international banks* (London, 1935).

Baum, Julian, "Chafing at the bit: Taiwanese firms dismiss warnings about China," *Far Eastern Economic Review* (November 7, 1996), pp. 90–91.

Baum, Richard, "China after Deng: ten scenarios in search of reality," *China Quarterly*, no. 145 (March 1996), pp. 153–75.

Berger, Suzanne, and Lester, Richard K., eds., *Made by Hong Kong* (Hong Kong, 1997).

Bergère, Marie Claire, *The golden age of the Chinese bourgeoisie, 1911–1937* (Cambridge, 1989).

Bernhardt, Kathryn, *Rents, taxes, and peasant resistance: the lower Yangzi region, 1840–1950* (Stanford, CA, 1992).

Block, Fred, "The roles of the state in the economy," in *The handbook of economic sociology*, edited by Neil J. Smelser and Richard Swedberg (Princeton, NJ, 1994), pp. 691–710.

Brauchli, Marcus W., "China stays a magnet for overseas money: economists expect inflows, like other indicators, to slow," *Wall Street Journal* (January 14, 1997), p. A14.

"Hong Kong is sold on Beijing's merchant: success makes China Resources a stock-market darling," *Wall Street Journal* (February 26, 1997), p. A12.

"Speak no evil: why the World Bank failed to anticipate Indonesia's deep crisis," *Wall Street Journal* (July 14, 1998), pp. A1, A10.

Brauchli, Marcus W., and Biers, Dan, "Green lanterns: Asia's family empires change their tactics for a shrinking world," *Wall Street Journal* (April 19, 1995), pp. A1, A8.

Brown, E. H. Phelps, "The Hong Kong economy: achievements and prospects," in *Hong Kong: the industrial colony*, edited by Keith Hopkins (Hong Kong, 1971), pp. 1–20.

Brown, Raj, "Chinese business and banking in South-East Asia since 1870," in *Banks as multinationals*, edited by Geoffrey Jones (London, 1990), pp. 173–90.

Burt, Ronald S., *Structural holes: the social structure of competition* (Cambridge, MA, 1992).

Butcher, John, and Dick, Howard, eds., *The rise and fall of revenue farming: business elites and the emergence of the modern state in Southeast Asia* (New York, 1993).

Cain, P. J., and Hopkins, A. G., *British imperialism: innovation and expansion, 1688–1914* (London, 1993).

Census and Statistics Department, *Annual review of Hong Kong external trade*, Government of the Hong Kong Special Administrative Region (Hong Kong, various years).

Employment & vacancies statistics (detailed tables) in transport, storage & communication, financing, insurance, real estate & business services, community, social & personal services, 1984, Government of the Hong Kong Special Administrative Region (Hong Kong).

Employment & vacancies statistics (detailed tables), series A (services sectors), 1997, Government of the Hong Kong Special Administrative Region (Hong Kong).

Hong Kong annual digest of statistics, Government of the Hong Kong Special Administrative Region (Hong Kong, various years).

Hong Kong review of overseas trade, Government of the Hong Kong Special Administrative Region (Hong Kong, various years).

Chai, Joseph, "Industrial co-operation between China and Hong Kong," in *China and Hong Kong: the economic nexus*, edited by A. J. Youngson (Hong Kong, 1983), pp. 104–55.

Chan, Anthony B., *Li Ka-shing: Hong Kong's elusive billionaire* (Hong Kong, 1996).

Chan, Christine, "Shanghai Industrial dismisses talk of Seapower Securities bid," *South China Morning Post* (February 19, 1998), Business Post, p. 3.

"Tianjin lures ex-Merrill director," *South China Morning Post* (April 4, 1998), Business Post, p. 2.

Chan, Wellington K. K., *Merchants, mandarins, and modern enterprise in late Ch'ing China* (Cambridge, MA, 1977).

Chandler, Alfred D., Jr., *The visible hand: the managerial revolution in American business* (Cambridge, MA, 1977).

Chandler, Alfred D., Jr., and Redlich, Fritz, "Recent developments in American business administration and their conceptualization," *Business History Review*, 35 (Spring 1961), pp. 1–27.

Chang, Chung-li, *The income of the Chinese gentry* (Seattle, 1962).

Chang, Hsin-pao, *Commissioner Lin and the Opium War* (Cambridge, MA, 1964).

Chao, Kang, *Man and land in Chinese history: an economic analysis* (Stanford, CA, 1986).

The development of cotton textile production in China (Cambridge, MA, 1977).

Chapman, Stanley, *Merchant enterprise in Britain: from the Industrial Revolution to World War I* (Cambridge, 1992).

The rise of merchant banking (London, 1984).

Chen, Edward K. Y., "The impact of China's four modernizations on Hong Kong's economic development," in *China and Hong Kong: the economic nexus*, edited by A. J. Youngson (Hong Kong, 1983), pp. 77–103.

Chen, Edward K. Y., and Ho, Anna, "Southern China growth triangle: an overview," in *Growth triangles in Asia: a new approach to regional economic cooperation*, edited by Myo Thant, Min Tang, and Hiroshi Kakazu (Hong Kong, 1994), pp. 29–72.

Chen, Edward K. Y., and Li, Kui-wai, "Manufactured export expansion in Hong Kong and Asian-Pacific regional cooperation," in *Manufactured exports of East Asian industrializing economies: possible regional cooperation*, edited by Shu-chin Yang (Armonk, NY, 1994), pp. 103–34.

Chen, Kathy, "China fills cabinet with younger technocrats," *Wall Street Journal* (March 19, 1998), pp. A12, A14.

"China to test waters of political reform," *Wall Street Journal* (July 27, 1998), p. A11.

"Early birds: China businesses flock to Hong Kong ahead of Beijing's takeover," *Wall Street Journal* (November 29, 1996), pp. A1, A4.

Cheng, Chu-yuan, "The United States petroleum trade with China, 1876–1949," in *America's China trade in historical perspective: the Chinese and American performance*, edited by Ernest R. May and John K. Fairbank (Cambridge, MA, 1986), pp. 205–33.

Cheng, Joseph Y. S., and Lo, Sonny S. H., eds., *From colony to SAR: Hong Kong's challenges ahead* (Hong Kong, 1995).

Cheng, Siok-hwa, *The rice industry of Burma, 1852–1940* (Kuala Lumpur, 1968).

Cheong, W. E., *Mandarins and merchants: Jardine Matheson & Co., a China agency of the early nineteenth century*, Scandinavian Institute of Asian Studies, Monograph Series no. 26 (London, 1979).

Cheung, Lai-kuen, "CSRC chief meets with red chips," *South China Morning Post* (January 20, 1998), Business Post, p. 2.

"China: closer ties expected with Hong Kong," *China Business Information Network* (April 1, 1998).

"China Liaoning firm secures world market foothold," *China Daily* (March 9, 1998).

"China resumes control of Hong Kong, concluding 156 years of British rule," *New York Times* (July 1, 1997), various articles.

"China: Shenzhen to focus investment on housing for local, Hong Kong residents," *China Business Information Network* (April 7, 1998).

Chiu, Stephen W. K., and Lui, Tai-lok, "Hong Kong: unorganized industrialism," in *Asian NIEs & the global economy: industrial restructuring & corporate strategy in the 1990s*, edited by Gordon L. Clark and Won Bae Kim (Baltimore, 1995), pp. 85–112.

Chiu, Stephen, Ho, Kong-chong, and Lui, Tai-lok, "A tale of two cities rekindled: Hong Kong and Singapore's divergent paths to industrialism," *Journal of Developing Societies*, 11 (June 1995), pp. 98–122.

Chow, Peter C. Y., and Kellman, Mitchell H., *Trade – the engine of growth in East Asia* (New York, 1993).

Chung, Jae Ho, and Lo, Shiu-hing, "Beijing's relations with the Hong Kong Special Administrative Region: an inferential framework for the post-1997 arrangement," *Pacific Affairs*, 68 (Summer 1995), pp. 167–86.

"CICC chairman outlines success," *Xinhua News Agency* (April 27, 1998).

Clifford, Mark L., "The new Asian manager: First Pacific's mix of East and West may be a model for the future," *Business Week*, Asian edition (September 2, 1996), pp. 22–25.

Coates, P. D., *The China consuls: British consular officers, 1843–1943* (Hong Kong, 1988).

Coble, Parks M., Jr., *The Shanghai capitalists and the Nationalist government, 1927–1937* (2nd edn.; Cambridge, MA, 1986).

Cochran, Sherman, "Commercial penetration and economic imperialism in China: an American cigarette company's entrance into the market," in *America's China trade in historical perspective: the Chinese and American performance*, edited by Ernest R. May and John K. Fairbank (Cambridge, MA, 1986), pp. 151–203.

Coclanis, Peter A., "Distant thunder: the creation of a world market in rice and the transformations it wrought," *American Historical Review*, 98 (October 1993), pp. 1050–78.

Cohen, R. B., "The new international division of labor, multinational corporations and urban hierarchy," in *Urbanization and urban planning in capitalist society*, edited by Michael Dear and Allen J. Scott (London, 1981), pp. 287–315.

Cohen, Warren I., and Zhao, Li, eds., *Hong Kong under Chinese rule: the economic and political implications of reversion* (Cambridge, 1997).

Coleman, James S., *Foundations of social theory* (Cambridge, MA, 1990).

Cook, Karen S., Emerson, Richard M., Gillmore, Mary R., and Yamagishi, Toshio, "The distribution of power in exchange networks: theory and experimental results," *American Journal of Sociology*, 89 (September 1983), pp. 275–305.

Corbridge, Stuart, Thrift, Nigel, and Martin, Ron, eds., *Money, power and space* (Oxford, 1994).

Cuthbert, Alexander R., "A fistful of dollars: legitimation, production and debate in Hong Kong," *International Journal of Urban and Regional Research*, 15 (June 1991), pp. 234–49.

Dalzell, Robert F., Jr., *Enterprising elite: the Boston Associates and the world they made* (Cambridge, MA, 1987).

Danhof, Clarence H., *Change in agriculture: the northern United States, 1820–1870* (Cambridge, MA, 1969).

Daniels, P. W., *Service industries in the world economy* (Oxford, 1993).

Daniels, Peter W., and Moulaert, Frank, eds., *The changing geography of advanced producer services* (London, 1991).

Dean, Britten, *China and Great Britain: the diplomacy of commercial relations, 1860–1864* (Cambridge, MA, 1974).

Dixon, Chris, *South East Asia in the world-economy* (Cambridge, 1991).

Downs, Jacques M., "American merchants and the China opium trade, 1800–1840," *Business History Review*, 42 (Winter 1968), pp. 418–42.

Durkheim, Emile, *The division of labor in society*, translated by George Simpson (New York, 1964).

East Asia Analytical Unit, *Overseas Chinese business networks in Asia*, Department of Foreign Affairs and Trade, Commonwealth of Australia (Parkes, ACT, 1995).

Eiichi, Montono, "'The traffic revolution': remaking the export sales system in China, 1866–1875," *Modern China*, 12 (January 1986), pp. 75–102.

Elson, Robert E., "International commerce, the state and society: economic and social change," in *The Cambridge history of Southeast Asia*, vol. II: *The nineteenth and twentieth centuries*, edited by Nicholas Tarling (Cambridge, 1992), pp. 131–95.

Elvin, Mark, "The high-level equilibrium trap: the causes of the decline of invention in the traditional Chinese textile industries," in *Economic organization in Chinese society*, edited by W. E. Willmott (Stanford, CA, 1972), pp. 137–72.

Emerson, Richard M., "Power-dependence relations," *American Sociological Review*, 27 (February 1962), pp. 31–41.

Endacott, G. B., *A history of Hong Kong* (rev. edn.; Hong Kong, 1973).

Enright, Michael J., Scott, Edith E., and Dodwell, David, *The Hong Kong advantage* (Hong Kong, 1997).

Erickson, Bonnie H., "Culture, class, and connections," *American Journal of Sociology*, 102 (July 1996), pp. 217–51.

Fairbank, John K., *China: a new history* (Cambridge, MA, 1992).

"The creation of the treaty system," in *The Cambridge history of China*, vol. X, *Late Ch'ing, 1800–1911, part I*, edited by John K. Fairbank (Cambridge, 1978), pp. 213–63.

Trade and diplomacy on the China Coast (First published, 1953; Stanford, CA, 1969).

Fairbank, John K., Reischauer, Edwin O., and Craig, Albert M., *East Asia: tradition and transformation* (Boston, 1973).

Farnie, D. A., *East and west of Suez: the Suez Canal in history, 1854–1956* (Oxford, 1969).

Faure, David, "The rice trade in Hong Kong before the Second World War," in *Between East and West: aspects of social and political development in Hong Kong*, edited by Elizabeth Sinn (Hong Kong, 1990), pp. 216–25.

Feldwick, W., ed., *Present day impressions of the Far East and prominent & progressive Chinese at home and abroad* (London, 1917).

Feuerwerker, Albert, *China's early industrialization: Sheng Hsuan-huai (1844–1916) and mandarin enterprise* (Cambridge, MA, 1958).

Fforde, Adam, "The political economy of 'reform' in Vietnam – some reflections," in

The challenge of reform in Indochina, edited by Borje Ljunggren (Cambridge, MA, 1993), pp. 293–325.

Fletcher, Max E., "The Suez Canal and world shipping, 1869–1914," *Journal of Economic History*, 18 (December 1958), pp. 556–73.

Floud, Roderick, and McCloskey, Donald, eds., *The economic history of Britain since 1700*, vol. I: *1700–1860* (2nd edn.; Cambridge, 1994).

Fong, Pang Eng, "Staying global and going regional: Singapore's inward and outward direct investments," in *The new wave of foreign direct investment in Asia*, Nomura Research Institute and the Institute of Southeast Asian Studies (Singapore, 1995), pp. 111–29.

Forney, Matt, "Private party: Jiang Zemin consolidates his power and vows to overhaul the state sector," *Far Eastern Economic Review* (October 2, 1997), pp. 18–21.

Freeman, Linton C., "Centrality in social networks: conceptual clarification," *Social Networks*, 1 (1978/79), pp. 215–39.

French, Sara, "National Mutual sees surge in Asian assets: Hong Kong manager puts red chips at core of regional strategy," *South China Morning Post* (March 23, 1997), Money section, p. 14.

Freris, Andrew F., *The financial markets of Hong Kong* (London, 1991).

Fukuyama, Francis, *Trust: the social virtues and the creation of prosperity* (New York, 1995).

Fung, Victor K., "Evolution in the management of family enterprises in Asia," in *Dynamic Hong Kong: business & culture*, edited by Wang Gungwu and Wong Siu-lun (Hong Kong, 1997), pp. 216–29.

Furber, Holden, *John Company at work: a study of European expansion in India in the late eighteenth century* (Cambridge, MA, 1951).

Rival empires of trade in the Orient, 1600–1800 (Minneapolis, 1976).

Gardella, Robert P., "The boom years of the Fukien tea trade, 1842–1888," in *America's China trade in historical perspective: the Chinese and American performance*, edited by Ernest R. May and John K. Fairbank (Cambridge, MA, 1986), pp. 33–75.

Gereffi, Gary, and Wyman, Donald L., eds., *Manufacturing miracles: paths of industrialization in Latin America and East Asia* (Princeton, NJ, 1990).

Gilley, Bruce, "Darkness dawns," *Far Eastern Economic Review* (April 11, 1996), pp. 14–15.

"Man on the spot," *Far Eastern Economic Review* (December 19, 1996), pp. 14–16.

Glain, Steve, and Stein, Peter, "Worries remain after Asian markets stage a rebound," *Wall Street Journal* (October 30, 1997), p. A10.

Goad, G. Pierre, "Slew of dismal economic news in Asia positions more countries for recession," *Asian Wall Street Journal* (June 1, 1998), pp. 1, 4.

Godley, Michael R., *The mandarin-capitalists from Nanyang: overseas Chinese enterprise in the modernization of China, 1893–1911* (Cambridge, 1981).

Goldstone, Jack A., "The coming Chinese collapse," *Foreign Policy*, no. 99 (Summer 1995), pp. 35–52.

"Good scientific cooperation between Guangzhou, Hong Kong," *Xinhua News Agency* (January 12, 1998).

Gopalan, Nisha, "Analysts say red chip frenzy repeat unlikely," *South China Morning Post* (February 8, 1998), Sunday Money, p. 2.

Graham, Gerald S., *The China station: war and diplomacy, 1830–1860* (Oxford, 1978).

Graham, S., "Cities in the real-time age: the paradigm challenge of telecommunications to the conception and planning of urban space," *Environment and Planning A*, 29 (January 1997), pp. 105–27.

Granovetter, Mark, "Economic action and social structure: the problem of embeddedness," *American Journal of Sociology*, 91 (November 1985), pp. 481–510.

"The strength of weak ties," *American Journal of Sociology*, 78 (May 1973), pp. 1360–80.

Gras, N. S. B., *An introduction to economic history* (New York, 1922).

Greenberg, Michael, *British trade and the opening of China, 1800–1842* (Cambridge, 1951).

Griffin, Eldon, *Clippers and consuls: American consular and commercial relations with eastern Asia, 1845–1860* (Ann Arbor, MI, 1938).

"Guangdong: the factory at the back," *China Economic Review* (November 1997), p. 19.

"Guanxi maximizing," *Wall Street Journal* (January 31, 1997), p. A18.

Guoli, "Strategic center for China operations established by BASF," *China Chemical Reporter* (July 13, 1998), p. 7.

Guyot, Erik, "Bank of China arm scores in Hong Kong," *Wall Street Journal* (May 5, 1997), p. B7D.

"China firm to cut Hong Kong holding, casting shadow over other 'red chips,'" *Wall Street Journal* (June 3, 1997), p. A18.

"Citic Pacific reaffirms link to China as net doubles," *Asian Wall Street Journal* (March 27, 1997), pp. 1, 24.

"Fears rise in Hong Kong over credit," *Wall Street Journal* (August 14, 1998), p. A13.

"Hong Kong dollar captive of real estate," *Wall Street Journal* (August 12, 1998), p. A11.

"HSBC's growth is fettered by Asian turmoil," *Wall Street Journal* (November 10, 1997), p. A19.

"Red-chip investments lose their allure," *Wall Street Journal* (September 26, 1997), p. A18.

"'Red chip' IPOs whip Hong Kong into fever pitch," *Wall Street Journal* (May 28, 1997), p. A12.

"Scarlet fever: Hong Kong delirium grows for China-backed shares," *Asian Wall Street Journal* (March 21–22, 1997), pp. 1, 6.

Haig, Robert Murray, *Regional survey of New York and its environs*, vol. I: *Major economic factors in metropolitan growth and arrangement* (New York, 1927).

Hamilton, David P., "Japan's new premier launches image offensive as economic problems pile up almost daily," *Wall Street Journal* (August 3, 1998), pp. A10–11.

Hamilton, Gary, ed., *Business networks and economic development in East and Southeast Asia* (Hong Kong, 1991).

Hao, Yen-p'ing, "Chinese teas to America – a synopsis," in *America's China trade in historical perspective: the Chinese and American performance*, edited by Ernest R. May and John K. Fairbank (Cambridge, MA, 1986), pp. 11–31.

The commercial revolution in nineteenth-century China (Berkeley, CA, 1986).

The comprador in nineteenth-century China: bridge between East and West (Cambridge, MA, 1970).

Harding, James, "Arriving in a trickle," *Financial Times* (May 19, 1998), p. 3.
Harley, Charles K., "The shift from sailing ships to steamships, 1850–1890: a study in technological change and its diffusion," in *Essays on a mature economy: Britain after 1840*, edited by Donald N. McCloskey (Princeton, NJ, 1971), pp. 215–34.
Henderson, Jeffrey, *The globalisation of high technology production: society, space and semiconductors in the restructuring of the modern world* (London, 1989).
Hewison, Kevin, *Bankers and bureaucrats: capital and the role of the state in Thailand* (New Haven, CT, 1989).
Hiebert, Murray, "Luring new wealth," *Far Eastern Economic Review* (May 2, 1996), pp. 46–48.
"Singapore: in search of a boost," *Far Eastern Economic Review* (September 25, 1997), pp. 106–08.
Hinton, Harold C., *The grain tribute system of China, 1845–1911* (Cambridge, MA, 1970).
"HK Bank open its door in Beijing," *South China Morning Post* (August 4, 1995), Business Post, p. 1.
Ho, Ping-ti, *Studies on the population of China, 1368–1953* (Cambridge, MA, 1959).
Ho, Samuel P. S., "Decentralized industrialization and rural development: evidence from Taiwan," *Economic Development and Cultural Change*, 28 (October 1979), pp. 77–96.
Ho, Yan Ki, "The money market and the foreign exchange market," in *Hong Kong's financial institutions and markets*, edited by Robert Haney Scott, K. A. Wong, and Yan Ki Ho (Hong Kong, 1986), pp. 79–103.
Ho, Yin-ping, *Trade, industrial restructuring and development in Hong Kong* (Honolulu, HI, 1992).
Hodder, Rupert, *Merchant princes of the East: cultural delusions, economic success and the Overseas Chinese in Southeast Asia* (Chichester, 1996).
Hong Kong Government Information Services, *Hong Kong: annual report* (Hong Kong, various years).
"Hong Kong – international shipping center," *Xinhua News Agency* (June 30, 1998).
"Hong Kong, Macao, Guangdong explore ways to develop tourism," *Xinhua News Agency* (July 14, 1998).
Hong Kong Monetary Authority, *Annual report* (Hong Kong, various years).
Money and banking in Hong Kong (Hong Kong, 1995).
Monthly statistical bulletin, no. 45 (May 1998).
"Hong Kong service sector attracted to southern Chinese province," *Xinhua News Agency* (March 4, 1998).
"Hong Kong–Shenzhen express train services to open," *Xinhua News Agency* (January 14, 1998).
"Hong Kong should move into hi-tech: scholar," *Xinhua News Agency* (June 30, 1998).
Hopkins, Keith, ed., *Hong Kong: the industrial colony* (Hong Kong, 1971).
Hsiao, Liang-lin, *China's foreign trade statistics, 1864–1949* (Cambridge, MA, 1974).
Huang, Philip C. C., *The peasant family and rural development in the Yangzi Delta, 1350–1988* (Stanford, CA, 1990).
Huang, Yasheng, "Why China will not collapse," *Foreign Policy*, no. 99 (Summer 1995), pp. 54–68.

Huff, W. G., *The economic growth of Singapore: trade and development in the twentieth century* (Cambridge, 1994).

"Huge investment put by Kodak in business expansion in China," *China Chemical Reporter* (June 26, 1998), p. 5.

Hyde, Francis E., *Blue Funnel: a history of Alfred Holt and Company of Liverpool from 1865 to 1914* (Liverpool, 1956).

Hymer, Stephen, "The multinational corporation and the law of uneven development," in *Economics and world order: from the 1970's to the 1990's*, edited by Jagdish N. Bhagwati (New York, 1971), pp. 113–40.

Immigration Department, Government of the Hong Kong Special Administrative Region, unpublished tabulations.

Industry Department, *Hong Kong's manufacturing industries*, Government of the Hong Kong Special Administrative Region (Hong Kong, various years).

Survey of external investment in Hong Kong's manufacturing industries, Government of the Hong Kong Special Administrative Region (Hong Kong, various years).

Survey of regional representation by overseas companies in Hong Kong, Government of the Hong Kong Special Administrative Region (Hong Kong, various years).

Ingram, James C., *Economic change in Thailand, 1850–1970* (2nd edn.; Stanford, CA, 1971).

"Thailand's rice trade and the allocation of resources," in *The economic development of South-East Asia: studies in economic history and political economy*, edited by C. D. Cowan (London, 1964), pp. 102–26.

Inspectorate General of Customs, *China: Imperial Maritime Customs*, Statistical Series No. 6, Decennial Reports, 1882–91 (Shanghai, 1893).

Jackson, Stanley, *The Sassoons* (New York, 1968).

Janelle, Donald G., "Central place development in a time–space framework," *Professional Geographer*, 20 (January 1968), pp. 5–10.

"Spatial reorganization: a model and concept," *Annals of the Association of American Geographers*, 59 (June 1969), pp. 348–64.

Jao, Y. C., *Banking and currency in Hong Kong: a study of postwar financial development* (London, 1974).

"Financing Hong Kong's early postwar industrialization: the role of the Hongkong and Shanghai Banking Corporation," in *Eastern banking: essays in the history of the Hongkong and Shanghai Banking Corporation*, edited by Frank H. H. King (London, 1983), pp. 545–74.

"Hong Kong's role in financing China's modernization," in *China and Hong Kong: the economic nexus*, edited by A. J. Youngson (Hong Kong, 1983), pp. 12–76.

"The development of Hong Kong's financial sector 1967–92," in *25 years of social and economic development in Hong Kong*, edited by Benjamin K. P. Leung and Teresa Y. C. Wong (Hong Kong, 1994), pp. 560–601.

"The rise of Hong Kong as a financial center," *Asian Survey*, 19 (July 1979), pp. 674–94.

Jesudason, James V., *Ethnicity and the economy: the state, Chinese business, and multinationals in Malaysia* (Singapore, 1989).

Johnson, Arthur M., and Supple, Barry E., *Boston capitalists and Western railroads: a study in the nineteenth-century railroad investment process* (Cambridge, MA, 1967).

Johnson, Graham E., "Continuity and transformation in the Pearl River Delta: Hong Kong's impact on its hinterland," in *The Hong Kong–Guangdong link: partnership in flux*, edited by Reginald Yin-wang Kwok and Alvin Y. So (Hong Kong, 1995), pp. 64–86.

Johnson, Linda Cooke, *Shanghai: from market town to treaty port, 1074–1858* (Stanford, CA, 1995).

Jones, F. C., *Shanghai and Tientsin* (San Francisco, 1940).

Jones, Geoffrey, *British multinational banking, 1830–1990* (Oxford, 1993).

Jones, Stephanie, *Two centuries of overseas trading: the origins and growth of the Inchcape Group* (London, 1986).

Jones, Susan Mann, "Finance in Ningpo: the 'ch'ien chuang,' 1750–1880," in *Economic organization in Chinese society*, edited by W. E. Willmott (Stanford, CA, 1972), pp. 47–77.

"The Ningpo pang and financial power at Shanghai," in *The Chinese city between two worlds*, edited by Mark Elvin and G. William Skinner (Stanford, CA, 1974), pp. 73–96.

Jones, Susan Mann, and Kuhn, Philip A., "Dynastic decline and the roots of rebellion," in *The Cambridge history of China*, vol. X: *Late Ch'ing, 1800–1911, part I*, edited by John K. Fairbank (Cambridge, 1978), pp. 107–62.

Kahn, Joseph, "A tale of 3 cities: Beijing warily pushes Shanghai to resume role in global finance," *Wall Street Journal* (November 25, 1994), p. A1.

"Hong Kong draws in Chinese diaspora: as changeover nears, returnees find best of both worlds," *Wall Street Journal* (May 1, 1997), p. A12.

Kahn, Joseph, and Guyot, Erik, "Red-chip fever hits Hong Kong market," *Wall Street Journal* (May 21, 1997), p. A10.

Kalathil, Shanthi, "Hong Kong is feeling clout of red-chip banks," *Wall Street Journal* (September 5, 1997), p. B7B.

"Swire Pacific Ltd. goes courting in China: Hong Kong giant aggressively pursues new deals," *Wall Street Journal* (October 22, 1996), p. A18.

Kao, John, "The worldwide web of Chinese business," *Harvard Business Review* (March–April 1993), pp. 24–36.

Keswick, Maggie, ed., *The thistle and the jade* (London, 1982).

King, Frank H. H., *The Hongkong Bank in Late Imperial China, 1864–1902: on an even keel* (*The history of the Hongkong and Shanghai Banking Corporation*, vol. I) (Cambridge, 1987).

The Hongkong Bank in the period of imperialism and war, 1895–1918: wayfoong, the focus of wealth (*The history of the Hongkong and Shanghai Banking Corporation*, vol. II) (Cambridge, 1988).

The Hongkong Bank between the wars and the bank interned, 1919–1945: return from grandeur (*The history of the Hongkong and Shanghai Banking Corporation*, vol. III) (Cambridge, 1988).

The Hongkong Bank in the period of development and nationalism, 1941–1984: from regional bank to multinational group (*The history of the Hongkong and Shanghai Banking Corporation*, vol. IV) (Cambridge, 1991).

Knox, Thomas, "John Comprador," *Harper's New Monthly Magazine*, 57 (August 1878), pp. 427–34.

Kohli, Sheel, "Foreign banks slash exposure to Hong Kong," *South China Morning Post* (August 31, 1998), Business Post, p. 1.

Kraar, Louis, "The death of Hong Kong," *Fortune*, 131, no. 12 (June 26, 1995), pp. 118–38.

Krause, Lawrence B., "Hong Kong and Singapore: twins or kissing cousins?" *Economic Development and Cultural Change*, 36, no. 3, Supplement (April 1988), pp. S45–S66.

Kuhn, Philip A., *Rebellion and its enemies in Late Imperial China: militarization and social structure, 1796–1864* (Cambridge, MA, 1970).

"The Taiping rebellion," in *The Cambridge history of China*, vol. X: *Late Ch'ing, 1800–1911, part I*, edited by John K. Fairbank (Cambridge, 1978), pp. 264–317.

Kunio, Yoshihara, *The rise of ersatz capitalism in South-East Asia* (Singapore, 1988).

Kwan, Chan Wai, *The making of Hong Kong society: three studies of class formation in early Hong Kong* (Oxford, 1991).

Kwok, Reginald Yin-wang, and So, Alvin Y., eds., *The Hong Kong–Guangdong link: partnership in flux* (Hong Kong, 1995).

Landa, Janet T., "The political economy of the ethnically homogeneous Chinese middleman group in Southeast Asia: ethnicity and entrepreneurship in a plural society," in *The Chinese in Southeast Asia*, edited by Linda Y. C. Lim and L. A. Peter Gosling (2 vols.), vol. I: *Ethnicity and economic activity* (Singapore, 1983), pp. 86–116.

Landers, Peter, "Men and boys: big bang separates strong from the weak in Japan," *Far Eastern Economic Review* (September 25, 1997), p. 153.

Landes, David S., *The Unbound Prometheus: technological change and industrial development in Western Europe from 1750 to the present* (Cambridge, 1969).

Lane, Frederic C., *Profits from power* (Albany, NY, 1979).

Langdale, John V., "The geography of international business telecommunications: the role of leased networks," *Annals of the Association of American Geographers*, 79 (December 1989), pp. 501–22.

Latham, A. J. H., *The international economy and the undeveloped world, 1865–1914* (London, 1978).

Latham, A. J. H., and Neal, Larry, "The international market in rice and wheat, 1868–1914," *Economic History Review*, 2nd series, 36 (May 1983), pp. 260–80.

Lau, James H., Jr., Hawkins, John, and Chan, Benjamin, "Monetary and exchange rate management with international capital mobility: the case of Hong Kong," in *Monetary and exchange rate management with international capital mobility* (Hong Kong, 1994), pp. 105–18.

Le Fevour, Edward, *Western enterprise in late Ch'ing China: a selective survey of Jardine, Matheson & Company's operations, 1842–1895* (Cambridge, MA, 1970).

Lee, Pui-tak, "Chinese merchants in the Hong Kong colonial context, 1840–1910," in *Hong Kong economy and society: challenges in the new era*, edited by Wong Siu-lun and Toyojiro Maruya (Hong Kong, 1998), pp. 61–86.

Lee, Roger, and Schmidt-Marwede, Ulrich, "Interurban competition? Financial centres and the geography of financial production," *International Journal of Urban and Regional Research*, 17 (December 1993), pp. 492–515.

Lee, Sheng-yi, *The monetary and banking development of Malaysia and Singapore* (Singapore, 1974).

Leeming, Frank, "The earlier industrialization of Hong Kong," *Modern Asian Studies*, 9 (July 1975), pp. 337–42.

Leung, Chi Kin, "Locational characteristics of foreign equity joint venture investment in China, 1979–1985," *Professional Geographer*, 42 (November 1990), pp. 403–21.

"Personal contacts, subcontracting linkages, and development in the Hong Kong–Zhujiang Delta region," *Annals of the Association of American Geographers*, 83 (June 1993), pp. 272–302.

Li, Lillian M., *China's silk trade: traditional industry in the modern world, 1842–1937* (Cambridge, MA, 1981).

"The silk export trade and economic modernization in China and Japan," in *America's China trade in historical perspective: the Chinese and American performance*, edited by Ernest R. May and John K. Fairbank (Cambridge, MA, 1986), pp. 77–99.

Lieberthal, Kenneth, *Governing China: from revolution through reform* (New York, 1995).

Lieu, D. K., *The growth and industrialization of Shanghai* (Shanghai, 1936).

Lim, Chee Peng, Nooi, Phang Siew, and Boh, Margaret, "The history and development of the Hongkong and Shanghai Banking Corporation in peninsular Malaysia," in *Eastern banking: essays in the history of the Hongkong and Shanghai Banking Corporation*, edited by Frank H. H. King (London, 1983), pp. 350–91.

Linge, G. J. R., and Forbes, D. K., eds., *China's spatial economy: recent developments and reforms* (Hong Kong, 1990).

Liu, Kwang-ching, *Anglo-American steamship rivalry in China, 1862–1874* (Cambridge, MA, 1962).

Lockwood, Stephen Chapman, *Augustine Heard and Company, 1858–1862: American merchants in China* (Cambridge, MA, 1971).

Lorsch, Jay W., and Allen, Stephen A., III, *Managing diversity and interdependence: an organizational study of multidivisional firms* (Boston, 1973).

Lovelock, Peter, "Information highways and the trade in telecommunications services," *Telecommunications in Asia: policy, planning and development*, edited by John Ure (Hong Kong, 1995), pp. 111–46.

Lubbock, Basil, *The opium clippers* (Glasgow, 1933).

Lucas, Louise, "Hong Kong dilutes its free-market image as it grapples with financial crisis," *Financial Times* (September 1, 1998), p. 7.

Ma, Debin, "The modern silk road: a tale of the integration of the global market for raw silk, 1850–1930," (Unpublished paper, Department of Economics, University of North Carolina, Chapel Hill, 1995), 48 pp.

"The modern silk road: the global raw-silk market, 1850–1930," *Journal of Economic History*, 56 (June 1996), pp. 330–55.

McDermott, Darren, "U.S. banks say loan exposure in Asia is limited," *Wall Street Journal* (November 7, 1997), p. A15.

McElderry, Andrea Lee, *Shanghai old-style banks (ch'ien-chuang), 1800–1935: a traditional institution in a changing society* (Ann Arbor, MI, 1976).

McGurn, William, "Money talks," *Far Eastern Economic Review* (March 21, 1996), pp. 40–42.

MacKenzie, Compton, *Realms of silver: one hundred years of banking in the East* (London, 1954).

MacPherson, Kerrie L., and Yearley, Clifton K., "The 2½% margin: Britain's Shanghai traders and China's resilience in the face of commercial penetration," *Journal of Oriental Studies*, 25, no. 2 (1987), pp. 202–34.

Mann, Susan, *Local merchants and the Chinese bureaucracy, 1750–1950* (Stanford, CA, 1987).

Marriner, Sheila, *Rathbones of Liverpool, 1845–73* (Liverpool, 1961).

Marriner, Sheila, and Hyde, Francis E., *The senior John Samuel Swire, 1825–98: management in Far Eastern shipping trades* (Liverpool, 1967).

Marx, Karl, *Capital*, 3 vols. (New York, 1967).

Meyer, David R., "A dynamic model of the integration of frontier urban places into the United States system of cities," *Economic Geography*, 56 (April 1980), pp. 120–40.

"Change in the world system of metropolises: the role of business intermediaries," *Urban Geography*, 12 (September–October 1991), pp. 393–416.

"Emergence of the American manufacturing belt: an interpretation," *Journal of Historical Geography*, 9 (April 1983), pp. 145–74.

"Expert managers of uncertainty: intermediaries of capital in Hong Kong," *Cities*, 14, no. 5 (1997), pp. 257–63.

"The division of labor and the market areas of manufacturing firms," *Sociological Forum*, 3 (Summer 1988), pp. 433–53.

"The formation of a global financial center: London and its intermediaries," in *Cities in the world-system*, edited by Resat Kasaba (Westport, CT, 1991), pp. 97–106.

"The rise of the industrial metropolis: the myth and the reality," *Social Forces*, 68 (March 1990), pp. 731–52.

"The world system of cities: relations between international financial metropolises and South American cities," *Social Forces*, 64 (March 1986), pp. 553–81.

Michael, Franz, *The Taiping rebellion: history and documents* (3 vols.), vol. I: *History* (Seattle, 1966).

Miller, Matthew, "U.S. insurance giant in line as Mainland set to open prime market," *South China Morning Post* (May 4, 1998), Business Post, p. 1.

Miller, Matthew, and Chan, Peter, "Foreign pioneers hope for busy life in quiet Pudong," *South China Morning Post* (May 14, 1998), China Business Review, p. 8.

Mitchell, Katharyne, "Flexible circulation in the Pacific Rim: capitalisms in cultural context," *Economic Geography*, 71 (October 1995), pp. 364–82.

Montgomery, James D., "Weak ties, employment, and inequality: an equilibrium," *American Journal of Sociology*, 99 (March 1994), pp. 1212–36.

Morse, Hosea Ballou, *The chronicles of the East India Company trading to China, 1635–1834* (5 vols., Oxford, 1926–29).

The international relations of the Chinese empire (3 vols., Shanghai, 1910).

Moulder, Frances V., *Japan, China, and the modern world economy: toward a reinterpretation of East Asian development ca. 1600 to ca. 1918* (Cambridge, 1977).

Mui, Hoh-cheung, and Mui, Lorna H., *The management of monopoly: a study of the English East India Company's conduct of its tea trade, 1784–1833* (Vancouver, 1984).

Murphey, Rhoads, *Shanghai: key to modern China* (Cambridge, MA, 1953).
The outsiders: the Western experience in India and China (Ann Arbor, MI, 1977).
Murray, Martin J., *The development of capitalism in colonial Indochina (1870–1940)* (Berkeley, CA, 1980).
Naquin, Susan, and Rawski, Evelyn S., *Chinese society in the eighteenth century* (New Haven, CT, 1987).
Neal, Larry, *The rise of financial capitalism: international capital markets in the Age of Reason* (Cambridge, 1990).
Nicholls, William H., "The place of agriculture in economic development," in *Agriculture and economic development*, edited by Carl Eicher and Lawrence Witt (New York, 1964), pp. 11–44.
Nicolle, Anthony, "Hong Kong in a regional perspective," in *Monetary management in Hong Kong* (Hong Kong, 1993), pp. 81–89.
No, Kwai-yan and Chan, Quinton, "Hope for progress on other projects," *South China Morning Post* (May 31, 1996), p. 6.
Nontapunthawat, Nimit, "Hong Kong as a financial centre in the Asian Pacific region in the 21st century," in *Hong Kong's role in the Asian Pacific region in the 21st century*, Proceedings and Papers of the ASEAN-China Hong Kong Forum 1992 (Hong Kong, 1993), pp. 39–52.
Nyaw, Mee-kau, "The experiences of industrial growth in Hong Kong and Singapore: a comparative study," in *Industrial and trade development in Hong Kong*, edited by Edward K. Y. Chen, Mee-kau Nyaw, and Teresa Y. C. Wong (Hong Kong, 1991), pp. 185–222.
Office of the Commissioner of Banking, *Annual report of the Commissioner of Banking for 1989* (Hong Kong, 1989).
Osborne, Milton E., *The French presence in Cochinchina and Cambodia: rule and response (1859–1905)* (Ithaca, NY, 1969).
Oshima, Harry T., *Economic growth in Monsoon Asia: a comparative survey* (Tokyo, 1987).
"The transition from an agricultural to an industrial economy in East Asia," *Economic Development and Cultural Change*, 34 (July 1986), pp. 783–809.
Overholt, William H., "China after Deng," *Foreign Affairs*, 75 (May/June 1996), pp. 63–78.
"Hong Kong's financial stability through 1997," *Bankers Trust Research, Hong Kong* (March 20, 1996).
"Indonesia after Suharto," *Emerging markets research: Asia*, BankBoston, Singapore, May 25, 1998.
"Korea: to the market via socialism," *Emerging markets research: Asia*, BankBoston, Singapore, July 21, 1998.
"Thailand: the next phase of crisis," *Emerging markets research: Asia*, BankBoston, Singapore, June 4, 1998.
The rise of China: how economic reform is creating a new superpower (New York, 1993).
"Overseas investment soars in Shenzhen Economic Zone," *Xinhua News Agency* (June 19, 1998).
Owen, David E., *British opium policy in China and India* (New Haven, CT, 1934).
Owen, Norman G., "Economic and social change," in *The Cambridge history of*

Southeast Asia, vol. II: *The nineteenth and twentieth centuries* (Cambridge, 1992), pp. 467–527.

Peng, Foo Choy, "HK, China explore bond issue," *South China Morning Post* (June 1, 1996), Business Post, p. 1.

"Shanghai briefing: territory holds high ground as upstart market battles Beijing," *South China Morning Post* (June 7, 1996), Business Post, p. 4.

Perkins, Dwight H., *Agricultural development in China, 1368–1968* (Chicago, 1969).

Pesek, William, Jr., "Dis-oriented markets: a year into the Asian crisis, no light at the end of the tunnel," *Barron's* (July 27, 1998), pp. 17–22.

Pitkin, Timothy, *A statistical view of the commerce of the United States of America* (New Haven, CT, 1835).

Pitman, Joanna, "Merrill Lynch heads stampede into battle with Tokyo's titans," *The Times* (April 15, 1998).

Polachek, James M., *The inner Opium War* (Cambridge, MA, 1992).

Pomeranz, Kenneth, *The making of a hinterland: state, society, and economy in inland North China, 1853–1937* (Berkeley, CA, 1993).

Porter, Barry, "Monetary body strives for spot on world stage," *South China Morning Post* (June 23, 1998), Business Post, p. 8.

Porter, Glenn, and Livesay, Harold C., *Merchants and manufacturers: studies in the changing structure of nineteenth-century marketing* (Baltimore, 1971).

Pratt, John W., and Zeckhauser, Richard J., "Principals and agents: an overview," in *Principals and agents: the structure of business*, edited by John W. Pratt and Richard J. Zeckhauser (Boston, 1985), pp. 1–35.

Pred, Allan, *City-systems in advanced economies* (New York, 1977).

The spatial dynamics of U.S. urban-industrial growth, 1800–1914 (Cambridge, MA, 1966).

Pritchard, Earl H., *The crucial years of early Anglo-Chinese relations, 1750–1800* (first published, 1936; New York, 1970).

Purcell, Victor, *The Chinese in Southeast Asia* (2nd edn.; London, 1965).

Putterman, Louis, *Continuity and change in China's rural development: collective and reform eras in perspective* (New York, 1993).

Rawski, Thomas G., "Chinese dominance of treaty port commerce and its implications, 1860–1875," *Explorations in Economic History*, 7 (1969–70), pp. 451–73.

Economic growth in prewar China (Berkeley and Los Angeles, CA, 1989).

Redding, S. Gordon, "Societal transformation and the contribution of authority relations and cooperation norms in overseas Chinese business," in *Confucian traditions in East Asian modernity: moral education and economic culture in Japan and the four mini-dragons*, edited by Tu Wei-ming (Cambridge, MA, 1996), pp. 310–27.

The spirit of Chinese capitalism (Berlin, 1990).

Reed, Howard Curtis, *The preeminence of international financial centers* (New York, 1981).

Regnier, Philippe, *Singapore: city-state in South-East Asia* (first published, 1987; translated and updated by Christopher Hurst, Honolulu, HI, 1991).

Reid, Alan, "The steel frame," in *The thistle and the jade*, edited by Maggie Keswick (London, 1982), pp. 11–54.

Remer, C. F., *The foreign trade of China* (Shanghai, 1926).

Reynolds, Lloyd G., "Agriculture in development theory: an overview," in *Agriculture in development theory*, edited by Lloyd G. Reynolds (New Haven, CT, 1975), pp. 1–24.

Richburg, Keith B., "Shanghai or Hong Kong: which will be China's financial king?" *International Herald Tribune* (August 26–27, 1995), p. 1.

Riedel, James, *The industrialization of Hong Kong* (Tübingen, Germany, 1974).

Rimmer, Peter J., "Transport and communications in the Pacific economic zone during the early 21st century," *Pacific Asia in the 21st century: geographical and developmental perspectives* (Hong Kong, 1993), pp. 195–232.

Roth, Cecil, *The Sassoon dynasty* (London, 1941).

Rothenberg, Winifred Barr, *From market-places to a market economy: the transformation of rural Massachusetts, 1750–1850* (Chicago, 1992).

Rowe, William T., *Hankow: commerce and society in a Chinese city, 1796–1889* (Stanford, CA, 1984).

Rowley, Anthony, "Retreat to China: Bank of Japan sees change in Hongkong's role," *Far Eastern Economic Review* (March 12, 1992), p. 50.

Rueschemeyer, Dietrich, *Power and the division of labor* (Stanford, CA, 1986).

Rush, James R., *Opium to Java: revenue farming and Chinese enterprise in colonial Indonesia, 1860–1910* (Ithaca, NY, 1990).

Sapsford, Jathon, "Bankers skeptical Tokyo has power to force reform," *Wall Street Journal* (June 19, 1998), p. A11.

SarDesai, D. R., *British trade and expansion in Southeast Asia, 1830–1914* (Columbia, MO, 1977).

Sargent, A. J., *Anglo-Chinese commerce and diplomacy* (Oxford, 1907).

Sassen, Saskia, *The global city: New York, London, Tokyo* (Princeton, NJ, 1991).

Schenk, Catherine R., "The Hong Kong gold market and the Southeast Asian gold trade in the 1950s," *Modern Asian Studies*, 29 (May 1995), pp. 387–402.

Schiffer, Jonathan R., "State policy and economic growth: a note on the Hong Kong model," *International Journal of Urban and Regional Research*, 15 (June 1991), pp. 180–96.

Schlote, Werner, *British overseas trade from 1700 to the 1930s*, translated by W. O. Henderson and W. H. Chaloner (first published, Jena, 1938; Oxford, 1952).

Schran, Peter, "The minor significance of commercial relations between the United States and China, 1850–1931," in *America's China trade in historical perspective: the Chinese and American performance*, edited by Ernest R. May and John K. Fairbank (Cambridge, MA, 1986), pp. 237–58.

Scott, Robert Haney, "Unit trusts and insurance companies," in *Hong Kong's financial institutions and markets*, edited by Robert Haney Scott, K. A. Wong, and Yan Ki Ho (Hong Kong, 1986), pp. 123–32.

Security Branch, Government of the Hong Kong Special Administrative Region, unpublished tabulations.

Sender, Henry, "Guns for hire: U.S. investment bankers swarm to Hongkong," *Far Eastern Economic Review* (October 29, 1992), pp. 78–79.

"Local hero: Nomura begins to diversify its Japanese face," *Far Eastern Economic Review* (August 1, 1996), p. 48.

"Tarnished lustre: if Chinese connections guaranteed business success in Hong

Kong, Everbright's fortune would be made," *Far Eastern Economic Review* (May 23, 1996), pp. 52–54.

"The sun never sets," *Far Eastern Economic Review* (February 1, 1996), pp. 46–50.

Shapiro, Susan, "The social control of impersonal trust," *American Journal of Sociology*, 93 (November 1987), pp. 623–58.

"Shenzhen and Hong Kong enhance cooperation in financial sector," *Xinhua News Agency* (January 22, 1998).

"Shenzhen and Hong Kong enhance cooperation," *Xinhua News Agency* (April 3, 1998).

Shi, Peijun, Lin, Hui, and Liang, Jinshe, "Shanghai as a regional hub," in *Shanghai: transformation and modernization under China's open policy*, edited by Y. M. Yeung and Sung Yun-wing (Hong Kong, 1996), pp. 529–49.

Shiba, Yoshinobu, "Ningpo and its hinterland," in *The city in Late Imperial China*, edited by G. William Skinner (Stanford, CA, 1977), pp. 391–439.

Shih, Ta-lang, "The role of the People's Republic of China's Hong Kong-based conglomerates as agents of development," *Research in International Business and International Relations*, 3 (1989), pp. 215–30.

Silver, Alan, "Friendship in commercial society: eighteenth-century social theory and modern sociology," *American Journal of Sociology*, 95 (May 1990), pp. 1474–504.

Silverman, Gary, "A run for their money: The Royal Hong Kong Jockey Club looks set to stride through 1997 into the era of Chinese rule," *Far Eastern Economic Review* (January 25, 1996), pp. 46–50.

"Now the good news: after a two-year lull, Southeast Asian buyers are re-entering Hong Kong's property market," *Far Eastern Economic Review* (August 1, 1996), pp. 46–47.

Simkin, C. G. F., *The traditional trade of Asia* (London, 1968).

Sinn, Elizabeth, "A history of regional associations in pre-war Hong Kong," in *Between East and West: aspects of social and political development in Hong Kong*, edited by Elizabeth Sinn (Hong Kong, 1990), pp. 159–86.

"Emigration from Hong Kong before 1941: general trends," in *Emigration from Hong Kong: tendencies and impacts*, edited by Ronald Skeldon (Hong Kong, 1995), pp. 11–34.

Growing with Hong Kong: the Bank of East Asia, 1919–1994 (Hong Kong, 1994).

Power and charity: the early history of the Tung Wah Hospital, Hong Kong (Hong Kong, 1989).

Sit, Victor F. S., "Hong Kong's industrial out-processing in the Pearl River Delta of China," in *Industrial and trade development in Hong Kong*, edited by Edward K. Y. Chen, Mee-kau Nyaw, and Teresa Y. C. Wong (Hong Kong, 1991), pp. 559–77.

Sit, Victor Fung-shuen, Wong, Siu-lun, and Kiang, Tsin-sing, *Small scale industry in a laissez-faire economy: a Hong Kong case study* (Hong Kong, 1980).

Sito, Peggy, "Bank to centralise back-office functions: Hang Seng puts $1b into property," *South China Morning Post* (August 12, 1995), Business Post, p. 1.

"CITIC outbids all, wins Tamar site for $3.35b," *South China Morning Post* (August 4, 1995), Business Post, p. 1.

"Hongs speed up Guangdong plan for investment," *South China Morning Post* (August 7, 1997), p. 3.

Skeldon, Ronald, "Emigration and the future of Hong Kong," *Pacific Affairs*, 63 (Winter 1990–91), pp. 500–23.

"Emigration from Hong Kong, 1945–1994: the demographic lead-up to 1997," in *Emigration from Hong Kong: tendencies and impacts*, edited by Ronald Skeldon (Hong Kong, 1995), pp. 51–77.

"Hong Kong in an international migration system," in *Reluctant exiles? Migration from Hong Kong and the new Overseas Chinese*, edited by Ronald Skeldon (Armonk, NY, 1994), pp. 21–51.

"Migration from Hong Kong: current trends and future agendas," *Reluctant exiles? Migration from Hong Kong and the new Overseas Chinese*, edited by Ronald Skeldon (Armonk, NY, 1994), pp. 325–32.

"Singapore as a potential destination for Hong Kong emigrants before 1997," in *Crossing borders: transmigration in Asia Pacific*, edited by Ong Jin Hui, Chan Kwok Bun, and Chew Soon Beng (New York, 1995), pp. 223–38.

"Turning points in labor migration: the case of Hong Kong," *Asian and Pacific Migration Journal*, 3, no. 1 (1994), pp. 93–118.

Skinner, G. William, "Regional urbanization in nineteenth-century China," in *The city in Late Imperial China*, edited by G. William Skinner (Stanford, CA, 1977), pp. 211–49.

Chinese society in Thailand: an analytical history (Ithaca, NY, 1957).

Smelser, Neil J., and Swedberg, Richard, eds., *The handbook of economic sociology* (Princeton, NJ, 1994).

Smith, Adam, *An inquiry into the nature and causes of the wealth of nations* (first published, 1776; Chicago, 1952).

Smith, Craig S., "Chinese companies hope dual listings in Hong Kong will bolster their shares," *Wall Street Journal* (March 19, 1997), p. A15.

"Chinese regions chase Hong Kong cash: capitalist city becomes a window for financing," *Wall Street Journal* (December 2, 1996), p. A16.

"Municipal-run firms helped build China; now, they're faltering," *Wall Street Journal* (October 8, 1997), pp. A1, A13.

Smith, David A., and Timberlake, Michael, "Conceptualising and mapping the structure of the world system's city system," *Urban Studies*, 32 (March 1995), pp. 287–302.

Solomon, Jay, "Salim assailed for unit's shift to Singapore: Indonesia's biggest firm falls prey to ethnic tensions," *Wall Street Journal* (August 5, 1997), p. A15.

"South China's Shenzhen City to promote high-tech industry," *Xinhua News Agency* (February 20, 1998).

Spence, Jonathan, "Opium smoking in Ch'ing China," in *Conflict and control in Late Imperial China*, edited by Frederic Wakeman, Jr., and Carolyn Grant (Berkeley, CA, 1975), pp. 143–73.

Spencer, Herbert, *The principles of sociology* (3 vols., New York, 1897–1906).

Spindle, Bill, and Sapsford, Jathon, "Japan's investors fear a game with new rules," *Wall Street Journal* (November 10, 1997), p. A15.

State Statistical Bureau, *China statistical yearbook* (Beijing, various years).

Steele, Ian K., *The English Atlantic, 1675–1740: an exploration of communication and community* (New York, 1986).

Stein, Peter, "China's push into Hong Kong stocks raises concerns over insider trading," *Wall Street Journal* (August 8, 1997), p. A10.

Stigler, George J., "The division of labor is limited by the extent of the market," *Journal of Political Economy*, 59 (June 1951), pp. 185–93.

Stockwell, A. J., "Southeast Asia in war and peace: the end of European colonial empires," in *The Cambridge history of Southeast Asia*, vol. II: *The nineteenth and twentieth centuries*, edited by Nicholas Tarling (Cambridge, 1992), pp. 329–85.

Stopford, John M., and Wells, Louis T., Jr., *Managing the multinational enterprise* (New York, 1972).

Sung, Yun-wing, "'Dragon head' of China's economy?" in *Shanghai: transformation and modernization under China's open policy*, edited by Y. M. Yeung and Sung Yun-wing (Hong Kong, 1996), pp. 171–98.

Suzuki, Yoshio, "The future role of Japan as a financial centre in the Asian-Pacific area," *Global adjustment and the future of Asian-Pacific economy*, edited by Miyohei Shinohara and Fu-chen Lo (Kuala Lumpur and Tokyo, 1989), pp. 384–97.

Swedberg, Richard, ed., *Explorations in economic sociology* (New York, 1993).

Szczepanik, Edward, *The economic growth of Hong Kong* (London, 1958).

Szostak, Rick, *The role of transportation in the Industrial Revolution: a comparison of England and France* (Montreal and Kingston, Ontario, 1991).

Szulc, Tad, "A looming Greek tragedy in Hong Kong," *Foreign Policy*, no. 106 (Spring 1997), pp. 76–89.

Taaffe, Edward J., and Gauthier, Howard L., *Geography of transportation* (Englewood Cliffs, NJ, 1973).

Tai, Lawrence S. T., "Commercial banking," in *Hong Kong's financial institutions and markets*, edited by Robert Haney Scott, K. A. Wong, and Yan Ki Ho (Hong Kong, 1986), pp. 1–18.

Tarling, Nicholas, ed., *The Cambridge history of Southeast Asia*, vol. II: *The nineteenth and twentieth centuries* (Cambridge, 1992).

Tasker, Rodney, "The Thai, Indonesian and Burmese armies have long been entrenched in politics and business," *Far Eastern Economic Review* (January 18, 1996), pp. 20–21.

Taylor, Graham Romeyn, *Satellite cities: a study of industrial suburbs* (New York, 1915).

Taylor, Michael, *Community, anarchy and liberty* (Cambridge, 1982).

Tett, Gillian, "'Big bang' adds to pressure on Japanese banks," *Financial Times* (April 1, 1998), p. 1.

The Basic Law of the Hong Kong Special Administrative Region of the People's Republic of China (Hong Kong, 1992).

Theroux, Paul, "Letter from Hong Kong," *New Yorker*, 73 (May 12, 1997), pp. 54–65.

Thoburn, John T., Chau, Esther, and Tang, S. H., "Industrial cooperation between Hong Kong and China: the experience of Hong Kong firms in the Pearl River Delta," in *Industrial and trade development in Hong Kong*, edited by Edward K. Y. Chen, Mee-kau Nyaw, and Teresa Y. C. Wong (Hong Kong, 1991), pp. 530–58.

"Three MNCs awarded regional headquarters status," *Business Times* (Singapore) (May 29, 1998), p. 4.

Thrift, Nigel, "New urban eras and old technological fears: reconfiguring the goodwill of electronic things," *Urban Studies*, 33 (October 1996), pp. 1463–93.

Spatial formations (London, 1996).

"The fixers: the urban geography of international commercial capital," in *Global restructuring and territorial development*, edited by Jeffrey Henderson and Manuel Castells (London, 1987), pp. 203–33.

Tilly, Charles, *Big structures, large processes, huge comparisons* (New York, 1984).

Coercion, capital, and European states, AD 990–1990 (Cambridge, MA, 1990).

Tom, C. F. Joseph, *The entrepot trade and the monetary standards of Hong Kong, 1842–1941* (Hong Kong, 1964).

Tong, Amy, "Hong Kong plans to push for high technology science park project seen as way to boost manufacturing sector," *Nikkei Weekly* (March 23, 1998), p. 21.

"Red chips are down, but Hong Kong hopeful," *Nikkei Weekly* (June 29, 1998), p. 26.

Tonnies, Ferdinand, *Community and society (Gemeinschaft und Gesellschaft)*, translated and edited by Charles P. Loomis (East Lansing, MI, 1957).

Tornqvist, Gunnar, *Contact systems and regional development*, Lund Studies in Geography, Series B, Human Geography, no. 35 (Lund, Sweden, 1970).

Trocki, Carl A., *Opium and empire: Chinese society in colonial Singapore, 1800–1910* (Ithaca, NY, 1990).

Tsai, Jung-fang, *Hong Kong in Chinese history: community and social unrest in the British colony, 1842–1913* (New York, 1993).

Tu, Wei-ming, ed., *Confucian traditions in East Asian modernity: moral education and economic culture in Japan and the four mini-dragons* (Cambridge, MA, 1996).

Turnbull, C. M., *A history of Singapore, 1819–1988* (2nd edn.; Singapore, 1989).

The Straits settlements, 1826–67: Indian presidency to Crown colony (London, 1972).

Uzzi, Brian, "The sources and consequences of embeddedness for the economic performance of organizations: the network effect," *American Sociological Review*, 61 (August 1996), pp. 674–98.

Vance, James E., Jr., *The merchant's world: the geography of wholesaling* (Englewood Cliffs, NJ, 1970).

Viraphol, Sarasin, *Tribute and profit: Sino-Siamese trade, 1652–1853* (Cambridge, MA, 1977).

Vogel, Ezra F., *One step ahead in China: Guangdong under reform* (Cambridge, MA, 1989).

The four little dragons: the spread of industrialization in East Asia (Cambridge, MA, 1991).

Wakeman, Frederic, Jr., *Strangers at the gate: social disorder in South China, 1839–1861* (Berkeley, CA, 1966).

"The Canton trade and the Opium War," in *The Cambridge history of China*, vol. X: *Late Ch'ing, 1800–1911, part I*, edited by John K. Fairbank (Cambridge, 1978), pp. 163–212.

The fall of imperial China (New York, 1975).

Wang, Yeh-chien, *Land taxation in imperial China, 1750–1911* (Cambridge, MA, 1973).

"Secular trends of rice prices in the Yangzi Delta, 1638–1935," in *Chinese history in*

economic perspective, edited by Thomas G. Rawski and Lillian M. Li (Berkeley, CA, 1992), pp. 35–68.
Webb, Sara, "Chinese shares show 'guanxi' appeal," *Wall Street Journal* (September 29, 1997), pp. C1, C12.
Weber, Adna Ferrin, *The growth of cities in the nineteenth century: a study in statistics*, Columbia University Studies in History, Economics and Public Law, vol. XI (New York, 1899).
Weber, Max, *Economy and society*, edited by Guenther Roth and Claus Wittich (2 vols., Berkeley, CA, 1978).
Weidenbaum, Murray, and Hughes, Samuel, *The bamboo network: how expatriate Chinese entrepreneurs are creating a new economic superpower in Asia* (New York, 1996).
Welsh, Frank, *A borrowed place: the history of Hong Kong* (New York, 1993).
White, Harrison C., "Agency as control," in *Principals and agents: the structure of business*, edited by John W. Pratt and Richard J. Zeckhauser (Boston, 1985), pp. 187–212.
Whitley, Richard, *Business systems in East Asia: firms, markets and societies* (London, 1992).
Wilkins, Mira, "The impacts of American multinational enterprise on American–Chinese economic relations, 1786–1949," in *America's China trade in historical perspective: the Chinese and American performance*, edited by Ernest R. May and John K. Fairbank (Cambridge, MA, 1986), pp. 259–92.
Williamson, Oliver E., *Markets and hierarchies* (New York, 1975).
The economic institutions of capitalism: firms, markets, relational contracting (New York, 1985).
"The modern corporation: origins, evolution, attributes," *Journal of Economic Literature*, 19 (December 1981), pp. 1537–68.
Wilson, John C., "Hong Kong as regional headquarters," in *Hong Kong's role in the Asian Pacific region in the 21st century*, Proceedings and Papers of the ASEAN-China Hong Kong Forum 1992 (Hong Kong, 1993), pp. 105–11.
Wong, Gilbert, "Business groups in a dynamic environment: Hong Kong 1976–1986," in *Business networks and economic development in East and Southeast Asia*, edited by Gary Hamilton (Hong Kong, 1991), pp. 126–54.
Wong, Jesse, and Lee-Young, Joanne, "Hong Kong looks to Mainland for economic, political support," *Wall Street Journal* (July 1, 1998), p. A15.
Wong, Lin Ken, *The Malayan tin industry to 1914* (Tucson, AZ, 1965).
Wong, Siu-lun, "Business and politics in Hong Kong during the transition," in *25 years of social and economic development in Hong Kong*, edited by Benjamin K. P. Leung and Teresa Y. C. Wong (Hong Kong, 1994), pp. 217–35.
Emigrant entrepreneurs: Shanghai industrialists in Hong Kong (Hong Kong, 1988).
Wong, Teresa Y. C., "A comparative study of the industrial policy of Hong Kong and Singapore in the 1980s," in *Industrial and trade development in Hong Kong*, edited by Edward K. Y. Chen, Mee-kau Nyaw, and Teresa Y. C. Wong (Hong Kong, 1991), pp. 256–93.
World Bank, *The East Asian miracle: economic growth and public policy* (New York, 1993).

World development indicators (Washington, DC, 1998), CD-ROM.
World tables 1993 (Baltimore, 1993).
Wright, Arnold, ed., *Twentieth-century impressions of Hongkong, Shanghai, and other treaty ports of China* (London, 1908).
Wu, Bian, "Hong Kong emigrants return home," *Beijing Review*, 40 (March 24–30, 1997), pp. 13–15.
Wu, Yuan-li, and Wu, Chun-hsi, *Economic development in Southeast Asia: the Chinese dimension* (Stanford, CA, 1980).
Xu, Xueqiang, Kwok, Reginald Yin-wang, Li, Lixun, and Yan, Xiaopei, "Production change in Guangdong," in *The Hong Kong–Guangdong link: partnership in flux*, edited by Reginald Yin-wang Kwok and Alvin Y. So (Hong Kong, 1995), pp. 135–62.
Yamagishi, Toshio, Gillmore, Mary R., and Cook, Karen S., "Network connections and the distribution of power in exchange networks," *American Journal of Sociology*, 93 (January 1988), pp. 833–51.
Yamaguchi, Kazuo, "Power in networks of substitutable and complementary exchange relations: a rational-choice model and an analysis of power centralization," *American Sociological Review*, 61 (April 1996), pp. 308–32.
Yang, Mayfair Mei-hui, *Gifts, favors and banquets: the art of social relationships in China* (Ithaca, NY, 1994).
Yang, Shu-chin, ed., *Manufactured exports of East Asian industrializing economies: possible regional cooperation* (Armonk, NY, 1994).
Yeh, Anthony G. O., "Pudong: remaking Shanghai as a world city," *Shanghai: transformation and modernization under China's open policy*, edited by Y. M. Yeung and Sung Yun-wing (Hong Kong, 1996), pp. 273–98.
Yeung, Henry Wai-chung, *Transnational corporations and business networks: Hong Kong firms in the ASEAN region* (London, 1998).
Yeung, Yue-man, "Infrastructure development in the southern China growth triangle," in *Growth triangles in Asia: a new approach to regional economic cooperation*, edited by Myo Thant, Min Tang, and Hiroshi Kakazu (Hong Kong, 1994), pp. 114–50.
Yiu, Enoch, "Independent Peregrine crash probe considered," *South China Morning Post* (June 9, 1998), Business Post, p. 1.
Yuan, Chen, "Financial relations between Hong Kong and the Mainland," *Money and banking in Hong Kong* (Hong Kong, 1995), pp. 47–55.
"Monetary developments in China and monetary relations with Hong Kong," in *Monetary management in Hong Kong* (Hong Kong, 1993), pp. 69–78.
Yuan, Lee Tsao, "Hong Kong and Singapore: government and entrepreneurship in economic development," in *25 years of social and economic development in Hong Kong*, edited by Benjamin K. P. Leung and Teresa Y. C. Wong (Hong Kong, 1994), pp. 632–59.
Zheng, Zhang Zhi, "Promoting economic co-operation between Guangdong and Hong Kong," *Hong Kong's role in the Asian Pacific region in the 21st century*, Proceedings and Papers of the ASEAN-China Hong Kong Forum 1992 (Hong Kong, 1993), pp. 129–43.

Index

Cambridge Studies in Historical Geography

1 Period and place: research methods in historical geography. *Edited by* ALAN R. H. BAKER and M. BILLINGE
2 The historical geography of Scotland since 1707: geographical aspects of modernisation. DAVID TURNOCK
3 Historical understanding in geography: an idealist approach. LEONARD GUELKE
4 English industrial cities of the nineteenth century: a social geography. R. J. DENNIS*
5 Explorations in historical geography: interpretative essays. *Edited by* ALAN R. H. BAKER and DEREK GREGORY
6 The tithe surveys of England and Wales. R. J. P. KAIN and H. C. PRINCE
7 Human territoriality: its theory and history. ROBERT DAVID SACK
8 The West Indies: patterns of development, culture and environmental change since 1492. DAVID WATTS*
9 The iconography of landscape: essays in the symbolic representation, design and use of past environments. *Edited by* DENIS COSGROVE and STEPHEN DANIELS*
10 Urban historical geography: recent progress in Britain and Germany. *Edited by* DIETRICH DENECKE and GARETH SHAW
11 An historical geography of modern Australia: the restive fringe. J. M. POWELL*
12 The sugar-cane industry: an historical geography from its origins to 1914. J. H. GALLOWAY
13 Poverty, ethnicity and the American city, 1840–1925: changing conceptions of the slum and ghetto. DAVID WARD*
14 Peasants, politicians and producers: the organisation of agriculture in France since 1918. M. C. CLEARY
15 The underdraining of farmland in England during the nineteenth century. A. D. M. PHILLIPS
16 Migration in Colonial Spanish America. *Edited by* DAVID ROBINSON
17 Urbanising Britain: essays on class and community in the nineteenth century. *Edited by* GERRY KEARNS and CHARLES W. J. WITHERS
18 Ideology and landscape in historical perspective: essays on the meanings of some places in the past. *Edited by* ALAN R. H. BAKER and GIDEON BIGER
19 Power and pauperism: the workhouse system, 1834–1884. FELIX DRIVER
20 Trade and urban development in Poland: an economic geography of Cracow from its origins to 1795. F. W. CARTER
21 An historical geography of France. XAVIER DE PLANHOL
22 Peasantry to capitalism: Western Östergötland in the nineteenth century. GÖRAN HOPPE and JOHN LANGTON

23 Agricultural revolution in England: the transformation of the agrarian economy, 1500–1850. MARK OVERTON*

24 Marc Bloch, sociology and geography: encountering changing disciplines. SUSAN W. FRIEDMAN

25 Land and society in Edwardian Britain. BRIAN SHORT

26 Deciphering global epidemics: analytical approaches to the disease records of world cities, 1888–1912. ANDREW CLIFF, PETER HAGGETT and MATTHEW SMALLMAN-RAYNOR*

27 Society in time and space: a geographical perspective on change. ROBERT A. DODGSHON*

28 Fraternity among the French peasantry: sociability and voluntary associations in the Loire valley, 1815–1914. ALAN R. H. BAKER

29 Visions of empire: nationalist imagination and geographical expansion in the Russian Far East, 1840–1865. MARK BASSIN

Titles marked with an asterisk * are available in paperback.